Ecological Causal Assessment

New Book Series

Environmental Assessment and Management

Series Editor

Dr. Glenn W. Suter II
U.S. Environmental Protection Agency
Cincinnati, OH, USA

Published Titles

Ecological Causal Assessment
edited by Susan B. Norton, Susan M. Cormier, and Glenn W. Suter II

Multi-Criteria Decision Analysis: Environmental Applications and Case Studies
Igor Linkov and Emily Moberg

Ecological Causal Assessment

Susan B. Norton
Susan M. Cormier
Glenn W. Suter II

CRC Press
Taylor & Francis Group
Boca Raton London New York

CRC Press is an imprint of the
Taylor & Francis Group, an **informa** business

CRC Press
Taylor & Francis Group
6000 Broken Sound Parkway NW, Suite 300
Boca Raton, FL 33487-2742

First issued in paperback 2017

ISBN-13: 978-1-4398-7013-6 (hbk)
ISBN-13: 978-1-138-07393-7 (pbk)

Library of Congress Cataloging-in-Publication Data

Ecological causal assessment / editors: Susan B. Norton, Glenn W. Suter II, Susan M. Cormier.
    pages cm. -- (Environmental assessment and management)
  Includes bibliographical references and index.
  ISBN 978-1-4398-7013-6 (hardcover : alk. paper) 1. Ecological assessment (Biology) 2. Ecosystem health. 3. Causation. I. Norton, Susan B., 1959-, editor of compilation, author. II. Suter, Glenn W., II, editor of compilation, author. III. Cormier, Susan M. (Susan Marie), 1954- editor of compilation, author.

QH541.15.E22E337 2015
577--dc23   2014032795

Visit the Taylor & Francis Web site at
http://www.taylorandfrancis.com

and the CRC Press Web site at
http://www.crcpress.com

Contents

Part 1   Introduction and Philosophical Foundation

Part 2   Conducting Causal Assessments

Part 2A   Formulating the Problem

Part 2B   Deriving Evidence

Part 2C   Forming Conclusions and Using the Findings

Preface

Causal thinking is an everyday activity. We all are confronted with questions of causation, whether to figure out why the car is making a funny noise or why a toddler is running a fever. Our fascination with investigating causes is reflected in the enduring popularity of detective stories and in the frequency of investigative reports in the news.

Because causal inference is commonplace,* a book on ecological causal assessment may seem unnecessary. However, causes are not always easy to determine. Ecosystems are complex; the factors we can influence interact with natural factors, random processes, and initial conditions to produce the effects that are observed. Taking corrective action to remedy an environmental problem before knowing its cause could target the wrong thing, depleting scarce resources and missing an opportunity to improve environmental quality.

Formal processes for causal assessment, as described in this book, are particularly helpful when the situation is complex or contentious. A well-articulated process guides the analysis of available data and optimizes further collection efforts. A transparent process helps others replicate results and is more likely to convince skeptics that the true cause has been identified. A consistent process helps meet legal and regulatory standards for reasonableness and ensures that scientific information contributes to these decisions. Perhaps most importantly, formal methods help to eliminate biases that arise because of the all-too-human tendency to make and defend causal judgments too readily. As aptly articulated by the physicist Richard Feynman, "The first rule of science is not to fool yourself—and you are the easiest person to fool."†

We began this project with a practical purpose—to share useful methods and strategies for identifying causes of undesirable biological effects in specific places. Causal assessment is a challenging, often humbling, but endlessly fascinating endeavor. It begins with the intrigue of a good mystery—why did this effect happen? Success requires the persistence to figure things out and solid strategies for using the information that you have and getting more of the right kind of information that you need. We feel fortunate to have been involved with adapting existing methods and testing new approaches. It has led us to renewed study of our intellectual heritage of science and philosophy, the strengths and foibles of human cognition, and the underlying

---

* Even infants are capable of recognizing causal processes (e.g., Leslie, A. M. and S. Keeble. 1987. Do six-month-old infants perceive causality? *Cognition* 25:265–288).

† Feynman, R. 2001. Cargo cult science: Some remarks on science, pseudoscience, and learning how not to fool yourself. In *The Pleasure of Finding Things Out: The Best Short Works of Richard Feynman*, edited by J. Robbins. Cambridge, MA: Perseus.

assumptions of different sampling designs and analytical methods. It has also allowed us to provide scientific assessments and advice on some of the more complex ecological problems of our times.

We have drawn on our personal experiences and those of our colleagues to provide examples and to describe approaches for assessing causes of undesirable biological effects in ecological systems. Some of these effects have captured the public's attention and concern: collapsing fisheries and bee colonies; bleaching coral reefs; endangered species; dwindling stream life; and kills of fish, birds, and bats. Behind these reports are scientists who monitor our ecological systems and carefully document when something is amiss. In the past 20 years, biological monitoring has become an essential part of the environmental management tool kit. Causal assessment is the next essential tool. When we wonder why a condition has worsened, causal assessment finds the explanation.

We believe that this book provides sound advice for the near term. We hope that it will lead the way to future improvements in methods and applicable scientific knowledge. We also hope that our study of causal assessment in the context of environmental management advances the larger field of causal assessment and provides insights into how we all can improve our causal reasoning.

Acknowledgments

This book grew out of our work producing the Causal Analysis/Diagnosis Decision Information System (CADDIS: http://www.epa.gov/caddis) in the U.S. Environmental Protection Agency's National Center for Environmental Assessment. We have aimed to discuss causal assessment from a common and consistent viewpoint. For this reason, we have drawn most of our contributing authors from the CADDIS project team. We thank our co-authors Laurie Alexander, David Farrar, Michael Griffith, Scot Hagerthey, Michael McManus, Leela Rao, Pat Shaw-Allen, Kate Schofield, Lester Yuan, and Rick Ziegler. We are very pleased to have the expertise of Jeroen Gerritsen and Lei Zheng to guide the chapters on the observational studies and the Clear Fork case study; Joseph Culp, Bob Brua, Alexa Alexander, and Patricia Chambers for the chapters on field, laboratory, and mesocosm studies and the Athabasca River case study, and Tom O'Farrell for the kit fox case study.

We thank the U.S. Environmental Protection Agency's (U.S. EPA's) Office of Research and Development, and especially the managers in the National Center for Environmental Assessment, for their support for this work and book: Jeff Frithsen, David Bussard, Mary Ross, Michael Slimak, Michael Troyer, and Annette Gatchett.

We thank Bill Swietlik and Donna Reed-Judkins for initiating our involvement in this field with the Stressor Identification Guidance Document.

We thank our colleagues who participated in U.S. EPA's Causal Analysis Team where many of these ideas were germinated and fostered: Maureen Johnson, John Paul, Amina Pollard, Robert Spehar, Brooke Todd, Walter Berry, Robert Cantilli, Maeve Foley, Scott Freeman, Jeff Hollister, Evan Hornig, Susan Jackson, Phil Kaufman, Kathryn Kazior, Jan Kurtz, Chuck Lane, Phil Larsen, Suzanne Marcy, Matthew Morrison, Chris Nietch, Doug Norton, Keith Sappington, Victor Serveiss, Treda Smith, Debra Taylor, Sharon Taylor, and Joe Williams. Thanks to our colleagues from Australia's Eco Evidence project for their perspectives and encouragement: Richard Norris, Sue Nichols, Angus Webb, Michael Stewardson, Ralph Ogden, and Michael Peat.

We thank the many biologists and environmental scientists in the front lines of environmental protection in state and local governments, and in the consulting firms that support them. They have encouraged us to develop methods, provided us with data, ideas, and constructive criticism, and have provided many of the stories that bring this book alive. In particular, we thank Christopher Bellucci, Dan Helwig, Scott Niemela, Mike Feist, Joel Chirhart, Craig Affeldt, Charlie Menzie, Danelle Haake, Tom Wilton, Thomas Isenhart, Susanne Meidel, Jeff Varricchione, Susan Davies, Kay Whittington, William Stephens, Matt Hicks, Chad Wiseman, Mike LeMoine, Ben Lowman, Jeff Bailey, Ken Schiff, David Gillette, Andy Rehn, Jim Harrington, Chris Yoder,

Ed Rankin, Art Stewart, David Eskew, Michael Barbour, Michael Paul, and Jerry Diamond.

We thank our reviewers for their insightful comments: Jackie Moya, Scot Hagerthey, Susan Jackson, Bruce Hope, Bob Miltner, Charlie Menzie, Piet Verdonschot, and Rick Racine. Additional reviewers of the CADDIS web-site have included Marty Matlock, Chuck Hawkins, Lucinda Johnson, Peter deFur, Rebecca Efroymson, Donald Essig, Roman Lanno, Mark Southerland, Daren Carlisle, Rob Plotnikoff, and Barbara Washburn.

Thanks for assistance in document production and graphics support from Bette Zwayer and Vicki Soto. Debbie Kleiser, Sandra Moore, Ashley Price, and Kathleen Secor, contractors with Highlight Technologies, Fairfax Virginia, and Stacey Herron, Thomas Schaffner, Linda Tackett and Lisa Walker, contractors with CACI, Arlington, Virginia also assisted in document production. Graphics assistance was provided by Katherine Loizos and Teresa Ruby, contractors from SRA International, Fairfax, Virginia. Susan Cormier designed the cover using a photograph by Doug Norton. Thanks also to the team at CRC Press: Irma Britton, Rachael Panthier, Florence Kizza, and Joselyn Banks-Kyle.

Thanks to the George Mason University, Department of Environmental Science and Public Policy and especially R. Christian Jones and Dann Sklarew for their encouragement at the beginning of this book project.

Finally, we thank our families who have inspired us to become better scientists and have exhibited tremendous patience during the long process of writing and editing. Thank you Doug Norton, Linda Suter, and Rick, Claire, Annie, and Lisa Racine.

Disclaimer: The views expressed in this book are those of the authors and do not necessarily reflect the views or policies of the U.S. Environmental Protection Agency or Environment Canada.

Susan B. Norton, Susan M. Cormier, and Glenn W. Suter II

Decision Information System (CADDIS) website and has applied these methods to identify and resolve causes of biological impairments in watersheds all over the country and influenced assessments worldwide. Her research and professional activity center on a recurring theme of generating and assessing scientific information to inform environmental management decisions.

Glenn W. Suter II is science advisor in the U.S. EPA's National Center for Environmental Assessment and Chairman of the Risk Assessment Forum's Ecological Oversight Committee. He has authored more than 200 publications including 3 authored books and 3 edited books over his 37-year career. He has received the SETAC Founder's Award and the AEHS Career Achievement Award, and is an elected fellow of the AAAS. His interests include ecological epidemiology, ecological risk assessment, and the conceptual bases for environmental science and decision-making. Since he left Oak Ridge National Laboratory for the U.S. EPA, it has been his pleasure and honor to work with Drs. Cormier and Norton on causal assessment.

Editors

Susan B. Norton is a senior ecologist in the U.S. Environmental Protection Agency's National Center for Environmental Assessment. Since joining the EPA in 1988, Dr. Norton has developed methods and guidance to better use ecological knowledge to inform environmental decisions. She is the lead author of the 2011 EPA review "The Effects of Mountaintop Mines and Valley Fills on Aquatic Ecosystems of the Central Appalachian Coalfields." She co-led the development of the EPA's Causal Analysis/Diagnosis Decision Information System (www. epa.gov/caddis) and contributed to many agency guidance documents including the 2000 Stressor Identification Guidance document, the 1998 Guidelines for Ecological Risk Assessment, the 1993 Wildlife Exposure Factors Handbook, and the 1992 Framework for Ecological Risk Assessment. She has published numerous articles and book chapters on ecological assessment and edited the book *Ecological Assessment of Aquatic Resources: Linking Science to Decision-Making*. Dr. Norton earned her BS in plant science from The Pennsylvania State University, her MS in natural resources from Cornell University, and her PhD in environmental biology from George Mason University.

Susan M. Cormier is a senior scientist with the U.S. EPA Office of Research and Development, having held both scientific and managerial positions. Before joining the U.S. EPA, she was an assistant professor at Vassar College, New York, and University of Louisville, Kentucky. Dr. Cormier earned her BA from the University of New Hampshire, her MS from the University of South Florida, and her PhD from Clark University of Worcester, Massachusetts, with stints at the Marine Biological Laboratories of Woods Hole, Discovery Bay Marine Laboratory, Jamaica, and the Bermuda Biological Station for Research. She has authored more than 100 peer-reviewed publications making substantive contributions to the development of methods for biocriteria, water quality criteria, causal assessment, and ecological risk assessment. Dr. Cormier co-led the development of the U.S. EPA Stressor Identification Guidance for the Agency and the Causal Analysis/Diagnosis

Contributors

Alexa C. Alexander
Environment Canada
Canada Centre for Inland Waters
Burlington, Ontario, Canada

Laurie C. Alexander
Office of Research and
    Development
National Center for Environmental
    Assessment
U.S. EPA
Washington, DC

Robert B. Brua
Environment Canada
National Hydrology Research
    Centre
Saskatoon, Saskatchewan, Canada

Patricia A. Chambers
Canada Centre for Inland Waters
Burlington, Ontario, Canada

Susan M. Cormier
Office of Research and Development
National Center for Environmental
    Assessment
U.S. EPA
Cincinnati, Ohio

Joseph M. Culp
University of New Brunswick
Fredericton, New Brunswick,
    Canada

David Farrar
Office of Research and Development
National Center for Environmental
    Assessment
U.S. EPA
Cincinnati, Ohio

Jeroen Gerritsen
Tetra Tech, Inc.
Owings Mills, Maryland

Michael Griffith
Office of Research and Development
National Center for Environmental
    Assessment
U.S. EPA
Cincinnati, Ohio

Scot E. Hagerthey
Office of Research and Development
National Center for Environmental
    Assessment
U.S. EPA
Washington, DC

Michael G. McManus
Office of Research and Development
National Center for Environmental
    Assessment
U.S. EPA
Cincinnati, Ohio

Susan B. Norton
Office of Research and Development
National Center for Environmental
    Assessment
U.S. EPA
Washington, DC

Thomas P. O'Farrell
WATASH, LLC
Boulder City, Nevada

Leela Rao
California Air Resources Board
El Monte, California

Kate Schofield
Office of Research and Development
National Center for Environmental
    Assessment
U.S. EPA
Washington, DC

Patricia Shaw-Allen
NOAA Fisheries
Office of Protected Resources
Silver Spring, Maryland

Glenn W. Suter II
Office of Research and Development
National Center for Environmental
    Assessment
U.S. EPA
Cincinnati, Ohio

Lester L. Yuan
Office of Water
Office of Science and Technology
U.S. EPA
Washington, DC

Lei Zheng
Tetra Tech, Inc.
Owings Mills, Maryland

C. Richard Ziegler
Office of Research and Development
National Center for Environmental
    Assessment
U.S. EPA
Washington, DC

Part 1

Introduction and Philosophical Foundation

Part 1 provides an introduction to causation and a solid foundation for performing a causal assessment. Chapter 1 introduces the causal assessment process using an example of low biological diversity in a stream and describes how scientific data and expertise are used to assess causation. Chapters 2 and 3 describe how the process for performing causal assessment is derived from the work of prior philosophers and scientists. Chapter 4 explains the characteristics of causation and introduces some commonly used types of evidence. Chapter 5 concludes Part 1 by discussing common errors and biases and ways of minimizing them.

1

Introduction

Susan B. Norton, Susan M. Cormier, and Glenn W. Suter II

What are these boxes? Each chapter begins with a text box that describes its contents and highlights.

This chapter provides an overview of the book and a brief example of a causal assessment. It describes the book's purpose: to show how scientific data and expertise can be used to reach credible, defensible conclusions about the causes of undesirable biological effects in ecological systems.

CONTENTS

It was a mystery. Chris Bellucci, a biologist with the State of Connecticut, had sampled the insects and other aquatic macroinvertebrates in the Willimantic River below the outfall from a publicly owned treatment works. He concluded that the macroinvertebrate assemblage did not meet the State's biological quality standard. This was not the mystery. The puzzle that confronted Bellucci was that macroinvertebrate samples taken upstream and above the influence of the treatment works were similarly degraded. Clearly, the treatment works was not the only reason that the macroinvertebrate assemblage did not meet the standard. But if effluent from the treatment works was not causing the upstream degradation, then who or what was the cause?

We have written this book to help investigate and solve environmental problems. In particular, the book is for scientists and engineers who are interested in finding the causes of undesirable ecological conditions, such as a fish kill, a decline in a population or assemblage, or increased incidence of disease or deformity. We describe an approach for assessing the causes of undesirable effects that includes developing a list of candidate causes of the observed effects, deriving evidence for or against each alternative, and identifying the best explanation by considering all of the evidence. We describe the philosophical and historical underpinnings of the approach and strategies for preventing common biases and blind spots. We hope the information and methods will provide you, our readers, with the tools and confidence needed to unravel tough environmental problems and help build the knowledge base for effective management solutions.

Causal assessments are not always easy to do, but their results are empowering. When one moves from identifying that a problem exists to understanding the causes of the problem, the stage is set for action. Even when one probable cause does not clearly emerge, a causal assessment can help narrow the field of possibilities, identify critical data needs, and provide the necessary impetus to collect the needed data that will reveal the cause. In the Willimantic River, chemicals from a broken sewer pipe in a tributary were eventually found to be the cause of the degraded assemblage. The State moved quickly to reroute the discharge. Over the next two years, they continued to monitor the macroinvertebrates. The stream assemblage recovered, verifying that the action was effective and the discharge was the cause.

1.1 What is an Ecological Causal Assessment?

Assessments can be broadly defined as technical support for decision-making.* Ecological causal assessments provide support for management decisions intended to solve environmental problems that adversely affect ecological systems and the biota that inhabit them. In this book, causal assessments are specific to a particular situation, system, or place. For example, was a particular fish kill caused by low dissolved oxygen levels from an algal bloom? Another kind of causal assessment evaluates

---

* This definition extends the definition of risk assessment in Suter (2007) to all types of assessments.

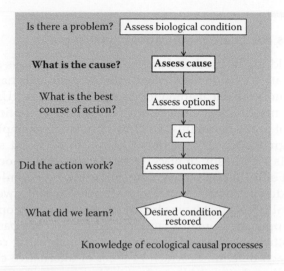

FIGURE 1.1
Causal assessment (shown in bold) is only one step in a series of activities needed to solve environmental problems. (Adapted from Norton, S. B., P. J. Boon, S. Gerould et al. 2004. In *Ecological Assessment of Aquatic Resources: Linking Science to Decision-Making*, Pensacola, FL: SETAC Press; Cormier, S. M., and G. W. Suter II. 2008. *Environ Manage* 42 (4):543–556.)

whether a factor is even capable of causing a specific effect. These types of assessments are prompted by questions such as "are algal blooms capable of causing fish kills?" and work out the scientific details describing how high algal biomass provides organic matter for bacterial decomposition, with its associated respiration and depletion of oxygen. Many of the approaches and specific methods we discuss in this book are useful in evaluating questions of general capability. But we will emphasize how to use this knowledge to investigate particular cases and solve problems at specific locations.

Causal assessments are usually undertaken as one in a series of activities used to identify and remedy environmental problems. Although many sequences and combinations are possible, one way that assessment activities can be linked together is through the following sequence of questions: "Is there an undesirable biological condition?," "What caused it?," "What is the best course of action?," and "Did the action work?" (see Figure 1.1). Each question is addressed using a specific type of assessment (e.g., of condition, cause, options, and outcomes). The sequence draws on and contributes to the knowledge foundation of ecological causal processes (depicted by the large gray box in Figure 1.1). Thus, the assessment sequence as a whole provides valuable information that can be used to improve future causal assessments, management actions, and our understanding of how ecosystems work.

1.2 Strategies for Ecological Causal Assessment

This book would not be needed if one method could always clinch a case and prove causation. Randomized, controlled experiments have long been held up as the most reliable method for determining cause.* Unfortunately, randomized, controlled experiments are not usually an option to investigate the types of problems addressed by this book. For example, we cannot randomly assign wastewater treatment plants to different streams—the treatment plants are already in place. Other factors that co-occur with the treatment plant effluent (e.g., stream flow) will not be randomly distributed upstream and downstream from the outfall. Additionally, we are investigating biological effects that have already occurred, so any opportunity to prospectively apply stressors in a randomized fashion has passed.

We need another way. Instead of proving and disproving causes one by one, our approach determines which of a set of alternative causes is best supported by all of the available evidence. The overall objective is to produce a coherent explanation of why some causes are likely and others are implausible.

This strategy is aligned with the way scientific research progresses, that is, not from a single experiment or fact. And as in all science, the explanation is only the best supported explanation based on the evidence available at the time. Even incremental knowledge can be useful for our ultimate goal of improving the environment. For example, reducing the list of candidate causes can focus further investigations on the remaining candidates or provide enough information to guide action.

The book is divided into three major parts: "Introduction and Fundamentals," "Conducting Causal Assessments," and "Case Studies." An overview of each of these sections follows. Readers interested in the philosophical and practical underpinnings of our approach should start with Part 1. Readers who are currently beginning a causal investigation may prefer to begin with Chapter 6 (our approach), one of the case studies in Part 3, and then focus on Part 2, which describes detailed methods and approaches for implementing the overall strategy. We do not review basic ecological, toxicological, and statistical principles and methods. Rather, our intent is to show how these methods and principles are used to investigate causes. We provide references to additional resources including those on the CADDIS website (U.S. EPA, 2012a) (Box 1.1) for readers interested in particular topics.

---

* In a randomized, controlled experiment, a stressor is randomly assigned and applied to a different experimental unit. The objective is to minimize the chance that other variables will influence the response. Randomization ensures that although other factors may introduce error into the results, they will not bias them. It enables scientists to conclude that the observed effects were caused by the stressor being manipulated.

BOX 1.1 CADDIS

The Causal Analysis/Diagnosis Decision Information System (CADDIS, available at www.epa.gov/caddis) was developed by the U.S. EPA to help scientists and engineers conduct causal assessments in aquatic systems. As of this writing, CADDIS contains a guide to the U.S. EPA's Stressor Identification process (originally documented in U.S. EPA, 2000a), information on commonly encountered stressors in aquatic systems, case examples, data analysis advice and tools, and literature databases. The method described in this book is a generalization of the more prescriptive method in CADDIS.

1.2.1 Part 1: Introduction and Fundamentals

Our approach builds on definitions and concepts about causation developed by the philosophers, scientists, and cognitive psychologists who have preceded us. The first part of the book reviews these foundations.

Chapters 2 and 3 begin by reviewing the ways that scientists and philosophers have thought about causes and their identification. In short, causes bring about their effects. They make things happen. In most ecological assessments, causes can be thought of as an addition of something harmful that was not there before (e.g., ethanol in a stream) or a removal of a required resource (e.g., gravel for spawning). Causes can be described as an event (e.g.., what happened?), a thing (e.g., what did it?), or a process (e.g., how did it transpire?). Different ways of describing causes can be used to develop clearly defined alternatives that are considered and eventually compared.

Causal relationships exhibit several basic characteristics (see Chapter 4) useful for suggesting ways that a cause-and-effect relationship can be observed and documented. We expect that (1) causes precede their effect in time, (2) there is a process or mechanism by which the cause and the biota can interact, (3) there is the opportunity for this interaction to occur, (4) the interaction is sufficient to produce the effect, (5) the interaction alters biota in specific ways, and (6) the causal event takes place within a larger web of causal events.

Chapter 4 also introduces types of evidence. Evidence can be thought of as associations and predictions that demonstrate (or alternatively refute) that a result expected of a causal relationship is obtained in the case. The data used to develop evidence can come from many different sources discussed further in Part 2.

Causal assessments are conducted by people. Along with the expertise, skills, and insights we all bring to an investigation, we may bring cognitive tendencies that can lead an investigation astray. Chapter 5 reviews common biases and blind spots and discusses strategies that can be used to prevent errors.

1.2.2 Part 2: Conducting Causal Assessments

Our approach is described in Chapter 6 (see Figure 1.2). The process provides the structure for Part 2 and has three major steps: (1) formulate the problem, (2) derive evidence, and (3) form conclusions. The product of the assessment identifies the cause or causes best supported by the evidence and those that lack support.

1.2.2.1 Part 2A: Formulate the Problem

Causal assessments are typically prompted by the observation of an undesirable effect. The effect could be diseased or dead organisms, such as coral bleaching or plants that fail to grow; a decline in a population such as a sport fish or endangered species; or a change in an assemblage of biota, such as

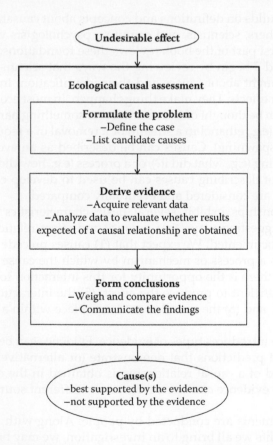

FIGURE 1.2
The causal assessment process, shown within the bold box, is typically prompted by an observation of an undesirable effect. The product identifies the cause or causes that are best supported by the evidence as well as those that are not.

the macroinvertebrates and fish that are frequently used to monitor stream conditions (see Figure 1.3).

Typically, the concerns that prompt a causal investigation must be further defined to support the assessment process. Formulating the problem to be investigated by defining its frame and focus greatly influences which causes will be considered and how data will be analyzed. The first part of problem formulation, discussed in Chapter 7, is the operational definition of the subject of the causal assessment (i.e., the case). The case definition describes the undesirable effects and the geographic and temporal dimensions of the investigation. It identifies places and times where undesirable effects have occurred and also identifies places or times that can be used for comparisons where effects either have not occurred or have occurred in a different way. The second major part of problem formulation is the development of the list of

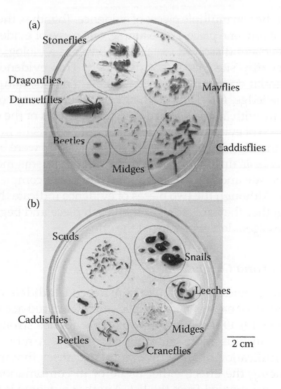

FIGURE 1.3

Example of stream biological monitoring samples from a high quality stream (a) and a stream receiving water from a stormwater drain and parking lot (b). The sample in (a) has many more organisms from sensitive taxonomic groups, such as mayflies, stoneflies, and caddisflies. The sample in (b) has many more organisms from tolerant taxonomic groups such as midges, scuds, and snails. Results like these can be used by biologists to judge the quality of water. In addition, these results can provide important clues to the causes affecting the biota living in the stream. (Courtesy of Thomas J. Danielson, Maine Department of Environmental Protection.)

candidate causes that will be investigated. Chapter 8 discusses strategies for developing the list, options for managing multiple causes, and the use of conceptual model diagrams to visualize hypotheses and organize information.

1.2.2.2 Part 2B: Derive Evidence

The core of a causal assessment is the evidence that can be used to argue for or against a candidate cause. Evidence is derived by analyzing data, which can come from many different sources, including observations at the site, regional monitoring studies, environmental manipulations, and laboratory experiments. Chapters 9–18 review different sources of information and the methods used to develop evidence. Each chapter points out the strengths and limitations of different approaches and provides referrals to more in-depth material.

Our strategy relies on multiple pieces of evidence. Together, they can mitigate the limitations of any one piece. For example, one piece of evidence may show that predawn levels of dissolved oxygen are lower at a biologically degraded site than at a nearby site that is not degraded. This evidence is uncertain because of potential errors in the oxygen measurements, natural variability, and lack of knowledge, for example, the likely co-occurrence of other factors that may co-occur with and possibly disguise the effects of the dissolved oxygen. Another piece of evidence may show that test organisms in the laboratory cannot survive the dissolved oxygen concentrations that were observed in the field. This evidence is uncertain because the test organisms may be different from those at the site and the test conditions will never completely match the field conditions. Although the two pieces of evidence indicate that low oxygen is the cause, together they are stronger than each piece and begin to build the argument that oxygen depletion caused the effects.

1.2.2.3 Part 2C: Form Conclusions

After the available evidence for and against each candidate cause is developed, the evidence for each one is weighed and compared across the alternatives (see Chapter 19). Optimally the available evidence strongly supports a candidate cause and discredits all other candidates. A more common outcome is the identification of all of the candidate causes that may be playing a role in producing the effect, either alone or in combination. Causes that lack support are winnowed from the list. Another outcome is the generation of new or refined alternative causes based on the first iteration of analyses. Even when one cause is not definitively identified, results can be useful by reducing the list of candidate causes that need further consideration and by pointing to fruitful directions for further data collection.

An explicit system for weighing evidence helps ensure that each cause is treated fairly and that all evidence is considered. In addition, an explicit system makes the basis for conclusions transparent and enables review.

However, an equally important part of the process is presenting the findings to others (see Chapter 20). Narrative explanations, tables, and diagrams that summarize the evidence communicate the conclusions of the causal assessment to decision-makers and stakeholders.

1.2.2.4 After the Causal Assessment: Using the Findings

The last chapter of Part 2 discusses what comes after causes are identified. As discussed above, a causal assessment is only one activity in a series of assessments typically undertaken to solve environmental problems. In some cases, the most effective management action will be obvious after the probable cause has been identified. In many cases, however, the investigation must identify sources and apportion responsibility among them. This task can be just as difficult as identifying the cause in the first place (e.g., quantifying the sources of fine sediment in a large watershed or deciding where to begin remediation at a large hazardous waste site). Identifying and implementing management options can also be a complex process that requires stakeholder involvement and additional analyses (e.g., economic comparisons, engineering feasibility). Chapter 21 discusses how the products of causal assessment can inform the activities that follow, for example, by helping define the goals and targets for management action and setting expectations for recovery.

1.2.2.5 A Brief Example Case: Causal Assessment in the Willimantic River, CT

The overall process is summarized with a synopsis of the Willimantic River investigation adapted from Bellucci et al. (2010).

1.2.2.5.1 The Undesirable Ecological Effect

The causal assessment in the Willimantic River was prompted by macroinvertebrate monitoring results used by the Connecticut Department of Environmental Protection to evaluate water quality. The monitoring results indicated that the biota did not meet state standards for healthy macroinvertebrate assemblages.

1.2.2.5.2 The Causal Assessment Process

1.2.2.5.2.1 *Formulate the Problem* The investigators used the macroinvertebrate monitoring results to home in on the decline in a sensitive group of insects that spend most of their lives in streams: the EPT taxa (Box 1.2). The investigators defined the spatial extent of the effects by mapping where the macroinvertebrate assemblage did not meet standards. They listed other sites for comparison within the watershed where standards were met. They listed the following six candidate causes: (1) toxic substances, (2) low dissolved oxygen, (3) altered habitat, (4) elevated temperature, (5) high flows, and (6) altered food resources. They developed conceptual model diagrams that hypothesized linkages between sources, stressors, and the observed effects.

BOX 1.2   BIOLOGICAL MONITORING IN STREAMS AND EPT TAXA

Biological monitoring programs use observations of biota as indicators of pollution and habitat quality. Scientists have developed different indicators for different environmental settings, for example, the extent of eel grass beds in coastal systems and floristic composition in terrestrial systems. Many water quality programs that monitor streams and rivers sample algae, fish, and invertebrates. The biological assemblages that are monitored have the advantage of reflecting exposure to many types of human-induced changes. They also directly represent biological resources that are valued and protected under laws like the U.S. Clean Water Act. Among other ecological services, organisms in these assemblages filter water, decompose organic matter, and form the base of aquatic-dependent food webs, feeding birds, bats, and fish.

The insects and other macroinvertebrates that live on the bottom of streams and rivers have been widely used as indicators (see Figure 1.3). Although all of the organisms are potentially useful, many monitoring programs have tracked the occurrences of three orders of insects: mayflies (E, for Ephemeroptera), stoneflies (P for Plecoptera), and caddisflies (T for Trichoptera). EPT taxa as a group respond to many different types of pollution and habitat change. Their responsiveness is valued by programs that document biological condition.

Changes in EPT taxa were used in the Willimantic River case study and are used in many of the other examples described throughout this book. However, the use of composite metrics like EPT taxa abundances makes it more difficult to distinguish the relative contributions of different stressors. For this reason, we expect that future biological and causal assessments will evolve toward disaggregating EPT metrics to evaluate whether individual genera or species show distinctive responses.

1.2.2.5.2.2 Derive Evidence   In the first iteration of the Willimantic River investigation, the scientists were able to develop several pieces of evidence. Levels of the different candidate causes were compared between the degraded sites and less-degraded comparison sites to establish whether causes occurred at the location where biota were affected. The levels of different candidate causes were associated with the level of effects to evaluate whether the direction of influence was consistent with expectation. Levels of different stressors were compared with results from laboratory tests and other field studies to evaluate whether the stressors reached levels sufficient to have produced effects in other situations.

1.2.2.5.2.3 Form Conclusions Investigators compared the evidence across the candidate causes and concluded that depleted oxygen, increased ammonia, and forceful flows were unlikely causes. Although the first iteration of assessment did not confidently identify a likely cause, the evidence pointed investigators to a reach of the stream where the effects seemed to begin. There, while resampling the stream, Bellucci and his team found the discharge emanating from a raceway into which a broken pipe was releasing waste from a textile mill.

1.2.2.5.3 After the Causal Assessment

The State moved quickly to reroute the discharge. Three years after rerouting the illicit discharge in the tributary to the Willimantic River, the impaired site reached acceptable biological condition as defined by the State's Department of Environmental Protection. These findings have given confidence to the state agency to apply causal assessment to other rivers and demonstrate that scientific information can be presented in a way that results in management action that improves the environment.

1.2.3 Part 3: Case Studies

Case examples of causal assessments are used throughout the book to illustrate the use of specific methods and approaches (see Table 1.1). The last section of the book provides four examples in greater depth. Three case studies (Long Creek, Clear Fork, and the kit fox) are described in detail to show how the overall approach is implemented to develop evidence and reach conclusions. The application of experimental approaches is highlighted in the case study from the Athabasca River in Canada. Our case studies and examples are admittedly biased toward our work and interest in streams and rivers. However, the principles can be adapted for other systems and places.

A necessary caveat is that all of these case studies are imperfect, reflecting the reality of performing causal assessments under deadlines and with the data that are available or obtainable—the best explanation with the available evidence. However, each of them improved the understanding of how human activities have affected the biota in these ecosystems. Many of them revealed the influence of unexpected factors and suggested directions for management action. We hope they will inspire further work to improve methods, to apply the ideas in new ways, to identify causes in additional ecological systems, and ultimately, to resolve environmental problems.

1.3 Summary

This book describes a strategy and methods for identifying the causes of undesirable biological effects. The strategy identifies the best supported

TABLE 1.1

Case Examples of Ecological Causal Assessments[a]

Location	Undesirable Effect	Likely Causes	Citations
Adirondack Lakes, New York, USA	Fish mortality	Monomeric aluminum	Overview of research provided by Jenkins et al. (2007)
Arkansas River, Colorado, USA	Altered benthic invertebrate assemblage	Metals	Clements (2004)
Athabasca River, Alberta, Canada	Altered benthic invertebrate assemblage	Primarily nutrient enrichment	Chambers et al. (2000); see also Chapter 24
Bogue Homo, Mississippi, USA	Altered benthic invertebrate assemblages	Primarily decreased dissolved oxygen (DO) and altered food resources	Hicks et al. (2010)
California Gulch, Colorado, USA	Altered plant community	Metals and acidification	Kravitz (2011)
Clear Fork, West Virginia, USA	Altered benthic invertebrate assemblage	Sulfate/conductivity, organic and nutrient enrichment, acid mine drainage, residual metals (particularly aluminum) at moderately acidic pH, excess sediment, and multiple stressors	Gerritsen et al. (2010); see also Chapter 23
Coeur d'Alene River basin, Idaho, USA	Kills of swans and other waterfowl	Lead in sediment	URS Greiner, Inc. and CH2M Hill (2001)
Elk Hills, California, USA	Decline in abundance of the endangered San Joaquin kit fox	Predation and accidents	Suter and O'Farrell (2008); see also Chapter 25
Florida Keys, Florida, USA	Coral disease	Human pathogen	Sutherland et al. (2010, 2011)
Groundhouse River, Minnesota, USA	Altered benthic invertebrate assemblage	Deposited and eroding sediment	U.S. EPA and MPCA (2004) and MPCA (2009)
Kentucky River, Kentucky, USA	Fish kill	Deoxygenated water from a bourbon spill	Cormier and Suter (2008); See also Chapter 6

Little Floyd River, Iowa, USA	Altered invertebrate assemblage	Excess sediment and nutrients	Haake et al. (2010a,b)
Little Scioto River, Ohio, USA	Altered benthic invertebrate assemblage	Altered habitat, PAHs, metal, and ammonia toxicity in different stream segments	Norton et al. (2002a) and Cormier et al. (2002)
Long Creek, Maine, USA	Altered benthic invertebrate assemblage	Primarily decreased dissolved oxygen and altered flow regime; secondarily decreased large woody debris, increased temperature, and increased toxicity due to salt	Ziegler et al. (2007a); see also Chapter 22
Pigeon Roost Creek, Kentucky, USA	Altered benthic invertebrate assemblage	Primarily nutrient enrichment and sediments	Coffey et al. (2014)
Salinas River, California, USA	Altered benthic invertebrate assemblage	Primarily increased suspended sediments and altered physical habitat as influenced by suspended sediments	Hagerthey et al. (2013)
Touchet River, Washington, USA	Salmonid decline, altered invertebrate assemblage	Excess sediment	Wiseman et al. (2010a,b)
Willimantic River, Connecticut, USA	Altered benthic invertebrate assemblage	Primarily a toxic effluent; secondarily sediment, altered food resource, increased temperature, mixed urban runoff	Bellucci et al. (2010)
Wilsonville, Oregon, USA	Bee death	Neonicotinoid pesticide	See Chapter 4

[a] This table summarizes case studies used in this book, listed in alphabetical order by location. Additional examples can be found on CADDIS (U.S. EPA, 2012a) and by using the term "Stressor Identification" in on-line searches.

cause or causes by weighing evidence for and against each candidate cause among a set of alternatives.

Our aim with this book is to show how scientific data and expertise can produce credible, defensible conclusions about causation. A thoroughly implemented causal assessment can direct resources and data collection toward the most important questions, increase confidence in management actions, and help communicate the rationale for those actions to the public.

Any strategy only supplements substantive knowledge. The subject areas of environmental science, biology, ecology, toxicology, and statistics provide the foundation for hypothesizing how effects could be caused and for judiciously interpreting results from sampling programs, toxicity tests, and other studies. By observing events through a causal lens, we can improve our understanding of how the world's natural systems operate, how they are degraded by human actions, and how they can be better protected and restored.

2

What Is a Cause?

Glenn W. Suter II

This chapter reviews how concepts of causality have been defined throughout history and how they have influenced the approaches and methods described in this book.

CONTENTS

Everyone, it seems, has an opinion about the best way to assess causes. Colleagues may throw around unfamiliar terms such as counterfactual or mention philosophers like David Hume. This chapter shows that causation is a surprisingly diverse concept that can be legitimately addressed in various ways. If you are already familiar with some of the controversies concerning causation, this chapter will show you how our understanding of the issues has led us to our methodology. This conceptual history may help you to think more deeply about causation and form your own opinions.

This book presents a historical overview of causation in two parts. This chapter reviews concepts primarily from philosophy concerning questions like "What do we mean that something caused something else?" (metaphysical questions) and "In what sense can we say that causes exist?" (ontological questions). The historical review is continued in Chapter 3, with concepts primarily from practitioners and encompasses questions like "How do we know something is a cause?" (epistemological questions). These causal concepts are presented in the order in which they appeared in the literature and their chief advocates and important critics are cited. Readers with a purely practical interest can skip to the methodological Chapter 6. And those intrigued by where they fit in the lineage as causal assessors can get an introduction to the major contributors in Tables 2.1 and 2.2 and more information in CADDIS.

It is important and surprisingly difficult to nail down what is meant by causation. The major causal concepts are presented in this section and defined in Table 2.3. You are likely to find all of these concepts to be relevant to some use of the word "cause" in your scientific endeavors or your daily life.

2.1 Causes as Agents

Prior to the development of materialistic natural philosophy by the ancient Greeks, people believed that things are caused by conscious agents (gods, humans, spirits, animals, etc.) (Mithen, 1998). Hence, causal explanation was a matter of assigning responsibility (Frankfort, 1946). Cognitive scientists refer to this tendency, which is still with us, as agency detection. Although Aristotle, Plato, and other Greek philosophers addressed causation more formally, they were still primarily concerned with metaphysical questions. In particular, they were concerned with not only the agent that induced the effect (the efficient cause), but also the purpose (*teleos*) which is the final cause (Mittelstrass, 2007). Agent causation is still an important concept (only things can affect other things), but the purposeful, teleological version is now the domain of psychology, theology, and criminal law (which seeks evidence of a motive as well as evidence of means and opportunity).

2.2 Causes are Necessary and Sufficient

Galileo Galilei provided the first modern and scientific concept of causation. He wrote in the *Dialogues Concerning Two New Sciences* (1638), "That and no other is to be called cause, at the presence of which the effect always follows, and at whose removal the effect disappears." He was arguing that a cause is *necessary* and *sufficient*—never E without C, and always E when C. This is a physicist's concept of causality and applies to the sort of simple systems, such as weights applied to levers, that Galileo investigated. For example, an

TABLE 2.1

An Abbreviated Time Line of the Philosophy and Science of Causality

When	Who	What
>500 BCE	Pre-ancient philosophers	Believed that causation was a matter of identifying the agent responsible (person, god, spirit, etc.)
427–347 BCE	Plato	Wrote that things are as they are because they participate in a form, but the forms are eternal and uncaused. Plato relied on reason to explain why things come into being and pass away
384–322 BCE	Aristotle	Classified causes as material (the statue is of marble), formal (to resemble an athlete), efficient (it was carved by a sculptor), and final (to earn a fee). Together, they explain what caused something to be the way it is
300 BCE–1500 CE	Roman and Medieval philosophers	Provided commentaries on Plato and Aristotle
1564–1642 CE	Galileo Galilei	Defined the first scientific theory of causation. Included necessary and sufficient conditions and manipulationist causation
1561–1626 CE	Francis Bacon	Described a scientific theory based on inference from positive and negative instances and, particularly, from elimination by failed predictions
1642–1727	Isaac Newton	Believed that causes must be *verae causae*, known to exist in nature (i.e., based on evidence independent of the phenomena being explained)
1632–1704	John Locke	Founded empiricism and the empirical epistemology of causation (causation is something we perceive rather than an ideal or entity); followed by Berkeley and Hume
1711–1776	David Hume	Stated that logic and evidence cannot prove causation. We accept causation based on observed patterns of association and the assumption that the future will be like the past
1724–1804	Immanuel Kant	To bridge the gap between rationalism and empiricism, Kant posited that human perception is filtered through innate categories of ideas. Hence, "Every event is caused" is a synthetic a priori truth that we apply to perceptions
1749–1827	Pierre-Simon Laplace	Deterministic causality: if we knew the state of the universe at a moment and had sufficient knowledge of natural laws and sufficient computational capability, we could predict all future states
1792–1881	John Herschel	Proposed five characteristics of causal relations: (1) Invariable antecedent of the cause and consequence of the effect, (2) invariant negation of the effect with the absence of the cause, (3) increase or diminution of the effect with the increased or diminished intensity of the cause, (4) proportionality of the effect to its cause, and (5) reversal of the effect with that of the cause

continued

TABLE 2.1 (continued)

An Abbreviated Time Line of the Philosophy and Science of Causality

When	Who	What
1806–1873	John Stewart Mill	The first to argue that only by manipulation (experiments) can causation be differentiated reliably from association. Two methods for identifying causes: Method of agreement—what is present in all cases of the effect? Method of difference—what distinguishes cases of the effect from other cases?
1843–1910	Robert Koch	Developed Koch's postulates, a set of criteria for determining the pathogen causing a disease. (1) The microorganism must be shown to be consistently present in diseased hosts. (2) The microorganism must be isolated from the diseased host and grown in pure culture. (3) Microorganisms from pure culture must produce the disease in the host. (4) Microorganisms must be isolated from the experimentally infected host, grown in culture, and compared with the microorganisms in the original culture. (The fourth step is often considered optional.)
1857–1936	Karl Pearson	Probabilistic causation—all knowledge of causation is captured by correlation
1872–1970	Bertrand Russell	Argued that physical laws make the concept of causation unnecessary
1890–1962	Ronald Fisher	A falsificationist who provided a method for probabilistically rejecting a null hypothesis in experiments. He allowed acceptance of a causal hypothesis by assuming that if the null hypothesis is rejected, and there is only one causal alternative it can then be accepted. He never accepted the causal link between smoking and lung cancer
1894–1981	Jerzy Neyman	Formalized Peirce's concept of confidence and confidence intervals and, with Egon Pearson, developed hypothesis testing by contrasting null and alternative hypotheses
1889–1988	Sewall Wright	Published the first causal network model and developed path analysis to quantify it
1897–1991	Austin Bradford Hill	Presented nine "considerations" for causation. He stated that they answer the question: "What aspects of this association should we especially consider before deciding that the most likely interpretation of it is causation?" His considerations are still commonly used and are called Hill's criteria. They are an expansion of the criteria developed by the U.S. Surgeon General's Advisory Committee on Smoking and Health
1902–1994	Karl Popper	Strong falsificationist—one can only tentatively accept the causal hypothesis that has withstood the strongest tests

TABLE 2.1 (continued)

An Abbreviated Time Line of the Philosophy and Science of Causality

When	Who	What
1891–1953	Hans Reichenbach	Developed the common cause principle—a correlation between events E_1 and E_2 indicates that E_1 is a cause of E_2, or that E_2 is a cause of E_1, or that E_1 and E_2 have a common cause. The principle has been abbreviated as "no correlation without causation"
1905–1997	Carl Hempel	Formalized the covering law concept of causal explanation—a phenomenon requiring an explanation is explained by premises consisting of at least one scientific law and suitable facts concerning initial conditions
1965	J. L. Mackie	Developed a formal theory of multiple causation—C is a cause of E if and only if: (1) C and E are both actual, (2) C occurs before E, and (3) C is an INUS condition, where INUS conditions are Insufficient but Necessary parts of Unnecessary but Sufficient set of conditions
1970	Patrick Suppes	Formalized the probability raising theory of causality in which C is identified as a *prima facie* cause if: (1) C precedes E, (2) C is real [i.e., $P(C) > 0$], and (3) C is correlated with E or raises the probability of E [i.e., $P(E \mid C) > P(E)$]. In addition, the relationship must be nonspurious. The theory holds for well-designed experiments
1973	David Lewis	Formalized and promoted the counterfactual theory of causation (had C not occurred, E would not have occurred)
1973	Mervyn Susser	Modified and clarified Hill's criteria and added a scoring system
1974	Donald Rubin	Developed the potential outcomes theory of causality for observational studies in which the effect is defined as the difference between results for two or more treatments of a unit, only one of which is observed. Various statistical techniques are used to estimate those differences, based on the observed outcomes and covariates
1977	Fredrick Mosteller and John Tukey	In their classic text on regression analysis, recognized that regression models do not demonstrate causation. They suggested that the following ideas are needed to support causation: (1) consistency, (2) responsiveness, and (3) mechanism
1979	J. D. Hackney and W. S. Linn	Adapted Koch's postulates to diseases caused by chemicals
1986	Kenneth Rothman	Argued that epidemiology cannot identify causes by statistics or criteria. Sufficiency of evidence should be identified by expert panels or by decision-makers
2000	Judea Pearl	Popularized causal analysis based on directed acyclic graphs. "Y is a cause of Z if we can change Z by manipulating Y" in a graphical network model

continued

TABLE 2.1 (continued)

An Abbreviated Time Line of the Philosophy and Science of Causality

When	Who	What
2003	Nancy Cartwright	Advocated causal pluralism—no theory of causation accounts for all uses of the concept
2004	A. M. Armstrong	Espoused singularist causation—we can identify causes in cases but not in general
2005	P. S. Guzelian	Created criteria for specific causation: (1) General causation, (2) dose–response, (3) temporality, (4) alternative cause (no confounders), and (5) coherence
2007	Phillip Wolff	Based on psychological experiments, Wolff argues that people infer causation from apparent physical interaction, not regular association
2007	Frederica Russo and Jon Williams	Pointed out that only two of Hill's criteria are actually used in most epidemiological studies: (1) Consistency of association and (2) plausible mechanism

Source: Suter, G. W., II. 2012. *A Chronological History of Causation for Environmental Scientists.* http://www.epa.gov/caddis/si_history.html (accessed February 1, 2014).

Note: The order is based on the dates of an author's major contribution. More contributions and greater details can be found in the CADDIS causal history.

electric current is necessary to illuminate a light bulb and every time that current passes through the tungsten filament, it is sufficient to light it.

Necessity and sufficiency have been largely set aside as a definition of causation because most effects can be caused by many things and because sufficiency is context-specific. Mill (1843) recognized that problem and argued that a necessary and sufficient cause is ideal but often unattainable. Mackie (1965, 1974) recognized that the problem came from the fact that many effects have multiple causes (plural causality) and each cause may have multiple components (complex causes). To describe this situation, he developed the concept of INUS (Insufficient but Necessary parts of Unnecessary but Sufficient) set of conditions. (A simpler acronym used in legal argument is NESS, a Necessary Element of a Sufficient Set.) For example, stream invertebrates are killed by hydrocarbons in storm water (an insufficient condition) that are activated by UV light (the sufficient set). The set would not kill without the hydrocarbons, so they are a necessary part. However, this set is unnecessary to kill stream invertebrates because other sets of conditions also can kill them.

Mackie recognized that we will not specify all members of the sufficient set; some must be treated as background. He called those unspecified conditions the "causal field." The INUS formulation is intuitively appealing and heuristically useful but can, like other versions of the necessary and sufficient definition of causation, lead to logical failures in some cases (Cartwright, 2007; Pearl, 2009).

How does this relate to ecological causal assessments? The idea that causes are sufficient and that some causes are necessary is helpful

TABLE 2.2

Applied Ecologists and Causation

When	Who	What
1979	Walter Westman	Pioneered the application of path analysis and multivariate statistics to complex ecological causation involving pollutants
1987	James Woodman and Ellis Cowling	Adapted Koch's postulates to effects of air pollution on forests
1990	Peter Chapman	Developed the sediment quality triad, a combined condition and causal assessment based on three standard types of evidence, to determine whether contaminants are causing adverse biological effects
1991	Robert Peters	Argued against causal analysis in ecology and in favor of predictive empirical modeling
1991	Glenn Fox	Advocated the use of Susser's causal criteria in ecology
1993	Glenn Suter	Adapted Koch's postulates to pollution effects, in general, and applied qualitative scoring to types of ecological evidence when Koch's postulates could not be met. Applied the approach to contaminated sites
2000	U.S. EPA (S. M. Cormier, S. B. Norton, and G. W. Suter)	Developed the *Stressor Identification Guidance* to determine the causes of specific biological effects in aquatic ecosystems. It includes three inferential methods: elimination, diagnosis, and strength of evidence. The strength of evidence method was inspired by Susser but highly modified. It has been further modified and expanded in the CADDIS technical support system
2002	Michael Newman	Argued that "belief in a causal hypothesis can be determined by simple or iterative application of Bayes' theorem"
2002	Valery Forbes and Peter Calow	Proposed seven causal criteria for ecosystems applied as a sequence of yes/no questions
2004	Wayne Landis	Advocated his relative risk model (subjective ranking of links in a network model) for ecological causal analysis to replace Hill's criteria and Chapman's triad
2005	Dick de Zwart and Leo Posthuma	Demonstrated a screening causal analysis using multivariate linear statistical models for a river basin to diagnose the causes of individual taxon abundances at specific sites with habitat variables and toxicity as the possible causes
2008	IPCC	Concluded that climate is the cause of an effect if: (1) the trend is consistent with that expected if temperature were the cause, (2) the change spatially co-occurred with increases in temperature, and (3) alternative causes are eliminated
2010	Susan Cormier	Described how the Hill/U.S. Surgeon General considerations are a mixture of causal characteristics, sources of information, quality of the information, and inference. She developed a system of separate characteristics, sources, and qualities
2012	Richard Norris and colleagues	Described a method for synthesizing the results of multiple studies to evaluate the degree of support for questions of cause and effect. Individual study results are weighted based on study design and replication. Results are combined using a system derived from Hill's and the U.S. Surgeon General 's considerations

TABLE 2.3

Concepts Related to Causation

Concept	Definition
Agent	Causes are things that act upon other things
Necessity and sufficiency	A cause of the effect is whatever is necessary (i.e., the effect never occurs without the cause) and sufficient (i.e., the effect always occurs when the cause occurs)
Covering law	Effects occur as a result of natural laws
Regular association	Causes occur before and in association with their effects
Events	Causes are events that induce other events
Characteristics	Whatever, in a particular set of circumstances, displays the characteristics of causation is the cause
Manipulation	A cause is something that, if manipulated, will change the effect
Probability raising	The cause is whatever increases the probability of an effect
Process connection	A cause is a process that induces the effect
Counterfactuals	Had cause C not occurred, effect E would not have occurred, therefore, C must be a cause of E
Pluralism	Causes are different things depending on the nature of the relationship and evidence

when weighing evidence that is equally compelling for several causes. It reminds us that some causes must act jointly in order to be sufficient to cause of the effect.

2.3 Causes as Natural Laws

Newton extended Galileo's causal concepts by providing the basis for the covering law theory of causation. That is, the natural laws that he developed seemed to reliably encompass natural causal phenomena. If an observation such as a falling object is covered by Newton's law of gravitation, then gravity is the cause. However, he did not see his laws as causes. In the *Principia* (1687), he argued that causes must be *verae causae*, known to exist in nature (i.e., based on evidence independent of the phenomena being explained). His advice against posing hypotheses (*Hypotheses non fingo*) had inordinate influence, leading physical scientists to largely abandon causes in favor of mathematically formulated laws.

Newton's contemporary, Gottfried Leibnitz, argued against Newton's theory of gravitation which was all mathematical law and no physical mechanism. Leibnitz stated "The fundamental principle of reasoning is nothing without cause" (Gleick, 2003). However, physical scientists and philosophers of science are generally content with covering laws. Mill (1843) considered natural laws to be the highest category of causal explanation. Bertrand Russell argued, based on the nature of physical laws, that there was no need for causality in science or philosophy. He wrote in *Mysticism and Logic* (1912) that "The law of causality … like much that passes muster among philosophers, is a relic of a bygone age, surviving like the monarchy, only because it is erroneously supposed to do no harm." Hempel (1965) formalized the idea that causation is simply the action of a natural law in relevant circumstances, that is, we infer causal explanations from one or more laws and one or more factual circumstances. This is the covering law or deductive-nomological model of causation.

How does this relate to ecological causal assessments? We believe that events in nature do follow predictable laws and therefore natural laws are useful causal constructs. However, laws are seldom available to causally explain events in the environment—except in trivial cases (e.g., the polluted water flowed between points A and B because of gravitation [the law] and the slope between the points [the fact]). However, natural laws are used in the development of mechanistic environmental models which are used in conjunction with site information to form evidence.

2.4 Causes as Regular Associations

The empirical philosophers of the British enlightenment believed that knowledge comes from experience. Beginning with John Locke and epitomized by David Hume, they developed the associationist theory of causation. They argued that people believe a relationship to be causal based on constant conjunction and lively or vivid impression. Hume's terminology is not consistent, but he expressed causal criteria as: *contiguity*, *priority*, and *constant conjunction*. Hence the definition "an object, followed by another, and where all the objects similar to the first are followed by objects similar to the second" from *A Treatice of Human Nature* (1739). In that way, Hume replaced the concept of necessity in causation with regularity. Furthermore, he made the argument that the cause of a unique event cannot be determined, because there can be no consistent conjunction. Hence, singular causal events must be instances of a general causal relationship.

How does this relate to ecological causal assessments? The cause and effect must be associated, and regular association is evidence. In fact, most of the quantitative analyses performed in causal assessments involve the application of statistics to quantify the regularity of associations. Sometimes regular association can be demonstrated in a case. For example, every time an orchard is sprayed with insecticide, a fish kill has occurred in the stream that flows through it. However, the particular association in a case often does not involve repeated instances. In such cases, an association in the case may be shown to be an instance of regular associations in similar situations.

2.5 Causes as Events

Another view is that causation is the result of an association between events, rather than between an agent and an affected entity. In modern philosophy, Hume's event causation has largely replaced agent causation. The philosopher of science, Bunge (1979) wrote that "the causal relation is a relation among events." For example, the striking of a window by a brick (causal event) caused the breaking of the window (effect event).

How does this relate to ecological causal assessments? Often, events such as oil spills or treatment failures are considered the causes of environmental effects. Using a perspective of events as causation can be useful for developing causal pathways (a series or network of cause–effect relationships leading to the effect of interest) and depicting them in a conceptual model. This perspective can also be useful in resolving the problem once the cause is discovered because the detail in describing events suggests options for reducing or eliminating them. If the series of events is divided into numerous mechanistic steps, event causation becomes a discrete version of process causation (see Section 2.9).

2.6 Causes as Whatever has the Necessary Characteristics

One practical concept for recognizing a causal relationship is that it is whatever, in a particular set of circumstances, displays the characteristics of causation (i.e., the attributes that distinguish a causal relationship). John Herschel, in *A Preliminary Discourse on the Study of Natural Philosophy* (1830), defined five "characteristics" of causal relations:

1. Invariable antecedent of the cause and consequence of the effect, unless prevented by some counteracting cause.
2. Invariate negation of the effect with the absence of the cause, unless some other cause be capable of producing the same effect.
3. Increase or diminution of the effect with the increased or diminished intensity of the cause.
4. Proportionality of the effect to its cause in all cases of *direct unimpeded* action.
5. Reversal of the effect with that of the cause.

Herschel believed that these characteristics were necessary attributes of a true cause but did not prove causation because of the possibility of confounding or interfering agents.* "That any circumstance in which all the facts without exception agree, *may* be the cause in question, or, if not, at least a collateral effect of the same cause ...." Various other lists of causal characteristics have been developed since Hershel. Lists of criteria for judging that a relationship is causal, such as Koch's postulates for identifying pathogens responsible for diseases and Hill's considerations for identifying causes in epidemiology (Hill, 1965), are useful guides to causal inference, although they do not define characteristics of causation. However, lists of characteristics of causation may also be used as guides for inferring causation (Russo and Williamson, 2007; Cormier et al., 2010).

How does this relate to ecological causal assessments? We believe that evaluation of the evidence in terms of a set of considerations, characteristics, or criteria is generally the best method for organizing and weighing multiple pieces of evidence (see Chapter 3).

2.7 Causes as Manipulations

Another perspective is that a cause is something that, if manipulated, will change the effect. Further, in cases of a network of multiple factors that jointly affect E, a manipulationist says that the cause is the thing that is manipulated. Symbolically, we distinguish interventional probabilities $P(E|\text{do } C)$ from the simple conditional probability $P(E|C)$ (Pearl, 2009). John Stuart Mill described how evidence is combined in logical arguments and is the founder of the manipulationist theory of causation since he was the first philosopher of science to clearly argue the priority of experiments over uncontrolled observations: "... we have not yet proved that antecedent to be the cause until we have reversed the process and produced the effect by means of that antecedent artificially, and if, when we do, the effect follows, the induction is complete ...." from *A System of Logic, Ratiocinative and Inductive* (1843).

Fisher (1937) made experimentation a more reliable means of identifying causal relationships by introducing the random assignment of treatments to replicate units, to minimize confounding. However, when we extrapolate from the experimental results to the uncontrolled real world, we run into the same problem of inferring from instances identified by Mill. That is, we have no reliable basis for assuming that the causal relationship seen

---

* A confounding variable is an extraneous variable that is correlated with both the cause and the effect. An interfering agent blocks the effects of an otherwise sufficient cause.

in an experiment will hold in a real-world case. In fact, the uncertainty is greater because experimental systems are usually simplifications of the real world. In addition, because of the complexity of ecological systems, the manipulations themselves may be confounded. For example, some experiments to determine whether biological diversity causes stability have actually revealed effects of fertility levels, nonrandom species selection, or other "hidden treatments" (Huston, 1997).

Contemporary philosophers have avoided the charge that manipulationist theories are circular by treating manipulation as a sign or feature of causation rather than a definition and by allowing natural manipulations and even hypothetical manipulations such as interventions in models (Pearl, 2009; Woodward, 2003).

How does this relate to ecological causal assessments? Our goal is to identify causes that may be manipulated to restore the environment, so our causes are at least potentially the types of causes recognized by manipulationists. Further, manipulations (both experiments and uncontrolled interventions) can provide particularly good evidence of causation. However, we do not require evidence from manipulations to identify the most likely cause.

2.8 Causes as Probability Raisers

A cause can also be viewed as anything that raises the probability of an effect. Although some prior philosophers recognized the importance of chance, Karl Pearson presented the first probabilistic theory of causation in *The Grammar of Science* (1911). Pearson took Galton's concept of "co-relation," developed it as a quantitative tool, and made causation, at most, a subset of it. For Pearson, everything people can know about causation is contained in contingency tables. "Once the reader realizes the nature of such a table, he will have grasped the essence of the concept of association between cause and effect." By this definition, causation is probabilistic consistency of association, and a cause is anything that raises the probability of an effect. Clearly, this definition is unreliable due to confounding and symmetry (the causal relationship is one-way, but correlations are symmetrical, so they do not indicate which of a pair of variables is the cause and which the effect). However, correlation is the most common basis for causal inference in environmental sciences.

How does this relate to ecological causal assessments? Anyone doing causal assessments needs to understand the maxim, "correlation is *not* causation," while simultaneously recognizing the value of correlation as a fundamental tool for exploring data and generating evidence of causation.

2.9 Causes as Process Connections

Ironically, although Russell famously opposed the idea of causality, he attempted to develop a scientifically defensible theory of causation (Russell, 1948). He defined causation as a series of events constituting a "causal line" or process. However, he did not distinguish between causal processes and non-causal processes (Reichenbach, 1956; Salmon, 1984).

The modern process theory of causation was developed by Salmon (1984, 1994, 1998) and Dowe (2000). This view of causation involves an exchange of invariant or conserved quantities such as charge, mass, energy, and momentum. However, causation in many fields of science are not easily portrayed as exchanges of conserved quantities (Woodward, 2003). Numerous philosophers have published variants and presumed improvements on Salmon's and Dowe's process theory. Some psychologists and psycholinguists have adopted a version of the physical process theory of causation and argue based on experiments that people inherently assume that a process connection (their terms are force dynamics or the dynamics model) is involved in causal relationships (Pinker, 2008; Wolf, 2007).

How does this relate to ecological causal assessments? All in all, this is a very mechanistic way of thinking about causation and is satisfying when enough data and knowledge exist to describe the processes. For the most part, in environmental and epidemiological investigations, there are not enough data and the data typically relate to states, not processes. However, it is helpful to develop conceptual models of processes that could cause the effect, including processes that generate, move, and transform the causal agent and those that determine the susceptibility of affected organisms. This proves to be a useful tool in considering how evidence might be generated and how the overall case can be presented. Furthermore, some of the cutting edge research for analytical methods is inspired by this desire to richly describe at least several steps in a causal sequence and that sequence could be described as a process sequence.

2.10 Causes as Counterfactuals

Counterfactual causation consists of the argument that had C not occurred, E would not have occurred; therefore, C must be a cause of E. Although Hume and Mill described counterfactual arguments, the concept did not catch on until formalized by Lewis (1973). For example, if the daphnids had not been exposed to high concentrations of copper, the daphnids would have lived. It is popular with philosophers because it seems to have fewer logical problems than regular association as an account of causation (Collins et al., 2004). However, there are conceptual objections as well as practical ones.

One problem with Lewis's original alternative worlds approach is that it requires hypothesizing possible worlds in which C did not occur and demonstrating that in every one E did not occur. Clearly, defining an appropriate set of possible worlds presents difficulties because, in general, a world without C would differ in other ways that are necessary to bring about the absence of C, which would have other consequences. Hence, Lewis developed the concept of similarity of worlds and of the nearest possible world. Also, counterfactual accounts of causation can result in paradoxes involving preemption (an intervention that blocks a cause), overdetermination (more than one sufficient cause acting in a case), and loss of transitivity (a relation is transitive if whenever A is related to B and B is related to C, then A is related to C) (Cartwright, 2007). An example of overdetermination follows. If two chemicals are spilled into a stream resulting in lethal concentrations of each, neither one is the counterfactual cause of the subsequent fish kill, because even if one was absent, the other would still have killed the fish. The counterfactual argument is not true; if one chemical was absent, the other would have killed the fish.

How does this relate to ecological causal assessments? This concept is seldom useful for determining a cause, but might be useful for setting up an experiment or thinking about multiple causes. Counterfactuals are the inspiration for controlling confounding by techniques such as propensity score analysis and trimming a data set.

2.11 Causal Pluralism

Since the late 1980s, many philosophers, led by Nancy Cartwright (2003), came to believe that no attempts to reduce causation to a particular definition (counterfactual, probability raising, etc.) could succeed. Therefore, they proposed causal pluralism which has been reviewed and shown to have two distinct meanings (Campaner and Galavotti, 2007; Hitchcock, 2007). (1) The idea of plural causes presented in this chapter is that there are multiple types of causes and of causation (ontological pluralism). That is, causation is a cluster of distinct types of relationships that happen to share a common name. (2) An epistemological view, more relevant to Chapter 3, is that causation can be approached from multiple, potentially legitimate and useful perspectives given different questions, bodies of evidence, and contexts (conceptual or epistemic pluralism). That is, we cannot provide a satisfactory definition of causality that is useful in all instances of causation, but we can identify a practical concept of causality for any instance. Russo and Williamson (2007) argue that epistemic pluralism applies to the health sciences and also subsumes ontological pluralism: "The epistemic theory of causality can account for this multifaceted epistemology, since it deems the relationship between various types of evidence and the ensuing causal claims to be constitutive of causality itself. Causality just is the result of this epistemology."

How does this relate to ecological causal assessments? Causal pluralism does not provide a way to define a cause or a causal relationship. But, it does justify choosing the approach that works best because it recognizes some utility in all of the concepts of causation.

2.12 Summary

None of the concepts of causation adequately describes all of the relationships that people think of as causal, and philosophers who have devoted their careers to causation cannot agree on a definition. Philosophers may enjoy debating the fundamental nature of causation, but what is a practical minded person to do?

We suggest a pragmatic approach that aggregates concepts into a useful view of causation for environmental scientists. Although we draw on all of these philosophical views, we are not simply causal pluralists. Rather, we have a composite view of causation that is useful for environmental problem solving. Causality depends on a relationship between events involving a process connection between a causal agent and an affected entity. The connection is a physical interaction that can be characterized by a mechanism acting at a lower level of organization. A description of a causal relationship is the best explanation that accounts for the evidence. Therefore, when we gather and weigh evidence, our goal is to arrive at the best explanation and then decide whether the evidence is strong enough to establish that causal relationship and predict that our actions will be beneficial.

3

How Have Causes Been Identified?

Glenn W. Suter II and Susan M. Cormier

Many methods for identifying causes have been defined and used by philosophers and scientists. This chapter describes these different methods and how they have contributed to the approach recommended in this book: identifying the causes that are best supported by the evidence.

CONTENTS

No individual analytical technique or inferential method can be used to reliably identify causes in all cases. For this reason, the approach described in this book combines many different methods into an overall approach that identifies the cause or causes that are best supported by the evidence.

This chapter describes the many ways people have identified causes in specific cases, from enumerating associations, to conducting experiments, to comparing model fits. Each method has strengths and limitations, which are better understood by reviewing their origins. Each of the methods has informed our overall approach by suggesting different ways that evidence of a causal relationship can be derived, synthesized, and compared. Our approach employs many methods so that many types of evidence can be derived for many different types of causes in many different settings. Evidence is synthesized using qualitative weights so that all relevant types of evidence can be included. The candidate causes are compared with respect to the full body of evidence so that the cause or causes that are best supported can be identified.

3.1 Identification and Enumeration of Associations

According to Hume (1748), association is the fundamental evidence from which people infer causation. Association in space and time is a requirement of causation because the affected entity must be exposed to the causal agent. Hence, the cause must at least have co-occurred with the biological effect. If the biological effect recurs, the association should occur in each case (unless there are multiple causes of the effect). Further, if the same cause occurs in other cases, then the same effect should occur.

This logic of causal inference from regular associations was formalized by Mill (1843). His method of agreement stated that effects always occur with their causes. His method of difference states that where the effect does not occur the cause also does not occur. This applies to individual cases (specific causation) as well as to general causes. For example, if the concentration of copper in a stream is elevated at locations with few mayfly taxa, that association is evidence that copper is a cause of that effect (method of agreement). That copper is not elevated at unaffected upstream sites is also evidence that copper is a cause in the stream (method of difference).

An association is derived from a set of measurements and their spatial relationship to the effect. Associations can be quantified by counting the numbers of co-occurrences in a single case over space or time. For example, copper concentration is elevated at the affected location every year, while the mayflies continue to be depauperate. More powerfully, independent cases may be enumerated. For example, in 24 cases in which a salmon spawning river is dammed, the associated population declines and that regular association is evidence of causation. That regular association constitutes a general model that can be applied to infer that the association of a salmon decline with damming of a particular river was causal and not just coincidental.

It is tempting to infer causation from only a single vivid association (e.g., the impaired reach begins below a wastewater outfall or dead birds are found on a golf course the day after a pesticide application). We all know that association does not prove causation because coincidences happen. However, in many cases, interpretation of that single vivid association as causal would be correct, even if not absolutely defensible.

How is this related to ecological causal assessment? Association is essential and fundamental evidence. Documenting the occurrence of an association between a candidate cause and an effect may simply involve co-occurrence in the single instance of an effect or the enumeration of instances in which the two are associated. The strength of the association is judged by the number of instances or the magnitude of the difference in the level of a candidate cause between affected and unaffected sites. Lack of association can refute a cause. However, association is weak, positive evidence, particularly in specific cases, because of the possibility that other potential causes are also associated with the effect.

3.2 Probabilistic Associations

Causal associations in complex systems are not invariant, so most causal assessments involve statistical analysis of the relative frequency of spatial and temporal associations between a candidate cause and its putative effect. That is, the strength of a causal relationship is expressed by the probability that the cause and effect are associated. The simplest and most generally useful expression of these associations is the contingency table (see Table 3.1). These frequencies may be converted into probabilities, but frequencies convey the actual basis for the evidence (the number of occurrences of each type of association) and most people find probabilities to be less easily interpreted (Gigerenzer and Hoffrage, 1995; see Chapter 12).

In this hypothetical example, a contingency table is formed from 100 site observations using channelization as the candidate cause and the effect at the study site, ≤3 species, as the contingent effect. This table serves as a general model* of the probability of the effect with and without channelization. The resulting evidence is that the probability is 0.90 of there being ≤3 species for any channelized location including the channelized study site. Therefore, this evidence strongly supports channelization as the cause of ≤3 species at a specific channelized site.

TABLE 3.1

A Contingency Table for the Association between Channelization of Streams and Degraded Biological Condition, Defined as Three or Fewer Fish Species

	>3 Species	≤3 Species	Total	Probability of ≤3 species
Channelized	5	45	50	0.90
Natural channel	30	20	50	0.40
Total	35	65	100	

---

* The term "general model" does not imply that the model is applicable outside its data set parameters, but rather that it describes the capability of a cause to produce an effect (i.e., general causation) rather than an instance of specific causation.

Many candidate causes, such as pollutant concentrations, are not categorical as in the channelized and natural channel categories of Table 3.1. For these, the degree of association is generally analyzed by some form of correlation or regression analysis. Correlation is the basic statistical measure of the degree of association of variables. However, regression lends itself more readily to prediction and verification than does correlation. For example, a certain proportion of genera may be predicted to be affected using the exposure level at the specific site based on a regression model. That predicted effect may or may not be consistent with the effect of interest.

Alternatively, Bayes' theorem could be used to calculate the conditional probability of degraded condition (i.e., ≤3 species) given channelization. Note that the probabilities in Table 3.1 are already conditional on channelization by the design of the study. However, Bayes' theorem provides the conditional probability of the association even in undesigned data. The Bayesian approach is particularly appropriate when there is good prior information such as from previous studies of the effects of channelization on fish species richness. Then the new data update the prior information rather than standing alone.

Many epidemiologists and others who use empirical inferential approaches for causal analysis limit themselves to calculating probabilistic associations. Some ecologists have disparaged the weighing of multiple lines of evidence and advocated Bayesian probabilities or multivariate generalized linear models as, in themselves, adequate and appropriate expressions of causation (de Zwart et al., 2009; Newman et al., 2007). However, correlation is still not causation. Further, restricting the analysis of causation to quantification of the consistency of association, as useful as it is, leaves out many important types of evidence.

How is this related to ecological causal assessment? Analysis of probabilistic associations enables assessors to evaluate the strength of associations and estimate the degree of confounding. Often there are insufficient data to form probabilistic associations using only data from the case, and a wider geographic area is needed to derive a general model.

3.3 Experimentation

Since Mill (and particularly since Fisher), experimental science has been considered the most reliable means of identifying causal relationships. Through random assignment of replicated systems to alternative treatments, confounding can be eliminated and variance among treatments can be differentiated from variance among systems. However, extrapolation from experimental results to the uncontrolled real world introduces problems. No reliable basis exists for assuming that the causal relationship we see in an experiment will hold in a real-world case. Applying experimental results involves extrapolating from the simplified experimental conditions to conditions of the real world. Omitted conditions may affect the outcome. In addition, because of the complexity of ecological systems, the manipulations themselves may alter more than just the cause being investigated (see Section 2.7).

Replicated and randomized experiments cannot be performed on the affected system itself, but it is possible to manipulate the system and observe the results. Most commonly, manipulations not only consist of attempted remediation or restoration actions, but they may also include "natural experiments" such as a temporary shutdown of an industry or quasi-experiments such as artificially shading a stream reach or introducing caged fish. Although confounding and random effects are possible in these unreplicated and unrandomized manipulations, the evidence from manipulations is likely to be stronger than from mere observations.

Experimental approaches that investigate mechanisms of action in a controlled laboratory setting are also important sources of information for causal assessments. Laboratory studies of media from the impaired site can be used in experiments to determine their toxicity and mode of action. Symptoms or a disease can be identified to provide evidence of an interaction with a pathogen.

How is this related to ecological causal assessment? Laboratory experiments are a common source of evidence. For example, we infer from a toxicity test the concentration required to kill 50% of the test organisms (the median lethal concentration, LC_{50}). Based on this, we infer that concentrations at a specific site greater than LC_{50} were high enough to have caused a fish kill. Field experiments are less common but can potentially fill the gap between unrealistic laboratory tests and uncontrolled field observations.

3.4 Statistical Hypothesis Testing

Statistical hypothesis tests are a quantitative technique developed for experimental data to determine whether a hypothesis can be rejected. Some examples are t-tests and analysis of variance. Most commonly, a hypothesis of no effect is tested by determining whether data, as extreme as those obtained in an experiment or more extreme, would occur with a prescribed low probability given that the null hypothesis is true (the agent does not cause the effect). Statistical hypothesis testing was developed by Fisher (1937) to test causal hypotheses, such as does fertilizing with sulfur increase alfalfa production, by asking whether the noncausal hypothesis is credible given experimental results. Neyman and Pearson (1933) improved on Fisher's approach by testing both the noncausal and causal models, but their approach is seldom used.

Fisher's probabilistic rejection of hypotheses became even more popular as Popper's rejectionist theory of science caught on in the scientific community.* Statistical hypothesis testing came to be taught in biostatistics courses as the standard form of data analysis. As a result, Fisher's tests have been applied indiscriminately to test causal hypotheses in inappropriate data sets, including those from environmental monitoring programs.†

Fundamentally, statistical hypothesis testing does not prove a cause or indicate the strength of evidence for a cause. The results are expressed as the *probability of the data* given the absence of a cause rather than the *probability of the cause* given the data. Numerous critiques of statistical hypothesis testing have demonstrated its failings (Anderson et al., 2000; Bailar, 2005; Germano, 1999; Johnson, 1999; Laskowski, 1995; Richter and Laster, 2004; Roosenburg, 2000; Stewart-Oaten, 1995; Suter, 1996; Taper and Lele, 2004). However, many scientists have chosen to ignore or are unaware of these failings. Many are lured with the false comfort of statistical significance. As a consequence, the misuse of statistical hypothesis testing persists.

In field studies, statistical hypothesis tests can be misleading for several reasons. Assumptions of tests usually are not met, as treatments are not replicated or randomly applied (e.g., sewage outfalls are not randomly placed on different streams). Very large sample sizes can find statistical significance in a small, biologically meaningless difference. Small sample sizes or high sampling error may cause a biologically relevant difference to not be statistically significant. An illustrative example is provided in Box 3.1.

How is this related to ecological causal assessment? Statistical hypothesis testing is applicable only to experimental studies in which independent replicate systems are randomly assigned to treatments (e.g., toxicity tests). However, even in those cases, statistical hypothesis test results do not provide the needed exposure–response model. Many cause–effect relationships are unimodal, that is, there is an optimum rather than a monotonic relationship. Hypothesis testing identifies a statistically significant level without elucidating the full range of increasing and decreasing responses. Even at tested levels, statistical hypothesis tests do not indicate the nature or magnitude of effects, only that an effect is or is not "statistically significant."

Observational data, such as those from natural experiments and environmental monitoring studies, are inappropriate for statistical hypothesis testing because treatments are seldom replicated and are not randomly located. Replicate samples from the same location are pseudoreplicates and cannot be used to evaluate the effect of the treatment (i.e., the candidate cause). Pseudoexperimental designs such as Before-After Control-Impact (BACI)

---

* Karl Popper argued that scientific hypotheses can be rejected but not accepted.
† One of the conceptual flaws in this practice was pointed out by Hurlbert (1984), who invented the term "pseudoreplication" to describe the practice of treating multiple samples from a system as if they were from replicate systems. Pseudoreplicates test whether the sampling locations are statistically different, not whether the effects of treatments are different.

BOX 3.1   EXAMPLE OF BEING MISLED BY STATISTICAL HYPOTHESIS TESTING

Consider the application of hypothesis testing to specific causation in two different streams, River A and River B (the names have been withheld but the data are real). Two sites are measured on each stream, one above and the other below a point source (see table below). At both downstream locations, the fish are reduced in number and diversity relative to the upstream location. Data collected from the point sources have high biological oxygen demand, so one candidate cause is low dissolved oxygen (DO) at the downstream locations. Which scenario presents a stronger case for DO causing adverse effects? What can be inferred from each scenario?

River A, Scenario 1	River B, Scenario 2
• DO measured upstream and downstream over 9 months	• DO measured upstream and downstream over 3 months
• Upstream = 9.3 mg/L	• Upstream = 7.9 mg/L
• Downstream = 8.4 mg/L	• Downstream = 4.2 mg/L
• Difference significant at $p < 0.05$	• Difference *not* significant at $p < 0.05$

In Scenario 1: The only thing that can be said is that DO at the downstream site is lower than that at the upstream site. Any good fishery biologist would tell you that the difference between 9.3 and 8.4 mg/L is just not enough to explain the phenomenon, statistically significant or not. Hypothesis tests alone never show biological relevance.

In Scenario 2: There is no significant difference between 7.9 and 4.2 mg/L in this data set, so classical hypothesis tests tell us nothing about causation. In this data set, DO concentrations are more variable at River B than at River A, so a standard *t*-test shows no statistical difference. However, if the average DO at the downstream sites is 4.2, the DO had to be even lower than 4.2 mg/L at times. Our fishery biologist would toss out the statistics and point to the data. From general knowledge, the downstream DO levels are sufficient to cause biological effects. The statistical test is misleading.

designs can reduce—but not eliminate—the likelihood that the study will be confounded (Stewart-Oaten, 1996).

Finally, testing the null hypothesis tells you little to nothing about the strength or likelihood that an association is causal, because all environmental variables considered in a causal assessment have some effect that would be "significant" if enough samples were taken. We are interested in determining the relationship between the cause and effect (e.g., estimating a concentration–response relationship from test data), not in determining whether the data set is sufficient to reject the hypothesis that the cause had no effect. Therefore, when deriving evidence, we find that descriptive statistics and statistical models are more useful than hypothesis tests.

3.5 Networked Associations

Network models represent causation graphically, with nodes representing entities or states connected by arrows that represent the direction of causal influence. In system analysis, nodes represent state variables and arrows represent models of individual causal processes or probabilities of the implied processes. The advantages of network models are that, unlike conventional equations, they convey directionality and make explicit the structure of interactions in multivariate causal relationships. Statistical methods for analyzing causal networks include path analysis, structural equation models, and Bayesian belief networks. A network can also be modeled mechanistically through mathematical simulation (e.g., systems of differential equations, also known as systems analysis (Bartell, 2007). Causal diagram theory provides a formal logic for analyzing network diagrams that can be used to analyze complex causal relationships, distinguish possible causes from noncausal associations, and identify potential confounders (Pearl, 2009; Spirtes et al., 2000; Greenland et al., 1999).

Quantitative analysis of causal networks began with Wright (1920, 1921), who developed path analysis (basically, a combination of directed graphs and regression analysis) to analyze the effects of genes and environment on phenotypes. It was first applied broadly by economists and social scientists, where data sets are often large and include quantification of multiple causal factors. However, the most important technical developments and the most influential texts on causal networks come from the field of artificial intelligence (Pearl, 2009; Spirtes et al., 2000). The quantitative implementation of these networks is performed using Bayesian analysis [Fenton and Neil (2013) provide an accessible introduction]. Statistical analysis in applied ecology has more often been performed by an extension of path analysis called structural equation modeling (see Shipley, 2000 for a clear presentation with biological examples). Network modeling has seldom been applied to the assessment of specific biological effects, because of inadequate volumes of data relative to the complexity of the causal networks.

How is this related to ecological causal assessment? Conceptual models that diagram ecosystem processes (see Chapter 8) provide the foundation for the construction and implementation of formal directed acyclic graphs. Such networks can be quantified to create exposure response models for whole systems using Bayesian belief networks or structural equation models. The diagrams can also help one to identify confounding variables and direct analyses to minimize the effects of confounding (see Chapter 13).

3.6 Diagnosis

Diagnosis is the identification of a cause by recognizing characteristic signs and symptoms. The diagnosis of a disease based on characteristic signs is as old as the practice of medicine. The first fully natural theory of disease and diagnosis comes from the Hippocratic treatises. The current practice of medicine is based on an approach, developed by William Osler in the late nineteenth century, in which a diagnosis is based on an algorithmic analysis of symptoms and the generation of signs through testing. Archibald Garrod extended diagnosis to include individual biochemical and genetic differences. In the last few decades, a theory of diagnosis has been developed within the field of artificial intelligence that is used in diagnostic expert systems (Reiter, 1987). In addition, new diagnostic symptoms are being developed based on genomics, metabolomics, and proteomics.

Diagnostic protocols for plants and nonhuman animals are available in the plant pathology, veterinary, wildlife, and fishery literatures. For example, Beyer et al. (1998a) developed reliable diagnostic criteria for lead poisoning in waterfowl as part of a study of the causes of waterfowl kills in the Coeur d'Alene basin (see Chapter 17).

How is this related to ecological causal assessment? Diagnostic sets of symptoms have been developed for many plant and animal pathologies. As the body of evidence grows, the breadth and reliability of diagnostic symptoms is improving for population and community-level effects (see Chapter 17). At present, diagnosis is seldom possible, but assessments of specific causation can be strengthened by using symptoms in conjunction with other types of evidence.

3.7 Analogy of the Cause

The idea of formal inference from similarities traces back to the ancient Greeks as *analogia*, a relationship between any two things or concepts. Analogy is an inference from the nonspecific principle that things that have a similar structure have a similar function. In modern times, inferences by analogy are used to infer attributes or modes of action of a candidate cause by relating it to a better characterized cause.

Analogy can be symbolically written as follows:

F is similar to *E*
E has attribute *A*
Therefore
F has attribute *A*

For example, because the molecular conformation of estrogens confers a feminizing capability, similarly shaped molecules should have feminizing effects. Therefore, we infer that a particular molecule that has an analogous shape will cause feminization.

Analogy appears as one of Hill's (1965) criteria for causation in epidemiology. However, it has been sharply criticized. "Whatever insight might be derived from analogy is handicapped by the inventive imagination of scientists who can find analogies everywhere. At best, analogy provides a source of more elaborate hypotheses about the association under study; absence of analogies only reflects lack of imagination or lack of evidence" (Rothman and Greenland, 1998). Assessors must be careful to give appropriate weight to inferences from analogies, because it is based on similarity not actuality.

How is this related to ecological causal assessment? Because the term analogy is used in so many ways in the vernacular, we have restricted its use to one type of evidence: inference of expected symptoms or effects based on modes of action from molecular structure or DNA sequences. However, other analogies may be appropriate in particular cases.

3.8 Case-Based Inference and Artificial Intelligence

Case-based reasoning is a formalized logic within the field of artificial intelligence. Like diagnosis, it is based on similarities of effects rather than analogy between similar causes. However, it includes all measured effects in a case and not just a simple standard set of signs and symptoms. Case-based reasoning follows the general process (Harrison, 1997):

Retrieve the most similar case(s) by comparing the new case to the
   library of past cases;
Use the retrieved cases to try to solve the current problem;
Revise and adapt the proposed solution if necessary; and
Retain the final solution in the library of cases.

This case-based technique relies heavily on updating information and models between steps 2 and 3. Examples include diagnostic systems that retrieve past medical cases with similar signs, symptoms, progression, patient characteristics, and other available data concerning cases. The technique also includes assessment systems that determine the values of variables by searching for similar implementations of a model. Case-based systems are very appealing for ecological epidemiology. For example, one might infer that if Stream A resembles all other streams found to be impaired by high temperatures, then Stream A is impaired by high temperatures.

How is this related to ecological causal assessment? To date, case-based reasoning systems for environmental problems do not exist partly because few formal and documented causal assessments have been made. Until a body of knowledge is developed and is accessible, this type of inference is not available as a formal tool for ecological causal assessment. However, in informal causal assessments, it is common for experienced field biologists to infer causation by similarity to prior cases, so the technique seems promising.

3.9 Comparison of Causal Models

Selection of the cause–effect relationship (empirical or mechanistic model) that best explains the data has been advocated by some authors as a means of choosing the best causal explanation (Josephson and Josephson, 1996; Taper and Lele, 2004). The Bayesian version has been called strongest possible inference (Newman and Clements, 2008). When applied to mechanistic models, the approach has been termed model-based inference (Anderson, 2008; Hillborn and Mangel, 1997). This approach develops models for each of the candidate causal relationships and compares them using sums of squares (simple goodness-of-fit statistics), likelihood ratios (conventional statistics), Bayesian probabilities (if you have good prior information or use subjective judgments), or Akaike's information criterion (the information theoretic approach). To be reliable, these methods require large data sets that include the appropriate metrics for all of the models to be compared. Model comparison can be done in cases such as models of harvesting and climate as causes of the decline in a fishery, because long time series of harvest and climate data are available. If data sets are not all reasonably large and of consistent quality, the comparisons may tell you more about the quirks of the data than about the explanatory power of the models or the causal hypotheses that they represent. Therefore, these methods are more often useful as contributors to a qualitative weighing of evidence than conclusive in themselves.

How is this related to ecological causal assessment? Quantitative abductive inference is the use of statistics to estimate the degree to which models associating each candidate cause and effect (including causal networks) are consistent with the data and use of that single piece of evidence as the decision criterion. This method is still subject to the objection that "correlation (or any other measure of association) is not causation." It is best thought of as a means of comparing models of candidate causal relationships, given a data set that is applicable to all of them. In practice, this approach is rarely feasible, because, except in simple cases, no one data set is complete enough to develop comparable models for all candidate causes.

3.10 Weighing Evidence

When multiple types and lines of evidence are available, causal inference requires weighting the individual pieces of evidence, combining them into an overall weight for each candidate cause and determining which alternative provides the best explanation (see Chapter 19). Weighing multiple pieces of evidence is a pragmatic method and an important tool of inference. C.S. Peirce, a founder of pragmatism, argued that any single line of reasoning, like a chain, is likely to fail due to a weak link, so science should be like a cable spun from multiple strands (Berstein, 2010). William James, pragmatism's other founder, introduced the term "pluralism" to the English language to describe the application of multiple methods to analysis of a case (Menand, 2001).

The weighing of evidence is particularly important for evaluating which candidate cause is best supported by a diverse body of evidence (Box 3.2). In many cases, it is not clear what causal explanation best accounts for the evidence. A cause may appear to explain the evidence because it is true or because of coincidence, technical errors, or some unknown factor. How applicable is that laboratory test to the field? Is the association in that field study representative of a general causal relationship? Are the results consistent with a more general theory and with theory in related fields as well as with the evidence (i.e., is it consilient—Whewell, 1858; Wilson, 1998)? Is

---

BOX 3.2   WEIGHTING AND WEIGHING

These terms can be confusing because weight is used as a noun and as a verb and it sounds like the verb to weigh. We refer to the process of scoring the importance of a piece or category of evidence as weighting. Evidence with higher weight exerts more influence on the final conclusion. After multiple pieces of evidence supporting or weakening a candidate cause have been weighted, the body of evidence is evaluated as a whole based on the constituent weights. The result of that weighing process is the weight of evidence. Our approach for weighting and weighing is described in Chapter 19.

there a mechanism that explains the results? Is it confirmed by independent investigations and is that confirmation insensitive to conditions that should be extraneous? This is where the various conceptual tools of science must be deployed to go beyond an individual experiment or observational project. These issues are discussed in detail in Chapter 19.

Weighting and weighing bodies of evidence are often performed implicitly using professional judgment (Weed, 2005) but is best done in some explicit and systematic fashion (Linkov et al., 2009). Weights may be numerical (Menzie et al., 1996) or symbolic (usually some number of +, 0, or − symbols depending on how strongly a line of evidence supports or discredits a candidate cause) (Fox, 1991; Susser, 1986). The pieces of evidence may be organized into types of evidence or causal characteristics for weighting and comparison (see Chapters 4 and 19). The weights assigned to evidence may be combined arithmetically, by ad hoc judgment or by judgment guided by standard considerations and logic (Suter et al., 2000).

How is this related to ecological causal assessment? Weighing evidence is a way of evaluating and synthesizing the evidence to arrive at the best available explanation for the cause of an environmental effect. There is no standard for defining enough evidential weight, but scientists can identify the weightiest of a set of alternative causal explanations (see Chapter 19).

3.11 Causal Criteria

Causes can be identified by determining whether the evidence meets certain criteria. For example, Koch's postulates (aka the Henle-Koch postulates) are a set of three or four criteria (depending on the version) that together constitute a standard of proof for infectious agents as causes of disease (see Table 2.1 and Box 3.3).

The Surgeon General's Committee and Austin Bradford Hill developed criteria to demonstrate that the body of evidence supported cigarettes as a cause of lung cancer (Hill, 1965; U.S. DHEW, 1964). Susser (1986) extended Hill's criteria and added a scoring system. Many other authors, particularly epidemiologists, have developed lists of criteria, but Hill's are the most often cited. Criteria have been adopted and adapted by ecologists for ecoepidemiology (Fox, 1991; Suter, 1990, 1998; U.S. EPA, 1998, 2000a–e). Hill considered his "criteria" to be only viewpoints for considering whether epidemiological associations are causal. Some have argued against the use of criteria as too subjective (Rothman et al., 2008), while others have argued for mandatory criteria (Guzelian et al., 2005). Criteria for assessing causation may be the demonstration of characteristics that are believed to define a causal relationship (see Chapter 4), but more often they are simply types or qualities of evidence (Cormier et al., 2010).

BOX 3.3   KOCH'S POSTULATES

Koch's postulates are a research plan to acquire scientific evidence required to perform a series of inferences that establish the cause of a disease. Koch's postulates were applied to the investigation of coral reef declines in the Florida Keys (Sutherland et al., 2010, 2011). Koch's postulates were satisfied: (1) characteristic lesions were identified, (2) the pathogenic bacteria were isolated from the field, and (3) the lesions were recreated under laboratory and field conditions. Koch's postulates have been adapted to causes other than pathogens, but they work best with a single clear cause (Woodman and Cowling, 1987; Yerushalmy and Palmer, 1959).

How is this related to ecological causal assessment? Clear inferential arguments are necessary for determining the best explanation for instances of specific causation. Hill's criteria are still conventional among human health assessors, but they have not been used to the same extent among ecological assessors. More fundamentally, Hill's criteria are a mixture of a few characteristics of causation (e.g., temporality), qualities of evidence (e.g., strength), and sources of information (e.g., experiment) (Cormier et al., 2010). However, criteria are useful as aids for maintaining consistency within the process of weighing evidence. Therefore, we have attempted to provide more complete and consistent sets of characteristics of causal relationships, types of evidence derived from different sources of information, and qualities of evidence for use in weighting the evidence (Chapters 4 and 19).

3.12 Our Conceptual Approach

Our approach to causal assessment draws on many of the methods discussed in this chapter and organizes them with other concepts into a flexible but powerful method that is the subject of Part 2. After defining the problem and listing candidate causes, the inferential methods described in this chapter are used to derive evidence, rather than to directly determine the cause. Each piece of evidence is weighted and the body of evidence for each candidate cause is weighed. The evidence is then compared across the candidate causes to determine which alternative is best supported. This approach uses all relevant evidence, shows which candidate cause is best supported by the evidence, and indicates how much is known about causation in the case. To facilitate the process, fundamental characteristics of causal processes (building on the concepts described in Chapters 2 and 3) are used to guide the derivation and organization of evidence.

4

Characteristics and Evidence of Causation

Susan M. Cormier, Glenn W. Suter II, and Susan B. Norton

This chapter has three sections. The first section provides a working description of a causal relationship in environmental applications. The second section describes the components and characteristics of causal relationships. The third section introduces common ways of deriving evidence of the causal characteristics.

CONTENTS

Causality is the principle that everything that happens has a cause. The cognitive capacity of humans for making causal connections is documented in infants (e.g., Michotte, 1946; Leslie and Keeble, 1987) and like other emergent properties of the brain is inherited and subject to evolutionary processes* (Lorenz, 1965, 2009; Campbell, 1982; Freeman, 2000; Ruse, 1989; Scarfe, 2012). People's

---

* Evolutionary epistemology refers to natural selection of cognitive mechanisms, such as the ability to form causal connections, and analogously, the evolution of scientific theories that survive selection by the scientific enterprise. Epigenetics is the causal mechanism by which genes bring about phenotypes, including the expression of cognitive abilities to form causal connections.

recognition of causality is reinforced and informed by repeated observations of specific instances of causation, the act of something causing an effect.

This chapter provides a convention for articulating expectations about causal relationships by providing explicit terminology for the elements and characteristics of causal relationships. Causal characteristics are used to suggest ways to develop and interpret evidence. They also provide a useful framework for organizing evidence for easier comparison and for recognizing pieces of evidence that are particularly compelling (see Chapters 19 and 20). Relating the evidence back to causal characteristics provides a useful way to sum up the conclusions of an assessment.

4.1 A Working Description of Causation

In this book, we define causation as a relationship between events involving a process connection between a causal agent (i.e., a stressor) and an affected entity (e.g., an organism).

Different aspects of causation—as an agent, event, or process—tend to be emphasized when causes are described. Agent causation is the simplest and uses nouns and adjectives to describe causes and effects. *Dissolved copper* caused *many* **dead fish**. The cause and effect are indicated by bold italics. Event causation links a preceding event with a subsequent event and emphasizes action rather than things. *Exposing* the fish to copper ions resulted in *killing* the fish. Process causation emphasizes mechanisms and modes of action, that is, one process leads to another ultimately resulting in the effect. In this book, mechanism is defined as a process that brings about the mode of action. The mode of action is the way that the mechanism ultimately affects the entity. For example, binding of copper ions to the gills *disrupts ionic regulation* resulting in low blood sodium and chloride concentrations, which affects blood viscosity (mechanism) which in turn causes *cardiac arrest* (mode of action) in the fish (Grosell et al., 2002).

In a causal relationship, both the agent and the entity are changed (see Figure 4.1). For example, when a parasitic lamprey adheres to a lake trout, the interaction (see arrows in Figure 4.1) of attaching and ingesting blood by the lamprey provides nutrients and energy to the lamprey, while the

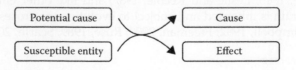

FIGURE 4.1

In a causal relationship, two entities interact in such a way that they are both changed. The interaction between the two entities that changes them is shown by the arrows. For practical reasons, one changed entity is treated as the cause and the other as the effect.

fitness of the trout is reduced. Even when both the lamprey and the trout are perceptibly changing, it is usually the change in the lake trout that is of greater interest or concern. For the purposes of this book, we call the trout before it is changed the *Susceptible Entity* and after it is changed, the *Affected Entity or Effect*. The cause (lamprey) is the thing we want to control so that future adverse events do not occur. We call the entity we want to control (the lamprey) a *Potential Cause* before the effect occurs. This implies that it has the potential to do harm, not that it has. It is called the *Cause* after the effect has occurred. In the diagram, the curved arrows depict the interaction between potential cause and the susceptible entity. After the interaction happens and the susceptible entity is affected, the cause and the effect are so named.

How does this relate to ecological causal assessments? Having common conventions and terminology makes it easier to describe expectations and to develop and evaluate evidence of specific causal relationships. We chose to dissect a causal relationship into its components and characteristics of causation. These characteristics can be used to define expectations of a causal relationship, which in turn are evaluated using evidence.

4.2 Characteristics of a Causal Relationship

We suggest six characteristics that we have found useful for assessing causes. They are co-occurrence, sufficiency, time order, interaction, alteration, and antecedence* (see Table 4.1).

These characteristics reflect systems for identifying causes developed by others, although the terminology varies. Most notably, they appeared in the 1964 report on smoking provided to the U.S. Surgeon General (U.S. DHEW, 1964), popularized by the transcription of an address by Sir Bradford Hill to the British Academy of Science (Hill, 1965). Hill listed nine considerations that are a mixture of types of evidence, sources of information, and types of inference, but we have simplified the list by focusing on the fundamental characteristics of a causal relationship and capturing the qualities of evidence when weighing the evidence (Cormier et al., 2010).

How does this relate to ecological causal assessments? The characteristics are used to articulate expectations of what would be observed if a causal relationship had occurred. Then, evidence is used to determine whether the expected results are obtained and to evaluate whether a relationship exhibits the characteristics of causation.

For example, large numbers of dead and dying bumble bees were reported to the Xerces Society in a commercial area south of Portland, Oregon, on June 18, 2013 (Figure 4.2). One might posit high or low temperatures, pesticides,

---

* Antecedence was called preceding causation in previous publications.

TABLE 4.1

Descriptions of Characteristics of Causal Relationships

Causal Characteristic	Description	What Evidence of a Characteristic Shows
Co-occurrence	The cause co-occurs with the susceptible entity in space and time	The presence of both the cause and the effect and the potential for exposure
Sufficiency	The intensity, frequency, and duration of the cause are adequate, and the entity is sufficiently susceptible to produce the type and magnitude of the effect	Enough of the cause and a sufficiently susceptible entity that can result in the level of the observed effect
Time order	The cause precedes the effect	Change in the entity after interaction with the cause and not before
Interaction	The cause interacts with the entity in a way that can induce the effect	Signs of initiation of the change by the causal agent such as contact or uptake
Alteration	The entity is altered by interacting with the cause	Changes in the entity attributable to or at least appropriate to the cause
Antecedence	The causal relationship is a result of a larger web of antecedent cause-and-effect relationships	Earlier events that led to the particular causal event

Source: Adapted from Cormier, S. M., G. W. Suter II, and S. B. Norton. 2010. *Hum Ecol Risk Assess* 16 (1):53–73.

FIGURE 4.2
Image of the high density of dead bumble bees on the pavement of a store parking lot in Wilsonville, Oregon. Between 25,000 and 50,000 bumble bees and 300 wild colonies were estimated to be killed. (From Rich Hatfield, Conservation Biologist, The Xerces Society, used with permission.)

chemical fumes, or other candidate causes, but for illustration purposes we have focused on one, a pesticide, which was the actual cause of the bumble bees' deaths.

If a pesticide were the cause of a massive number of bumble bee deaths, one might expect to find that a pesticide had been applied to flowering trees in the area (antecedence) and that bumble bee deaths occurred after contact with the flowers and not before (time order). Furthermore, one would expect that dead bumble bees would occur where the pesticide was sprayed and not where it was not sprayed (co-occurrence) and applied at levels known to cause death in insects (sufficiency). In fact, evidence of these characteristics was established and the pesticide was identified as a neonicotinoid with an exposure route through sap and nectar that causes paralysis and death of insects. As would be expected, the affected bees showed signs of impaired neurological function by falling from the trees and then dying on the pavement below (alteration). Binding of the pesticide to acetylcholine receptors was not measured in the bees but would be expected to occur (interaction). In this specific case, no one disputes that the systemic pesticide was the cause of the effect based on the strong evidence for most of the characteristics of a causal relationship. In presenting the findings, the characteristics did not need to be called out, just the evidence (Case, 2013; Black, 2013), but we might surmise that a general understanding of the characteristics of causation guided what evidence was collected and what was reported by several news groups.

To more fully explore each causal characteristic, the rest of this chapter uses a familiar experimental construct, a toxicity test (see Figure 4.3). A toxicity test is a controlled laboratory experiment in which organisms are

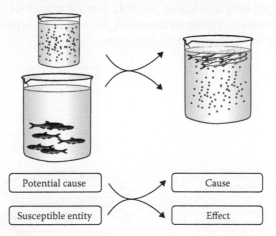

FIGURE 4.3
To illustrate the characteristics of causal relationships, we use a situation that is familiar to many biologists: a toxicity test. Although this book is dedicated to environmental assessments, the simplicity of the toxicity test is a useful illustration. The potential cause is dissolved copper ions depicted as dots. The susceptible entities are young fish in a beaker of water. The cause is copper ions. The effect or affected entity is dead fish floating at the surface.

exposed to a potentially toxic chemical. If the organisms are susceptible and the chemical is toxic at the tested level of exposure, the organisms die or exhibit sublethal effects such as reduced egg production or growth. A toxicity test allows a scientist to witness a causal event as it happens and potentially observe evidence of each causal characteristic. The second part of this book (especially Chapters 9–18) extends these concepts to a broad variety of sources of information and types of evidence.

4.2.1 Co-Occurrence

Because a potential cause and susceptible entity must have interacted to result in an effect, they must have co-occurred in space and time (Hume, 1748). In the laboratory toxicity test, copper that has been dissolved and added to the beaker co-occurs with the test fish (see Figure 4.4). During and after the death of the fish, the expectation is that copper ions in the water could be detected.

How does this relate to ecological causal assessments? When a stressor has caused the effect, we expect that the stressor will be present where and when the effect occurred and will not be present where and when the effect did not occur. The concept of co-occurrence does not require physical contact and may refer to co-occurrence with the absence of something. For example, absence of food, rather than the presence of food, is a cause of death in humans and other animals.

In most cases, measurements are available after the effect occurred, so stressor measurements reflect the presence of a causal agent rather than the potential cause. In the field, the copper measured in the water sample is not the specific copper ions that killed the fish, but we infer that it is representative of the conditions just prior to the effect. Most measurements are approximations of what we wish we could measure.

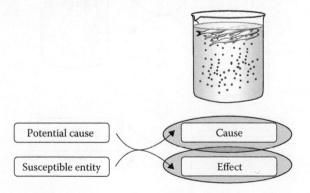

FIGURE 4.4
Co-occurrence: The cause and susceptible entity are collocated in space and time. The gray area in the diagram indicates parts of a causal relationship which usually provide the measurements used to develop evidence of co-occurrence. For example, fish are dead in the beaker (effect) and copper is present (dots).

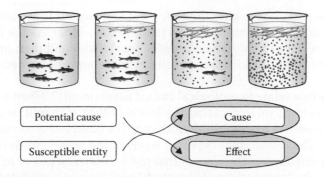

FIGURE 4.5
Sufficiency: The intensity, frequency, and duration of the cause are adequate to produce the magnitude of the effect, given the susceptibility of the entity. In the first beaker, at very low copper (dots) concentration, all fish appear healthy (gray fish). As copper concentration increases, fish begin to die (white fish) until all are dead at the highest concentration in the last beaker.

The expectation of co-occurrence is frequently extended to predict that stressor levels will be greater at locations where the effect occurred than at locations where it did not occur. This is because many causal agents (e.g., dissolved ions) occur naturally and many are necessary for living things. Naturally occurring chemicals are always present at least at very low levels.

Also, time lags and movements of organisms are important considerations when evaluating co-occurrence. One well-known example is the decline of birds of prey after the introduction of DDT. Bioaccumulation occurred over time in prey and then in adult birds, and populations declined later when reproduction failed and young did not replace their parents. Effects may be manifested at later times or at distant places from the original co-occurrence.

4.2.2 Sufficiency

Sufficiency indicates that the magnitude or duration of exposure was enough for the effect to occur. In laboratory toxicity tests, sufficiency is illustrated by a series of tests with increasing concentrations of a chemical (see Figure 4.5).

As a cause interacts with a susceptible entity more frequently, for longer amounts of time, or at greater concentrations or intensities, the magnitude or severity of the effect increases.* For example, dissolved copper at 54 µg/L

---

* Some causal relationships have stressor–response patterns that are not monotonic. For example, unimodal responses may increase and then decrease as stressor levels increase. The response of the mayfly genus *Isonychia* to salinity exhibits this type of pattern: as salinity increases, the capture probability of the mayfly genus *Isonychia* increases occurrence to a maximum then decreases and then no longer occurs (see Figure 12.5).

for 10 days is sufficient to kill salmon fry, but exposure at that concentration for a day is not (given conditions in Hansen et al., 2002). As the heights of dams increase, passage of fish becomes more difficult and then impossible. Less and less cover for kit foxes results in more and more predation (see Chapter 25).

How does this relate to ecological causal assessments? When a cause has produced an effect, we expect that the levels of the cause will be consistent with those known to be capable of producing the observed type and severity of the effect. The relationship between levels of the cause and the effect can be observed directly in experiments that use a dilution series or in field situations where the cause is diluted by natural processes (e.g., pollutant concentrations in a stream diluted below the confluence with a cleaner tributary). Alternatively, the expected levels of the cause that are capable of producing the effect can be derived using models developed from the results of laboratory experiments or field observational studies.

4.2.3 Time Order

Causes come before effects even when writing the phrase, "cause and effect." The characteristic of time order concerns the proper sequence of events. In the laboratory, the proper sequence of events is established when the investigator documents that fish die only after addition of copper to the beaker (see Figure 4.6).

At the spatial and temporal scales of human experience, time does not reverse and effects do not cause their own causes (Salmon, 1984; Dowe, 1991). This asymmetrical nature of our experience with time is termed directionality (Eddington, 1928). Because of directionality, causal relationships have the characteristic of time order. For example, dead fish do not come back to

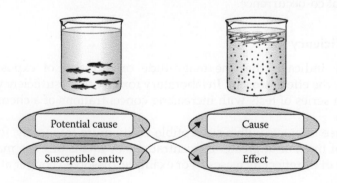

FIGURE 4.6
Time order: The cause precedes the effect. Gray ovals indicate the parts of a causal relationship which usually provide the measurements used to establish the temporal sequence of events. The left beaker contains a small amount of dissolved copper and the fish appear healthy. After more copper is added, the fish die.

life and copper released from them does not return to a jar on the laboratory shelf.

How does this relate to ecological causal assessments? When a cause has produced an effect, we expect that observations of the cause will be coincident with or precede observations of the effect. Developing evidence for time order requires knowledge about the agent and the susceptible entity before and after the event occurs. If conditions have not been observed before and after the event, then an association showing time order cannot be made. However, when available, information about time order is strong evidence, as in the case of the bumble bee deaths.

4.2.4 Interaction

The characteristic of interaction delves deeper beneath the concept of co-occurrence, to the mechanistic processes that initiate the effect. Observations from laboratory experiments can provide evidence of the actual interaction as biochemical markers or chemical accumulation in a target organ (see Figure 4.7). For example, Grosell and Wood (2002) measured copper binding to brachial ionic channels and changes in sodium and chloride concentrations in fish blood (Grosell et al., 2002). In field investigations that begin after an effect has appeared, investigators often have to settle for traces of an interaction that persists such as partially metabolized chemicals (Cormier et al., 2002; Norton et al., 2002a).

The concept of interaction is more than one entity touching the other, or even a mechanism that involves contact. Salmon (1984) described causal interactions as the "transmission or interruption of an invariant or conserved quantity (e.g., charge, mass, energy, and momentum) in an exchange between two causal events." Other authors (e.g., Wolf, 2007) have extended

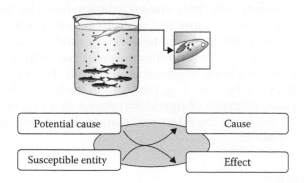

FIGURE 4.7

Interaction: The potential cause and the susceptible entity influence each other in a way that induces the effect. The gray oval indicates the part of a causal relationship which usually provides the measurements used to evaluate results based on expected interactions between the cause and the entity. In the copper toxicity test example, copper ions (black dots) are bound to the gill surface.

the physical concept of interaction to causal interactions such as information transfer (e.g., nucleic acid codes in viral infection), sensing of environmental conditions, or communication. For example, although abrasion of fish gills by suspended sediment is a physical interaction, suspended sediment in a stream which interferes with light penetration and sighting of prey is also a physical interaction.

How does this relate to ecological causal assessments? When a cause has produced the effect, then we expect to observe evidence of exposures and initiation of changes that are part of the process that leads to the effect of interest. As with alteration, the mechanism or mode of action of a candidate cause must be known in order to develop expectations concerning interaction. This knowledge comes from previous work in the laboratory, mesocosm, or field settings. Typical types of evidence that demonstrate that exposure has occurred include pathogens or body burdens of a chemical, molecular binding to receptors such as DNA, or tissue damage.

4.2.5 Alteration

Another characteristic of a causal relationship is that the susceptible entity is changed by the interaction with the cause. Causes change susceptible entities in different ways. Herein lies a means of generating expectations and evidence. When the expected kind of alteration does not occur or does not match the cause, that candidate cause is less likely or rejected. A jagged wound on a manatee is not caused by pesticide toxicity. It is caused by a sharp object like a propeller. During a laboratory toxicity test, specific effects might include fish gasping at the surface or displaying hemorrhagic gills (see Figure 4.8).

How does this relate to ecological causal assessments? When a cause has produced the effect, then we expect that specific manifestations of that effect will be consistent with those known to be produced by the cause. The knowledge of the expected signs and symptoms are typically developed based on laboratory experiments or field observational studies. For some causes, lists of known signs and symptoms have been compiled by experts (see Chapter 17). In this book, evidence of alteration shows changes in the organism that leads to the effect (e.g., paralysis in the bumble bee in Section 4.2). In comparison, evidence of interaction shows that exposure occurred and may or may not be in the effect pathway, such as bioaccumulation in nontarget tissues like fat.

4.2.6 Antecedence

Causes are caused by other causes (Pearl, 2009). That is, each causal relationship is a result of a larger web of preceding cause–effect relationships. Identifying sources and pathways to exposure is a way of putting the

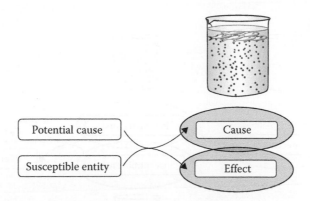

FIGURE 4.8
Alteration: The entity changes by interacting with the cause. Gray ovals indicate the parts of a causal relationship which usually provide the measurements used to develop evidence to evaluate whether expected specific effects are observed. Strong evidence of alteration is provided when a single cause is known to be associated with multiple specific effects, illustrated here by lack of melanin and opaque eyes, which are different from the set of effects caused by other candidate causes, such as bleeding gills.

potential cause and susceptible entity at the scene, so to speak. Such evidence increases confidence that the causal event actually occurred (e.g., that it was not a result of a measurement error or hoax) (Bunge, 1979).

There are typically multiple antecedent processes leading up to a causal interaction. In addition to leading to a proximate cause, antecedents also lead to an entity becoming more susceptible or likely to be exposed. For example, organisms subjected to freezing temperatures may be more susceptible to the effects of metals (Holmstrup et al., 2010). Reduced prey abundance increases time that kit foxes must hunt and are vulnerable to predation (see Chapter 25). Flowers in bloom attract bees.

In the laboratory experiment example, newly hatched fish, which are more susceptible to environmental contaminants than adult fish, were reared for the experiment, copper solutions were created, and then both were added to the beaker (see Figure 4.9). These steps preceded exposure of the fish to the copper.

Because aphids leave residues on parked cars, the store manager hired a contractor to eradicate the aphids. The contractor sprayed linden trees in bloom and foraging bees were attracted to the blooms. These steps preceded the bees drinking the neonicotinoid-laced nectar.

Relationships that precede the causal relationship of interest are causal in their own right and possess the characteristics of causation. Therefore, evidence of the causal events that produced the proximate cause can be based on any of the characteristics of co-occurrence, sufficiency, interaction, alteration, or time order. For example, if depleted oxygen is the cause of a fish kill,

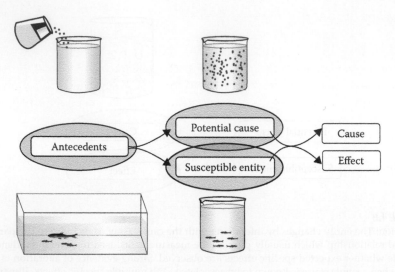

FIGURE 4.9
Antecedence: The specific causal relationship occurs as a result of a larger web of cause-and-effect relationships. Gray ovals indicate parts of a causal relationship which usually provide the measurements used to develop evidence of a causal progression. For example, copper release and then dissolution in water is an antecedent causal event. The copper solution is then in contact with the fish. Events also lead to an entity being susceptible to exposure where the interaction can occur, shown here as the selection of young fish. Evidence of antecedence usually demonstrates a causal progression of events leading to the potential cause and the susceptible life stage being in a situation where the interaction can occur.

evidence could demonstrate how the oxygen became depleted. Perhaps there were *insufficient* riffles to aerate the water. Perhaps an effluent containing organic matter *occurred* at the affected site and was decayed by bacteria and fungi depleting oxygen. Note that the lack of riffles or the organic matter did not harm the fish, but they led to the cause that did.

How does this relate to ecological causal assessments? When a cause has produced the effect, we expect to observe evidence of the causal events that led up to either the occurrence of the cause or increased susceptibility of the entity. Evidence of a more complete causal pathway strengthens the overall case. In some cases, measurements of the proximate cause are not available and evidence of an antecedent may be used as a surrogate for the cause such as organic matter as a surrogate for depleted oxygen in water. When there are appropriate models, the concentration and time of exposure can be estimated for a moving cause or migrating entity. In some cases, antecedents may trace all the way back to the source of the cause or illustrate a connection to a factor that made the entity more susceptible or the cause more toxic. Antecedents that lead to the presence of a cause, for example, emissions from a source, are of particular interest because steps in these pathways are frequently the subject of management action.

4.3 Characteristics and Evidence

The previous sections described several characteristics of causal relationships and how that construct can be used to hypothesize the kinds of results that are expected to be observed when a cause has produced an effect. This section provides a brief overview of how data from field studies, experiments, and models are used to derive evidence. Later chapters develop these ideas much more fully.

Evidence is information used to evaluate whether results expected of a causal relationship have been observed. If Cause X produced Undesirable Effect Y, we would expect to observe Z. The information used to judge whether Z in fact is observed is a piece of evidence. For example, because causes precede effects, when the effluent from a wastewater treatment plant is the cause of an undesirable effect, it is expected that the effect began after the effluent was first released and did not occur beforehand. Actual evidence of when the plant went into operation relative to the observation of the effect should confirm these expectations. When evidence does not match expectations, then these pieces of evidence weaken, or even refute, the effluent as the cause.

As in all scientific enterprises, it is better to establish in advance what kind of evidence would support a candidate causal relationship and what would not. When an effect is anticipated, data can be deliberately collected to develop evidence of any aspect of a causal relationship. This is not possible when the effect is not anticipated. Nevertheless, even if an investigation begins after the causal event and even after the collection of data, results that would support or weaken a candidate cause can be stated before deriving any evidence. Stating expectations is possible because general causation has been described for many environmental causes at least in terms of the type and direction of expected effects (e.g., U.S. EPA, 2012b). Assessors can first describe what is expected based on this more general knowledge, then use the site data to generate evidence, and evaluate whether that evidence is consistent or inconsistent with the expectation. Using this formal process reduces the chance that information will be molded into a false but interesting narrative.

4.3.1 Opportunities to Develop Evidence

Ecological causal assessments typically begin with at least one piece of information: the observation of the undesirable effect at the site under investigation (i.e., the affected site). Evidence is developed when that initial observation (i.e., the effect) is combined with additional data from the case or compared with causal associations developed outside of the particular case. Typical approaches include calculating the magnitude of an association at a site relative to locations without the effect, or comparing case observations with predictions of a model. Some evidence may be qualitative or descriptive in nature, for example, documenting the species of infected coral and symptoms such as the shape of lesions. How evidence is generated (e.g., a photograph, a multivariate analysis, an experiment) depends on the cause, effects, and causal characteristic that one is trying to evaluate, as well as, the source, kind, amount, and availability of data.

Organizing what is known about the case is a useful way to start to develop evidence. Common types of evidence can be derived using only data from the case. Then evidence can be made using data from the case related to more general associations. To do this, established associations about causal relationships are compared with data from the particular case. For example, when a set of symptoms is associated with a pathogen, those known symptoms should appear in the affected organisms in the particular situation if the effects are a result of that pathogen and not when they are the result of something else.

4.3.2 Types of Evidence

In ecological causal assessments, evidence is derived using qualitative and quantitative associations between measurements of effect and exposure. Measurements of effects include presence, absence, abundance, survival, and symptoms. Measurements of exposure include environmental concentrations and physical habitat attributes; biomarkers, body burdens, and physiologically relevant information. Sources of information include field observational studies, laboratory experiments, field tests, and models. The evidence is derived using statistical and logic-based analyses. The expectations prompted by different causal characteristics and the array of data result in a wide variety of possible ways of deriving evidence. For this reason, the types of evidence described in CADDIS are provided as a handy menu of options (see Table 4.2) (Suter et al., 2010a). These types of evidence reflect some of the more common ways evidence is developed, but not the only ways.

The types of evidence in Table 4.2 are organized by causal characteristic(s). It is not necessary to categorize evidence by characteristic or type; however, the two categories provide useful functions. The types of evidence emphasize the approaches used, for example, field observations versus laboratory experiments. The assessor may begin by going through the list of types and asking whether data are available for each type and whether a type is appropriate for the case. The characteristics emphasize the fundamental attributes of a causal relationship. The assessor may go through the list asking what evidence can be generated that illustrates each of the characteristics. Both types and characteristics serve as reminders of the many different ways evidence can be derived. It is not necessary to develop all these types of evidence to infer the best causal explanation for an effect.

TABLE 4.2

Types of Evidence Organized by Causal Characteristic

Type of Evidence	Expected Result If Relationship Is Causal	Example of Type of Evidence
Co-occurrence: The results are expected because potential causes and the susceptible entity both must be present for a causal process to happen		
Spatial/temporal co-occurrence	The cause is observed where and when the effect of interest is observed	Excess fine black sludge was intermittently observed in Buffalo Fork where macroinvertebrate assemblages were degraded and was not observed at reference sites in the watershed (Gerritsen et al., 2010; see also Chapter 23)
Covariation of the stress and effect from the affected site and nearby comparison sites[a]	More stressful levels of a cause are associated with a higher level or frequency of an effect	Where salt concentrations were higher, there were fewer EPT taxa in Long Creek (Ziegler et al., 2007a; see also Chapter 22)
Manipulation of exposure at the affected site	Field experiments or management actions that remove the cause eliminate the biological effect. Introduction of the cause results in the biological effect	In the Cuyahoga River, removal of the dam increased oxygenation and within a few years the fish community had dramatically improved (Tuckerman and Zawiski, 2007)
Time order: The results are expected because a potential cause must co-occur with a susceptible entity before an effect is produced		
Temporal sequence	Measurements demonstrate that the biological effect occurred after occurrence of the cause	Large numbers of fish died when an ethanol spill reached a new area but were unaffected beforehand (Cormier et al., 2010; see also Chapter 6)
Sufficiency: The results are expected because effects occur only when causes interact enough with the susceptible entity to produce a change		
Laboratory tests of site media	Controlled exposure in laboratory tests of contaminated or altered site media induce biological effects consistent with the effects observed in the field	In controlled greenhouse studies, adding lime to raise the pH of fluvial mine-waste deposits from 4.8 to 7.0 decreased exchangeable/extractable metal concentrations and increased growth of willow cuttings in greenhouse experiments (Fisher 1999, Fisher et al. 2000), supporting metals in soils with low pH as the cause of decreased plant growth in the flood plain of the Arkansas River, CO (Kravitz, 2011)

continued

TABLE 4.2

Types of Evidence Organized by Causal Characteristic

Type of Evidence	Expected Result If Relationship Is Causal	Example of Type of Evidence
Stressor–response relationships from laboratory studies	At the affected site, the cause is at levels that have been associated with similar biological effects in laboratory studies	Salmonids were absent from portions of the Touchet River where temperatures exceeded those reported in laboratory tests that caused stress and lethality in salmonids (Wiseman et al., 2010a, b)
Stressor–response relationships from field studies	At the affected site, the cause is at levels that have been associated with similar biological effects in field observational studies	Using a model of EPT richness as a function of percentage sands and fines developed from a database from West Virginia (Cormier et al., 2008), the amount of sand and fines in Pigeon Roost Creek, TN was shown to be sufficient to reduce EPT richness from 13 (baseline from reference site) to 4 at affected sites (Coffey et al., 2014)
Manipulation of exposure at the affected site	Field experiments or management actions at the affected site that increase (or decrease) exposure to a cause increase (or decrease) the biological effect	In stream microcosm studies (i.e., colonized rock trays), water simulating heavy metal concentrations in the affected reaches of the Arkansas River, CO decreased macroinvertebrate abundance (Clements, 2004)
Manipulation of exposure in field tests from other locations	At the affected site, the cause is at levels shown to be associated with similar effects in field experiments or management actions from other locations	After total suspended solids were reduced below a paper mill outfall in the Androscoggin River, the macroinvertebrate assemblage improved. This provided supporting evidence that total suspended solids were the cause of a degraded macroinvertebrate invertebrate assemblage community in the Presumpscott River below a similar pulp mill outfall (U.S. EPA, 2000a)
Stressor–response relationships from ecological simulation models	At the affected sites, the cause is at levels associated with effects simulated in mathematical models of the ecological processes	A population model of the kit fox showed that the observed high level of mortality from coyotes, particularly in young-of-the-year foxes, was sufficient to account for the 30% per year population decline and that reduced fecundity was not a contributing factor (Suter and O'Farrell, 2008; Chapter 25)

Interaction: The results are expected because potential causes affect the susceptible entity by physical processes. The emphasis is on initiation of the effect

| Evidence of exposure | Measurements of the biota show that relevant exposure to the cause has occurred | Metabolites indicative of exposure to polycyclic aromatic hydrocarbons were higher in the bile of white suckers sampled from the affected reach than an upstream comparison site of the Little Scioto River (Norton et al., 2002a; Cormier et al., 2002) |

Alteration: The results are expected because causes only change susceptible entities in a particular way. The emphasis is on the effect

Symptoms[b]	Biological measurements (often at lower levels of biological organization than the effect) can be characteristic of one or a few specific causes	Different corals are subject to different diseases that produce characteristic lesions (Sutherland et al., 2004)
Evidence of related effects[c]	Observations demonstrate that effects expected of the cause have occurred in addition to the effect that prompted the assessment	Coyote predation was a candidate cause of kit fox population declines. Abundances of jack rabbits, another prey species of coyotes, also declined during the same period (Suter and O'Farrell 2008; see also Chapter 25)
Evidence of mechanism or mode of action	Measurements showing that the relationship between the cause and biological effect are consistent with known principles of biology, chemistry, physics, as well as properties of the affected organisms and the receiving environment	Macroinvertebrate drift from deployed colonized rock baskets was strongly related to the concentration of heavy metals. These results indicated that emigration rather than death was the mode of action leading to depauperate benthic invertebrate communities (Clements, 2004)
Analogous stressors	Effects of agents similar to the candidate causal agent include the effects at the site. Usually, effects or symptoms based on mechanisms or modes of action predicted from molecular structure or DNA sequence are observed in affected systems	Deformities of Forster's tern and double crested cormorants in Green Eay exposed to high concentrations of polychlorinated biphenyls (PCBs) resembled chick edema disease associated with the structurally similar dioxin[d] (Gilbertson et al., 1991)

Antecedence: The result is expected because causes are produced by a preceding series of cause and effect relationships

Causal pathway	Measurements or models demonstrate the occurrence of steps in the pathways linking sources to the cause, increased susceptibility of the entity, or conditions permitting interaction of the cause and the entity	The causal pathway from waste water and pulp mill effluent, to nutrient enrichment, to algal abundance, to increased abundance of benthic invertebrates and fish biomass was shown in the Athabasca River, Canada (see Chapter 24)

[a] This type of evidence was called "stressor–response relationships from the case" in previous publications.

[b] Many physicians reserve the term "symptom" for subjective evidence of disease as observed by the human patient and not observed by the physician (e.g., a headache is a symptom, holding head and moaning is a sign). For simplicity, we are following the broader common language definition of symptom: a physical feature which is regarded as indicating a condition of disease.

[c] This type of evidence was called "verified prediction" in previous publications and has been revised because many types of predictions can be verified.

[d] 2,3,7,8-Tetrachlorodibenzo-p-dioxin (2,3,7,8-TCDD).

4.4 Summary

This chapter presents six fundamental characteristics of a causal relationship:

- The cause co-occurs with the unaffected entity in space and time (co-occurrence);
- The intensity, frequency, and duration of the cause are adequate for the susceptible entity to exhibit the type and magnitude of effect (sufficiency);
- The cause precedes the effect (time order);
- The entity is changed by the interactions with the cause (alteration);
- The cause interacts with the entity in a way that can induce the effect (interaction); and
- The causes and their effects are results of a web of causation (antecedence).

The characteristics of causation provide a conceptual basis for setting expectations and planning the collection of data to demonstrate whether or not any candidate causes produced the effect. These expectations are evaluated using evidence. Using causal characteristics as a framework for developing expectations and evidence also benefits the end of the causal assessment process by providing a logical structure for explaining the basis for causal conclusions.

Awareness of the logic involved in a causal assessment empowers an assessor to develop a complete set of relevant and interpretable evidence. Being able to clearly define the logic increases the credibility of both the evidence and the assessor. When a case is novel and routine approaches are lacking, the characteristics can help the deft assessor to move beyond past standard practices and adapt to the new challenge.

The evidence descriptions in this chapter provide many examples of results that would support expectations. However, evidence that weakens the case for a candidate cause can play a pivotal role in constraining the possible explanations and thereby a greater influence on the overall conclusions (Weed, 1988).

The process of deriving and interpreting evidence is a scientific endeavor that relies on technical skills and knowledge. Hence, it is a process conducted that is influenced by errors and biases introduced by the way many people view and process information. Strategies for minimizing these errors are described in the next chapter.

5

Human Cognition and Causal Assessment*

Susan B. Norton and Leela Rao

Understanding sources of errors and biases in causal assessments can suggest strategies for preventing mistaken conclusions.

CONTENTS

Human brains have evolved to process information in ways that influence how evidence is sought, generated, and used. Many recent books have summarized research findings on human cognitive strengths, limitations, and idiosyncrasies and how they apply to topics such as making decisions, judging information, and setting goals (Pinker, 1997; Dawes, 2001; Shermer, 2002; Gladwell, 2007; Lehrer, 2009; Thaler and Sunstein, 2009; Kahneman, 2011). This chapter describes how these findings can be applied to the process of causal assessment.

---

* Adapted from Norton et al. (2003).

Sections 5.1 and 5.2 provide an overview of the challenge. Section 5.1 discusses the human tendency to form initial judgments quickly based on our prior perceptions, rules of thumb, and information readily at hand. Once a judgment is formed, our tendency is to stick with it, filtering out any new information that may undermine initial conclusions. We form narratives to justify our conclusions that make us overconfident in judging accuracy.

Situations that prompt ecological causal assessments also have attributes that can exacerbate cognitive errors (see Section 5.2). Multiple sources and stressors are frequently encountered, many of which covary in time and space. In addition, an observed effect can be produced by different causes. For these reasons, the first or most obvious cause of an effect may not be the actual cause. Data sets are frequently small, preventing the use of statistical techniques to disentangle relationships. The costs and time required to implement management actions frequently prevent systematic experimentation and comparison of the results of different actions to better understand causes.

Section 5.3 describes how the overall approach and specific strategies described in this book help minimize errors and biases. Although the strategies take time and effort, the payoff comes by maintaining trust and confidence both within the investigative team and between the technical staff, managers, and the wider community.

5.1 Cognitive Tendencies that Contribute to Errors in Causal Assessments

Human cognitive tendencies evolved, in part, to efficiently process sensory input and make quick decisions. Quick processing does not always result in biases or errors. However, the quick processing can amplify biases in ways that can lead to high confidence in an erroneous conclusion. High confidence in a mistaken cause can result if a coherent story, is constructed from evidence that was selected either subconsciously or deliberately because it supported a mistaken first impression. Understanding how this sequence can transpire is addressed in the following sections by discussing first how initial judgments can be formed in error (see Section 5.1.1), followed by the development and selection of evidence (see Section 5.1.2), and ending with the formation of conclusions (see Section 5.1.3).

5.1.1 Mistaken First Impressions

Malcolm Gladwell's book *Blink* (2007) describes how quickly we can form opinions. But long before this book, Leo Tolstoy (2007) was well aware of the issue:

Man's mind cannot grasp the causes of events in their completeness, but the desire to find those causes is implanted in man's soul. And without considering the multiplicity and complexity of the conditions any one of which taken separately may seem to be the cause, he snatches at the first approximation to a cause that seems to him intelligible and says: "This is the cause!"

Jumping to conclusions is not a problem in itself. However, it is counterproductive when the wrong first impression delays or prevents understanding of causal processes. It is a part of human nature to form initial judgments without considering all of the information available. Initial perceptions are frequently overly influenced by evidence that is easily retrieved, worrisome, or dramatic, even when it is based on only a few samples (Nisbett and Ross, 1980). Too often, people neglect the overall rate of occurrence of a cause or forget that alternative causes can produce the same effect. Conclusions formed in these ways can start an investigation off in a biased direction (see Table 5.1).

TABLE 5.1

Ways that the Cognitive Tendency to Quickly Form Opinions Can Result in Biases and Errors

Description	Example
Anchoring: A dependency of belief on initial perceptions or estimates	A survey of citizens in the Chesapeake Bay watershed found that many of the respondents thought that stressors responsible for ecological effects in the past are still the predominant causes of degradation (Blankenship, 1994).
Easy representation: Explanations that are easy to envision are favored	An effect is more likely to be attributed to a cause that is discrete and observable (e.g., a point source discharge), than one that is diffuse and not readily apparent (e.g., atmospheric deposition of nitrogen). This may have contributed to public opinions, indicating that local industrial point sources are the sources of degradation of the Chesapeake Bay, rather than nutrient inputs from agricultural, residential areas, and atmospheric deposition (Blankenship, 1994).
Ignoring base rates: Evidence of a cause is considered without factoring in the overall probability of the cause's occurrence	The EPT taxa richness decreases with increasing copper contamination. However, an observed decline in EPT richness should not lead immediately to the conclusion that copper was the cause, because other causes such as siltation are more common.

Source: Descriptions summarized from Nisbett, R. and L. Ross. 1980. *Human Inference: Strategies and Shortcomings of Social Judgement.* Eaglewood Cliffs, NJ: Prentice-Hall; Dawes, R. M. 2001. *Everyday Irrationality.* Boulder, CO: Westview Press.

5.1.2 Filtering Evidence

Seeking out many different types of evidence is an effective way of reaching the right conclusion. Filtering evidence becomes problematic when one hypothesis is favored, evidence that runs counter to the hypothesis is ignored, or alternative explanations are neglected (see Table 5.2). Initial hypotheses greatly influence the search for evidence. After an opinion regarding a hypothesis is formed, investigators tend to collect information that supports their opinion, ignoring evidence to the contrary. Experts are apparently particularly adept at garnering evidence that supports their favored hypothesis (Shermer, 2002). At the extreme, intently focusing on one question can lead to observers completely missing events that are irrelevant to a hypothesis (Mack and Rock, 1998). Even when evidence accumulates to the contrary, a very human tendency is to continue to disregard it, holding on to an initial opinion. In the words of Nassim N. Taleb: "We treat ideas like possessions, and it will be hard for us to part with them" (Taleb, 2010).

TABLE 5.2

Cognitive Tendencies that Lead to Filtering Information

Description	Example
Hypothesis dependence[a]: The hypothesis dictates the types of evidence that will be accumulated to enhance or reduce belief	When investigators believe that biological effects observed in urban streams are due to flow extremes, they will tend to collect hydrologic data and not data for alternative causes such as road salt.
Inattentional blindness: Intently focusing on one event detracts from observing the occurrence of another	Previous studies focused on the effects of polycyclic aromatic hydrocarbons (PAHs) in the Little Scioto River missed the contribution of stream channelization to degraded fish assemblages (Norton et al., 2002a; Cormier et al., 2002).
Confirmation bias: The results of tests or observations that support a preferred theory or hypothesis are favored	Investigators tend to eliminate outliers because they must be wrong, or because they double-check them and find a reason to eliminate them. Data that fit the model are not scrutinized to the same degree.
Hypothesis tenacity[a]: Belief in the validity of a theory (i.e., causal explanation) despite accumulation of evidence to the contrary	Investigators who believed that the decline of peregrine falcons was due to shooting and collection by falconers dismissed evidence of effects of dichlorodiphenyltrichloroethane (DDT).

Source: Descriptions summarized from Kahneman, D. 2011. *Thinking, Fast and Slow*. New York: Farrar, Straus and Giroux; Dawes, R. M. 2001. *Everyday Irrationality*. Boulder, CO: Westview Press; Mack, A. and I. Rock. 1998. *Inattentional Blindness*. Cambridge, MA: MIT Press.

[a] Hypothesis dependency and tenacity are often referred to as theory dependency and tenacity in the literature. We use the word hypothesis here because most causal investigations evaluate specific hypotheses rather than overarching theories.

TABLE 5.3

Cognitive Tendencies that Lead to Unwarranted Confidence

Description	Example
Narrative fallacy: A set of facts that form a cohesive explanation leads to greater confidence despite the quality or abundance of information	Reports of an oil spill occurring just before a fish kill lead to the conclusion that the kill was caused by the spill.
What you see is all there is (WYSIATI): Conclusions are confidently formed based on the information that is available rather than on the information that is needed	Causal assessments often address only the candidate causes that have been measured or counted and included in available data sets.
Overconfidence: Subjective estimates of confidence exceed objective estimates of accuracy	Investigators who have extensively studied a particular cause (e.g., acid rain, toxic contamination) are more likely to be confident in attributing an effect to that cause.

Source: Descriptions summarized from Kahneman, D. 2011. *Thinking, Fast and Slow*. New York: Farrar, Straus and Giroux; Piattelli-Palmarini, M. 1994. *Inevitable Illusions: How Mistakes of Reason Rule our Minds*. Translated by M. Piattelli-Palmarini. Edited by K. Botsford. New York: Wiley. Dawes, R. M. 2001. *Everyday Irrationality*. Boulder, CO: Westview Press.

5.1.3 Overconfidence

Several cognitive tendencies can prevent an accurate assessment of the confidence in conclusions (see Table 5.3). Conclusions are frequently confidently made based on available information, even when that information is an imperfect fit to the question being posed (Kahneman, 2011). When the available information forms a cohesive story, confidence is increased further, even when there is little information or when it is of questionable quality. Most of us consistently overrate our accuracy in making estimates and conclusions. All of these errors can lead to a confident but incorrect conclusion.

5.2 Attributes of Ecological Causal Assessments that Contribute to Cognitive Errors

Ecological causal assessments are rarely conducted when the cause is obvious and the solution clear. Rather, ecological causal assessments are frequently conducted in situations where the effects of multiple, covarying stressors are difficult to discriminate. Conclusions may need to be reached based on sparse data. Furthermore, the time and cost of deploying management actions prevent rapid feedback that could be used to evaluate whether initial conclusions were correct.

5.2.1 Covarying Stressors and Plural Causes

Human activities frequently generate many candidate causes. For example, agricultural soil tillage can be associated with increased insecticides, fine sediments, and nutrients in nearby water bodies. All of these stressors would be expected to covary in time and space.

Distinguishing covarying stressors from causal relationships is important because management actions targeting a false cause will not result in environmental improvement. In some cases, remediating one candidate cause may fortuitously remediate the true cause (e.g., dredging sediments for one contaminant may remove others as well). However, in other cases, it may divert time and resources away from the true cause (e.g., building retention ponds to intercept storm flows in urban areas may not improve aquatic invertebrate taxa richness when increased temperature is the cause of declines).

Covariation can lead to mistaken causal attribution by influencing observed associations between variables (see Table 5.4). The analysis of associations is a cornerstone of causal inference methods. When stressors covary, an association with an effect may be observed even when a factor is not a cause. For example, both reduced base flows and increased stream temperature are expected to occur in urban areas because of reduced contribution to surface flows from groundwater. Effects on stream assemblages would be associated with temperature even when low base flows were the actual cause (see Figure 5.1a).

The situation is more complicated when each of the stressors that covary can cause the biological response (called plural causes, see Chapter 2). The

TABLE 5.4

Challenges from Plural Causes and Covarying Variables

Description	Example
Plural causes: Different causes can produce the same effects. Observing an effect that is typically associated with a particular stressor does not necessarily mean that the stressor is the cause	The richness of EPT taxa typically decreases with increasing metal contamination However, EPT richness also declines with other stresses, such as pesticides, nutrient enrichment, salinity, and habitat. degradation. Concluding that an observed decline in EPT richness was caused by metal contamination would be premature until other stressors that can cause EPT richness declines are also evaluated.
Covarying variables: An association between a candidate cause and an effect may be misinterpreted because of the influence of another factor that covaries with the cause	A biological response of macroinvertebrates may be associated with increased algal production in a set of streams, but the relationship disappears when corrected for stream size.

Source: Descriptions summarized from Dawes, R. M. 2001. *Everyday Irrationality.* Boulder, CO: Westview Press; Glymour, C. 2001. *The Mind's Arrows: Bayes Net and Graphical Causal Models in Psychology.* Cambridge, MA: MIT Press.

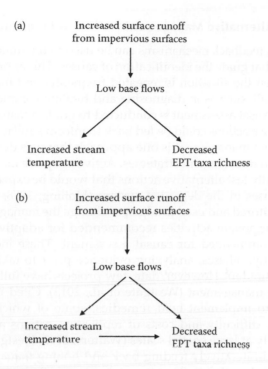

FIGURE 5.1
(a) Temperature will be correlated with EPT taxa richness when it increases with low base flows (the true cause in this hypothetical case). (b) The association between increased low base flows and decreased EPT taxa richness will be overestimated when increased temperature also contributes to decreased richness.

association between the biological response and each factor taken individually will be overestimated, because the estimate will include the influence of the covarying variable. In the example, both low base flows and high temperatures decrease EPT richness, so the association between low base flow and EPT richness will be overestimated (see Figure 5.1b). Natural gradients such as stream size, connectivity, and elevation can also covary with candidate causes. When the natural gradient also produces the effect of interest, the effect may be mistakenly attributed to the candidate cause.

5.2.2 Small Sample Sizes

Sample sizes for site-specific causal assessments are frequently small. Although the small sample size is not a problem in itself, a common cognitive tendency is to assume a small sample size is more representative than it really is. For example, temperate estimates made using a few samples will greatly underestimate the magnitude of temperature spikes. Estimating extreme values with small samples is even riskier for variables that tend to have skewed distributions, like many chemical concentrations.

5.2.3 Testing Alternative Management Actions is Frequently Infeasible

In everyday life, feedback mechanisms can be used to fine-tune intuitions and rules of thumb that guide the identification of causes. This approach is particularly useful when the situation is repeated frequently and the consequences of errors are small, such as in diagnosing and treating common illnesses like colds. When a causal assessment is conducted to guide a management action, the results of those actions could be fed back to inform similar situations.

Active adaptive management is one approach that was developed to provide feedback on management strategies. Active adaptive management projects systematically test alternative actions that would be expected to produce different responses in the degraded system (Holling, 1978; Walters, 1986). Results are monitored and used to adjust or change the management strategy. Many of the component activities recommended for adaptive management projects are recommended for causal assessment. These include developing alternative hypotheses, analyzing evidence prior to taking action, and involving stakeholders. However, date, few projects have fully implemented active adaptive management (Westgate et al., 2013). Cited reasons include the reluctance to implement bold remedies, some of which may be ineffective, and the difficulty and costs of replicating actions and monitoring results, especially at large spatial scales (Walters, 1997; Westgate et al., 2013). The difficulties in iteratively feeding back and honing management actions over time emphasizes the need to accurately identify the correct cause before management action is taken.

5.3 How Does This Relate to Ecological Causal Assessment? Strategies for Minimizing Errors

Awareness of our cognitive tendencies can go a long way toward minimizing errors and biases. One of the advantages of using a structured method for causal assessment, as described in this book, is that it can deliberately incorporate strategies to minimize errors and biases. This section introduces strategies that take advantage of our cognitive strengths while minimizing our limitations. By understanding how different techniques minimize errors, we can better implement them, augment them, and use them to improve the broader community of practice.

Identify and evaluate alternative candidate causes. Dawes (2001) has suggested that incomplete specification of the possible contributors to a problem is the primary reason for cognitive errors. Identification of a suite of candidate causes for evaluation guards against the tendency to selectively seek out information (see Chapter 8). For example, Winger et al. (2005) compared two causal assessments of a degraded stream community. One used the sediment quality triad approach and the other used the rapid bioassessment procedure. The two assessments gave different answers. The sediment quality triad approach examined only sediment contamination and identified it as the cause. The rapid bioassessment protocol examined only habitat disturbance and identified it as the cause. An assessment that addresses all candidate causes allows for a wider range of possibilities, and therefore, a greater opportunity to evaluate whether one cause is dominant or several are interacting to produce the effect. Even when a causal assessment must focus on only one candidate cause (e.g., for regulatory or legal purposes), considering other possible causes can help to prevent an erroneous conclusion.

Derive many different pieces and kinds of evidence. Errors associated with WYSIATI (what you see is all there is) can be combated by collecting additional data on candidate causes and by deriving many different pieces and kinds of evidence. Although any individual causal assessment will use only some types of evidence, familiarity with different ways evidence can be developed can suggest different options and directions for innovating new approaches.

This book discusses many different ways of developing evidence. Chapter 4 provides an overview of evidence from the perspective of causal characteristics. The types of evidence listed in Table 4.2 can provide a handy reference of options that have been used in past causal assessments. Chapters 9–18 provide many examples of evidence developed from case-specific observations, observational studies from other places, field tests, laboratory, and mesocosm experiments. Symptoms and simulation models that use data from many different sources are also discussed.

Conceptual models of causal relationships (see Chapter 8) document how human activities lead to different stressors and also stimulate ideas for generating evidence. Conceptual models articulate expected links in the causal webs of events and show where data can be used to provide evidence relevant to candidate cause. They are particularly useful for linking causes with effects that are distant in time and space, or when complex chains of events are involved. For example, unionid mussels may have low reproductive rates because their host fish species have been extirpated due to intermittent low dissolved oxygen caused by decomposing algae that have bloomed because of increased nutrient inputs. Evidence that establishes the presence or absences of any of these links could be developed.

Seek opportunities to isolate the effects of different causes. Frequently, the goal of site-specific causal assessments is to sort out the influences of multiple covarying stressors. Causal inference is easier when candidate causes vary independently or when only one cause changes at a time (Cheng, 1997).

There are many strategies for isolating the effects of different causes. Defining the problem narrowly in terms of space, time, and specificity of effect is discussed in Chapter 7. Investigators can seek to identify situations where the presence of each candidate cause is the only factor that varies between a comparison site and the site or unit under investigation (see

Chapter 10). Statistical approaches and data set trimming can be used to isolate the effects of a particular cause (see Chapter 13). Experiments (see Chapters 14–16) provide an opportunity to manipulate the factors that vary and to include a closely matched control.

In most causal assessments, it will not be feasible to completely isolate the effects of individual stressors. Conducting the analysis as a comparison of candidate causes provides an additional means for identifying when more than one cause is present and capable of contributing to the observed effects.

Use systems for organizing and tabulating results. Evaluating many pieces of information relating to a suite of candidate causes can quickly become a daunting effort. As the number of alternatives grows, so does the need for information (Churchland et al., 2008). However, human minds can only retain and process about seven pieces of information at a time (Miller, 1956). Breaking up information into pieces or chunks is an effective way to manage this complexity. Isolating each analysis helps prevent cognitive overloading and our tendencies to revert to information that is easily obtained.

The following chapters describe several strategies for organizing information evaluated during a causal analysis to support the fair evaluation of each piece of evidence. At the beginning of an assessment, the development of a conceptual model can provide an overarching structure for organizing the information relevant to each candidate cause and the data relevant to each (see Chapter 9). Data are analyzed and evidence is derived for each available information source (see Chapters 9–18). As each different piece of evidence is developed, results are recorded and placed aside. Evidence can be organized according to the causal characteristic it supports, the source of data, or a combination of the two (see Chapter 4).

An interesting and useful consequence of breaking up the analysis and evaluating each piece of information independently is that it can delay the synthesis of a final conclusion, thereby reducing biases associated with hypothesis dependency. In an experiment studying visual cognition, subjects who were prevented from forming early conclusions more accurately identified a picture's subject (Bruner and Potter, 1964). Isolating the analysis of each piece of evidence can prevent the conclusion reached from one analysis from influencing another, helping investigators fairly evaluate the evidence for each cause.

Evaluate and compare alternatives before developing the story. A high degree of confidence that a causal conclusion is correct is produced by using many different pieces of high-quality evidence that together provide the best explanation of the available observations. It can be useful to think of confidence and accuracy as the result of two different processes. Accuracy in causal conclusions is provided by synthesizing many different pieces of high-quality evidence. Confidence in conclusions is increased when the evidence forms a cohesive story. A cohesive story can be so compelling that it prevents a fair look at alternative explanations. The strategy recommended in this book is to weight and compare the evidence developed for all candidate causes (see Chapter 19) before assembling the evidence into a narrative (see Chapter 20).

Involve other people. No one thinks exactly the same way or shares exactly the same opinion. Involving other people with different viewpoints can increase the chance that an error will be detected. The scientific review process asks peers and experts to provide objective insights and detect unintended biases, faulty inferences, and overlooked issues. Because ecological causal assessments draw on so many scientific disciplines, outside perspectives can provide insights for a deeper knowledge of the patterns that have been recognized and point out patterns that were not detected. Without these colleagues, you must provide your own counterpoint and this is hard to do. Argument with a critic who also shares the same objectives or wants to solve the same problems can be an effective way to reveal unsupported assumptions and other weaknesses.

Displaying evidence and the logic behind conclusions reduces problems associated with hypothesis tenacity and over confidence. It makes it easier for investigators and their reviewers to see when a candidate cause is being fairly vetted. Displaying evidence makes data gaps obvious, decreasing the influence of hypothesis dependence on the collection of additional data.

The involvement of interested and affected parties (i.e., stakeholders) in causal assessment can vary from informal conversations to required consultations, depending on the decision context. Although describing strategies for effective engagement is beyond the scope of this book, we recognize that involving stakeholders in causal assessments has many benefits. In the early phases of the assessment, stakeholders may be able to suggest candidate causes and may know of sources of data. At the end of the process, discussing the findings of the assessment with the wider community increases understanding of the basis for action and is respectful to those affected by the decisions of the assessment. Involving stakeholders can increase the level of trust in conclusions. In turn, these stakeholders can provide the continuity of attention needed to see a management action through to fruition.

5.4 Summary

Recognizing the ways that human cognitive tendencies in typical causal assessment situations can lead to erroneous conclusions is the first step toward minimizing their effects. In short, the occurrence of covarying causes can lead to mistaken first impressions. The impressions can quickly become opinions that limit the range of evidence that is developed, producing explanations that appear to be cohesive but omit important pieces of evidence. The difficulty of iteratively conducting management experiments prevents feedback that could be used to adjust actions and improve the accuracy of causal conclusions.

Most causal assessments are conducted under circumstances that allow at least some deliberation. This provides the opportunity to follow a structured process and implement strategies for preventing mistakes. Overcoming biases and errors associated with our cognitive processes requires the discipline to prevent leaps to judgment, strategies to ensure that our minds are open to alternative explanations, and the diligence to ensure that all alternatives receive fair treatment and consideration.

Part 2

Conducting Causal Assessments

With the foundation laid in Part 1, Chapter 6 starts Part 2 with an overview of the process for ecological causal assessment. It is followed by chapters describing how to formulate the problem, how to derive evidence, and how to form conclusions. Part 2 concludes with chapters that discuss communicating the findings and using them to guide actions to address the identified cause.

Part 2 is divided into three subparts, corresponding to the three main steps of the Ecological Causal Assessment process.

Part 2A Formulating the Problem (see Chapters 7 and 8)

Part 2B Deriving Evidence

- Near-site data (see Chapters 9 and 10)
- Regional data (see Chapters 11–13)
- Experimental systems (see Chapters 14–16)
- Symptoms and simulation models (see Chapters 17 and 18)

Part 2C Forming Conclusions and Using the Findings (see Chapters 19–21)

6

Our Approach for Identifying Causes

Susan M. Cormier, Susan B. Norton, and Glenn W. Suter II

This chapter provides an overview of our approach to ecological causal assessment: list candidate causes, derive evidence for each, and identify which is best supported by the evidence. The approach is illustrated with a case of a fish kill in the Kentucky River.

CONTENTS

As we discussed in Chapter 5, figuring out a true cause is tricky because all of us tend to see causal relationships everywhere. People form opinions quickly and will give preference to information that supports initial opinions. Avoiding these pitfalls requires an investigative sense of what will work and awareness of where people are apt to go astray. A clearly described process charts the course.

The process is implemented in three main steps (within the bold box in Figure 6.1):

1. *Formulate the problem.* This includes defining the case that will be investigated and developing the list of candidate causes.

2. *Derive evidence.* Evidence is generated, using as many sources of data and methods as possible. The characteristics of a causal relationship introduced in Chapter 4 guide the search for data and information useful for generating evidence. A list of typical types of evidence serves as a reminder to be creative and inclusive when developing evidence.

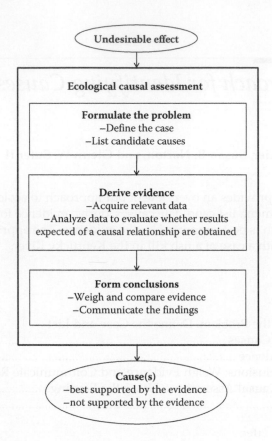

FIGURE 6.1
Steps and activities in the ecological causal assessment process (within the bold box). An undesirable effect prompts the assessment; the product is a description of the cause or causes that are best supported by the evidence as well as those that are not.

3. *Form conclusions.* The explanation that is best supported by the evidence is identified by weighing the evidence for each cause and comparing the body of evidence among the candidate causes. Conclusions are communicated using summary narratives, figures, or tables, supported by the documentation of the evidence.

The output of our approach is the identification of the cause or causes best supported by the evidence. Causes that are not supported by the evidence are also identified. The objective is to provide conclusions that will be used to inform management decisions and lead to actions that improve the environment. There are also ancillary benefits of those actions to improve or attempt to resolve the problem. Information about the assessment and subsequent actions provide important feedback to improve the assessment, the process of causal assessment, and our understanding of the causal processes that produce undesirable ecological effects.

As discussed in Chapter 5, our approach incorporates strategies to minimize errors and biases including considering a set of candidate causes, deriving many different kinds of evidence, using systems for breaking up complex assessments, and seeking other perspectives. Chapters 9–18 highlight additional strategies for taking advantage of the strengths and minimizing the limitations of different sources and types of evidence.

An overview of the approach was briefly described in Chapter 1 using the Willimantic River case study. For this chapter, we developed an additional example based on a massive fish kill that occurred after a warehouse fire and bourbon spill into the Kentucky River. Although the spilling of bourbon was very apparent, the explanation for the fish kill was not. The fish kill did not occur directly adjacent to where the bourbon flowed into the river and did not occur until 2.5 days after the spill. Several alternative causes seemed plausible. Nevertheless, it is a fairly straightforward example, and our intent is to expeditiously illustrate the process.

6.1 Formulate the Problem: Define the Case and List Candidate Causes

Problem formulation identifies the subject and scope of the assessment by defining: (1) the effects of concern, (2) the spatial and temporal extent of the problem, and (3) the candidate causes that will be explored. Even when the issue being addressed by the causal assessment seems obvious, it is worth taking a step back to carefully articulate the problem being investigated and the candidate causes under consideration.

The process of formulating the problem provides several excellent opportunities for minimizing biases and preventing mistakes. It is often tempting to delve immediately into analyzing data. Data analysis becomes more focused by taking the time to step back and carefully describe the effect and the spatial and temporal extent is before identifying a set of candidate causes. The all-too-human tendencies to jump to a conclusion and cling to a favorite hypothesis.

More details on problem formulation are provided in Chapters 7 and 8.

In the example case, the effect was a massive fish kill 22 km downstream from the site of a 20-million-liter bourbon spill into the Kentucky River. Candidate causes of the kill included ethanol narcosis, toxicity from other unknown chemicals, and asphyxiation in deoxygenated water. A conceptual model depicts the causal pathways from hypothesized sources, through mechanisms and modes of action, to the effect: a fish kill (see Figure 6.2).

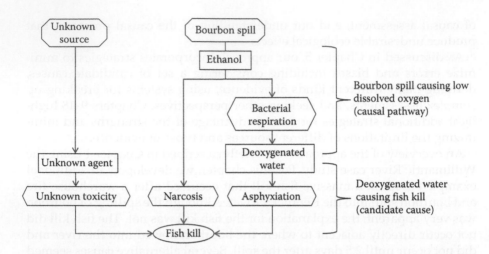

FIGURE 6.2
A conceptual model diagram depicting the relationships between the bourbon spill or an unknown source (octagons), candidate causes (rectangles), ecological and physiological processes (hexagons), and effect (oval). The brackets distinguish the causal pathway (upper) from the proximate causal relationship (lower) for low dissolved oxygen. (Adapted from Cormier, S. M., G. W. Suter II, and S. B. Norton. 2010. *Hum Ecol Risk Assess* 16 (1):53–73)

6.2 Derive Evidence

Evidence provides the substance of any explanation. Sleuthing out all the evidence is the challenging, detailed work of a causal assessment.

Evidence is more than data. It is information that is relevant to evaluating whether an apparent relationship is causal. In Chapter 4, we introduced approaches for seeking evidence and keeping it organized so that it is easy to recognize gaps in the overall arguments for each candidate cause and eventually to enable consideration of the whole body of evidence.

Evidence shows that results expected from a hypothesized causal relationship are (or are not) obtained. The characteristics of a causal relationship—co-occurrence, sufficiency, time order, interaction, alteration, and antecedence—provide a framework for articulating the results that would be expected if a causal relationship were occurring (see Chapter 4; Cormier et al. 2010). They provide a logical structure for evaluating whether a candidate cause exhibits the characteristics of a causal relationship.

Although there are six causal characteristics, there are many ways to develop evidence. Evidence can be produced using observations taken in and near the locations under investigation, or combining information from the site with information from other field studies, laboratory and field manipulations, and models. The inferential approaches and sources of data provide many different possible combinations for producing evidence. Some commonly encountered combinations are reflected in the types of evidence developed as part of U.S. EPA's CADDIS project (see Table 4.2; U.S. EPA, 2012a; Suter et al., 2002). These types of evidence serve as a handy list of some, but not all of the possibilities.

The Kentucky River fish kill provides examples of the variety of evidence that can be derived and how to compare it among candidate causes. For example, fish did not die at the maximum ethanol concentrations and did not exhibit the symptoms of ethanol poisoning; therefore, ethanol toxicity is an unlikely cause. In contrast, dissolved oxygen was 0–1 mg/L during and in the vicinity of the fish kill, providing evidence that deoxygenated water co-occurred with the kill. Dissolved oxygen levels less than 1 mg/L do not support most species of fish in Kentucky streams, and several species die in laboratory tests at levels below 5 mg/L (U.S. EPA, 1986a), demonstrating that the oxygen levels were sufficiently low and mortality would be expected. Additional evidence is summarized in Table 6.1.

6.3 Form Conclusions: Weigh Evidence and Communicate Results

The objective of a causal assessment is to produce a coherent explanation of why some causes are likely and others are implausible. This is done is by demonstrating that some explanations for the cause of the effects are supported by the body of evidence and others are not.

In our approach, we advocate that each piece of evidence be scored, with plusses indicating support for the candidate cause, minuses indicating that the argument for the candidate cause is weakened, and zeroes for ambiguous evidence. Evidence is given additional weight when it is considered especially strong or reliable. The body of evidence is weighed for each cause and compared among the alternatives. The summary is recorded. For complicated assessments with many stressors and pieces of evidence, we recommend tracking the evidence in tables. This bookkeeping strategy helps prevent information overload. It also provides a way to check that causes were treated fairly and for summarizing the overall body of evidence for each cause. More details of the theory and methods are described in Chapter 19.

Even though a causal relationship possesses all the causal characteristics, it is not likely nor necessary to have evidence for every characteristic or possible type of evidence (see Box 6.1). It may be enough to provide high-quality evidence that levels of the cause were sufficient and that the specific effects observed were consistent with those expected from the causal agent

TABLE 6.1

Summary of the Kentucky River Fish Kill Case

Formulate the Problem

On May 9, 2000, a 7-story warehouse housing 20 million liters of bourbon caught fire, and burning bourbon flowed into the Kentucky River in Lawrenceburg, KY. There was a 2.5-day delay between the spill and the effect, a massive fish kill. Candidate causes included (1) asphyxiation in deoxygenated water, (2) unknown toxicant, and (3) ethanol toxicity.

Derive Evidence

Characteristic *Type of Evidence*	Body of Evidence and Scores for Asphyxiation in Deoxygenated Water	Evidence and Scores for Other Candidate Causes
Antecedence *Causal Pathway*	Supported: Aged bourbon containing 53–56% ethanol spilled into the river 2.5 days before the fish kill, a sufficient time (2.5 days) for bacterial depletion of dissolved oxygen. The odor of bourbon was reported at the site of the fish kill. Hydrologic models accurately tracked movement of the spill. The area of influence measured about 6–8 km, bank to bank, surface to stream bed, moving at a rate of about 8 km per day (++)	Unknown toxicant weakened: No other toxic spills reported before or during the fish kill (−) Ethanol supported: Ethanol spilled into the river after the explosion (++)
Time order *Temporal Sequence*	Supported: Fish died (May 11) after the spill occurred and not before (May 9). An extreme fish kill was reported 22 km downstream (May 14, 2000). After the spill moved out of each area, dissolved oxygen returned to pre-spill concentrations as uncontaminated water flowed from upstream and fish moved into the area. The fish kill ceased when diluted by the Ohio River (+++)	Ethanol supported: The effect occurred 2.5 days after exposure to ethanol (+)
Co-occurrence *Spatial-temporal co-occurrence*	Supported: Dissolved oxygen was 0–1 mg/L during and in the vicinity of the fish kill. A diverse and abundant fish assemblage died during the event, and fish fleeing downstream were trapped behind dams and died (++)	Ethanol refuted: No kill for the first 22 km of the ethanol spill (−−)

Sufficiency *Stressor-response from field studies*	Supported: Dissolved oxygen levels less than 1 mg/L do not support most species of fish in Kentucky streams (+) (U.S. EPA, 1986a, b). The severity of the fish kill increased as the dissolved oxygen level decreased at each new location as the plume moved downstream	Ethanol weakened: No kill when or where the ethanol levels were at their maximum (−−)
Stressor-response from laboratory studies	Supported: Many fish species die in laboratory tests at levels below 5 mg/L (++) (U.S. EPA, 1986a, b)	No evidence
Alteration *Symptoms*	Supported: Flight is symptomatic of fish exposed to low dissolved oxygen. Fish fled downstream in advance of the spill and into tributaries where dissolved oxygen was greater. When trapped into deoxygenated water by barriers, fish were seen gasping at the surface and attempting to jump over dams (+)	Ethanol weakened: A key symptom of narcosis is lethargy rather than flight from the spill (−−)
Evidence of mechanism or mode of action	Supported: Acute lethality of fish and other organisms is consistent with rapid depression of dissolved oxygen	Ethanol weakened: The lag between the spill and effect is inconsistent with mechanisms for rapid onset of narcosis (−)
Evidence of related effects	Supported: Air breathing reptiles and amphibians were unaffected	No evidence
Interaction *Evidence of exposure*	No evidence: Blood oxygen levels were not measured	No evidence

Form conclusions

The evidence indicated that the fish died of asphyxiation from deoxygenated water: fish died after coming into contact with the deoxygenated water; similar levels of oxygen depletion have been associated with lethality in laboratory and other studies; and fish exhibited symptoms of asphyxiation. Ethanol was refuted because the fish kill did not co-occur with the highest levels and fish did not exhibit symptoms of narcosis. There was no evidence of other toxicants

After the causal assessment

Immediate action: On May 18, 2000 U.S. EPA contractors and the U.S. Coast Guard began aeration of the affected area from six barges equipped with large oxygen compressors and trailing submerged perforated piping. Dissolved oxygen increased to 0.8 mg/L, which was insufficient to avert fish kills as the plume moved downstream.

Subsequent action: New methods for dealing with ethanol spills drew from the experience at the Kentucky River. Now guidance for responding to an ethanol spill recommends containment of ethanol, aerating contaminated water before DO gets low, and controlled burning of contained ethanol (MDEP, 2011).

Source: Adapted from Cormier, S. M., G. W. Suter II, and S. B. Norton. 2010. *Hum Ecol Risk Assess* 16 (1):53–73.

> **BOX 6.1   CAUSAL CHARACTERISTICS AND THE FELLING OF A TREE**
>
> If a tree falls in the forest and no one is there, it did make a sound. If you later see it intact but uprooted on the forest floor, it did not fall by means of a chainsaw. Why is it that we are so sure? We know how the effect of uprooting is different from the effects of sawing wood. We do not need evidence for every causal characteristic of wind felling a tree. But variety and abundance of evidence increase our confidence, especially if it is consistent. Would you not be more convinced if it were known that the day before documentation of the downed tree there had been sustained winds in excess of 70 miles per hour in the area? The more relevant, good-quality evidence there is, the greater the confidence that the real cause was identified, because it could have been a bulldozer and not the wind at all.

or to provide results from a well-conducted field study that combines time order with spatial co-occurrence. Ultimately, many pieces of evidence, consistently pointing to one cause, strengthen the case.

The final conclusions explain why some causes are implausible and others are so plausible that there is a willingness to change what is being done and make or recommend a decision (see Chapter 20).

In the Kentucky River fish kill example (see Table 6.1), the evidence consistently supported the candidate causes of deoxygenated water: fish died after coming into contact with the deoxygenated water; similar levels of oxygen depletion have been associated with lethality in laboratory and other studies; and fish exhibited symptoms of asphyxiation. The other alternatives lacked support. Ethanol was refuted because fish did not die when the ethanol concentrations were at their maximum. In addition, fish did not die within the expected time of onset of narcosis, and symptoms of narcosis were not observed. There were no reports of other toxicant spills.

6.4 After the Causal Assessment: Using the Findings

The immediate goal of our causal assessment approach is to provide useful and defensible causal explanations to decision-makers, to improve the practice of causal assessment, and to increase our understanding of the environment. Our ultimate goal is to improve the environment. A causal assessment that leads to a management decision that improves the environment provides evidence that our understanding of the causal processes, while never perfect, are correct enough to guide effective management actions. For example, dramatic recovery of fish and macroinvertebrates was observed after removing a dam that reduced oxygen in an impounded section of the Cuyahoga River (Tuckerman and Zawiski, 2007; Cormier and Suter, 2008).

In the Kentucky River fish kill example, attempts were made to aerate the water even while the fish mortality was continuing. On May 18, 2000, U.S. EPA contractors and the U.S. Coast Guard began aerating the affected area from six barges equipped with large compressors and trailing submerged perforated piping. Dissolved oxygen increased to 0.8 mg/L, which was inadequate to prevent fish kills as the plume moved downstream. In the end, only dilution with the Ohio River was sufficient to "treat" the 20-million-liter volume of the spill.

Even when management action is not successful, a causal assessment can help explain why. The clear evidence in the Kentucky River case allowed the Kentucky Department of Natural Resources to collect damages. In the years that have passed, many species have recolonized from nearby tributaries and the Ohio River. Nationally, the frequency of large ethanol spills associated with fire has increased owing to shipping of denatured ethanol for automobile fuel. Other states have drawn from the experience of the Kentucky River to develop guidance for dealing with ethanol spills. New recommendations include containing the ethanol as a key immediate site response, aerating contaminated water before dissolved oxygen gets low, and controlled burning of the contained ethanol (MDEP, 2011).

6.5 Summary

The objective of our approach to ecological causal assessment is to identify the cause(s) of undesirable ecological effects that is best supported by the evidence. It is implemented in three steps: (1) formulate the problem by defining the case and listing candidate causes, (2) derive evidence by analyzing data, and (3) form conclusions and communicate results. The goal is to provide a sound basis for management actions, which in turn provide a means to evaluate the assessment's conclusions.

Evidence is the foundation of any good causal assessment. However, only repeated experience hones someone's ability to figure out causes. A single fabulous experiment, statistical analysis, or mathematical model may provide evidence, but it will not form the conclusion or drive the decision. It is human beings weighing a body of evidence that leads to a credible explanation that is likely to change behavior or policy.

Evidence is never as complete as we would like. Our philosophy is pragmatic: decisions should be based on the best explanation available when a decision needs to be made. Completing the steps of causal assessment can strengthen the scientific basis for decisions made to improve biological conditions. Evaluating and sharing the results of actions can help improve the entire community of practice.

Part 2A

Formulating the Problem

An ecological causal assessment is done for a reason: there is a problem and someone wants it to be rectified. The process of formulating the problem, described in Chapters 7 and 8, ensures that the data collection and analysis will support the goals of the assessment. Chapter 7 describes how to define the case being assessed in a way that is clear and directed toward informing an environmental management decision. Chapter 8 describes strategies for developing the list of candidate causes and managing multiple causes.

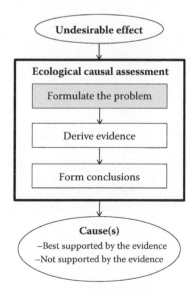

7

Defining the Case

Susan B. Norton and Glenn W. Suter II

The process of formulating the problem defines an assessment's focus and frame. It identifies the effects that will be investigated, the spatial and temporal scope of the assessment, and the candidate causes that will be considered.

CONTENTS

This chapter begins the discussion of the processes and activities involved with conducting a causal assessment. As described in Chapter 6, the assessment process consists of three major parts: (1) formulating the problem, (2) deriving evidence, and (3) forming conclusions.

Here and in the next chapter, we suggest strategies for formulating problems in causal assessments. We begin by describing how the broad and sometimes vague concerns that initiate a causal assessment can be sharpened to define the specifics of the case. The case definition identifies the specific changes or effects that are most striking and a first approximation of when and where they are occurring. A series of comparison sites are also identified that represent places where the effect is not occurring or is occurring in a different way. In Chapter 8, we describe strategies for developing the list of candidate causes that will be investigated before we move on to the chapters on deriving evidence and forming conclusions.

7.1 Formulating the Problem: An Overview

Problem formulation is the process by which the concerns that initiate an assessment are defined in a way that can be evaluated.* Problem formulation simplifies the world by defining an assessment's focus and frame. The focus of a causal assessment is defined by the effects that will be investigated. The frame describes the spatial and temporal extents of the assessment and the candidate causes that will be considered. As an investigation proceeds, its focus and frame may evolve with knowledge gained from preliminary results, or different options may be explored to extract new or different insights.

Problem formulation is one of the steps that is most influenced by an investigator's judgment, and it can have a large impact on conclusions. In the field of decision analysis, the way options are presented has been shown to heavily influence decisions and actions, especially in complex situations (Tversky and Kahneman, 1981; Thaler and Sunstein, 2009; Kahneman, 2011). In other words, the way an assessment problem is stated can influence the assessment's conclusions by directing which evidence should be pursued and presented. For example, an investigation into the causes of population declines of kit foxes initially defined the frame with a large spatial expanse and temporal extent. That assessment concluded that annual precipitation patterns were responsible for variance in abundance (Cypher et al., 2000). However, when the frame of the assessment was narrowed to examine the specific location and years of the severe decline, the assessment pointed to a different cause entirely: increased predation from coyotes (see Chapter 25).

Causal assessments are often conducted to support particular decisions. Decisions have their own focus and constraints that may not coincide with the optimal frame for the assessment. For example, the decision may be limited to actions relevant to only one source or type of cause, such as pesticides or toxic chemicals. Alternatively, a watershed partnership group may be deciding which recommendations to make to improve all of the water bodies within a catchment.

We recommend defining the case in a way that helps distinguish and isolate different causal processes. Identifying individual causal processes can lead to specific management actions that may be more feasible to implement than eliminating the source or human activity that initiated the chain of events. Urbanization may be identified as a cause, but it is unrealistic to replace a city with forest. However, when a causal assessment establishes that most of the biological degradation stems from inadequate groundwater

---

* Readers familiar with ecological risk assessment will recognize the term "problem formulation." Although the purpose of problem formulation is the same in causal and risk assessment, the components and sequence of activities differ.

inflow, the specific problem becomes tractable and can be mitigated by increasing recharge blocked by impervious surfaces.

7.2 Initiating the Process

Causal assessments of specific cases are initiated because undesirable biological conditions have been observed. Some examples include

- Kills of fish, invertebrates, plants, domestic animals, or wildlife
- Anomalies in any life form, such as tumors, lesions, parasites, or disease
- Changes in community structure, such as loss of species or shifts in species abundances (e.g., increased algal blooms, loss of mussel species, increases in tolerant species)
- Responses of indicators designed to monitor biological condition, such as the Index of Biotic Integrity (IBI) or the ratio of observed-to-expected taxa occurrences
- Changes in organism behavior
- Changes in population structure, such as population age or size distribution
- Changes in ecosystem function, such as nutrient cycles, respiration, or photosynthetic rates
- Changes in the area or pattern of different ecosystems, such as shrinking wetlands or increased sandbar habitats

Causal assessments can be a part of a set of investigations initiated for reasons other than an undesirable effect, such as concern over a source or a stressor. For example, the Northern River Basins Study (see Chapter 24) was initiated in part by concerns over pulp and paper mill discharges. In these cases, an assessment that characterizes the effects that are occurring is needed to provide the focus for the causal assessment.

The concerns that initiate a causal assessment may also reflect the decision context within which the assessment is being conducted. For example, in the United States, the nonattainment of a Water Quality Standard (WQS) due to an undesirable biological effect can lead to the listing of a stream reach as impaired under Section 303(d) of the Clean Water Act. States are then required to develop a plan for its improvement, either by developing a watershed management plan or a total maximum daily load for a pollutant of concern. Without an identified cause, such plans cannot be developed. An analogous process is being developed under the European Water Framework Directive.

No matter how they are prompted, causal assessments begin with at least some data that provided the basis for the initial concerns. The data could be biological survey measurements or anecdotal observations of kills. More than likely the data were collected for another purpose, for example, developing estimates of biological condition at the regional or national scale. One challenge in the early stages of a causal assessment is to use available data even though they are an imperfect match for the investigation's needs.

The following sections discuss defining the undesirable effects of concern (see Section 7.3) and the spatial and temporal extents of the assessment (see Section 7.4). These two topics are often explored simultaneously when an undesirable effect is recognized and a causal assessment is begun.

7.3 Defining the Undesirable Effect

The objective of this step is to define the undesirable effects to be investigated. Implicit in the concept of an undesirable effect is that some biological attribute has changed from a more desirable to a less desirable state. In some cases, the fact that an undesirable effect is occurring must be verified, and in most cases the effect must be more specifically described. In addition, defining the effect may require clearly articulating the desired condition. The desired condition may be obvious, such as fish with no spinal deformities or liver tumors. Other cases may require quantifying background frequencies of effects. For example, because the background frequency of intersex in smallmouth bass (*Micropterus dolomieu*) was not known, additional research was required to establish that the observed frequency of fish intersex in the Potomac River was indeed a cause for concern (Blazer et al., 2007; Hinck et al., 2009). In still other cases, the desired biological condition is similar to what is observed at high-quality reference sites (Bailey et al., 2004; Stoddard et al., 2006).

The undesirable effect that prompted the investigation would seem to provide an obvious starting point for a causal assessment's focus. However, undesirable effects that initiate an assessment are often too general in detail than is optimal for causal assessment. They may be couched in general terms like poor fish health or poor biological integrity or declining biodiversity. For example, causal assessments may be initiated by an observed decline in a multimetric index used by states or other government groups to evaluate biological condition. Multimetric indexes combine indicators that themselves are summaries of individual measurements such as taxa counts or abundance data. Frequently used metrics in stream invertebrate condition assessments include overall taxa richness, abundances of particular taxa, frequency of deformities, relative abundances of trophic groups, or nominal sensitivity of species. Multiple metrics are used because each presumably responds to different changes produced by stressors. As a whole, the index is intended to improve the ability to detect changes produced by many different types of causes. However, because the goal of causal assessment is to identify the different causal processes that are operating, it is usually most beneficial to focus on the biological measurements that are changing the most (see Box 7.1).

Defining the effects more specifically can help separate the signal of the causal relationship from the noise of other sources of variation. Associations between stressors and specific effects should be stronger than those between stressors and indexes that aggregate many different responses. A specifically defined effect also makes it easier to relate field observations to experimental results that typically measure specific responses.

In other cases, a causal assessment may have been prompted by a very conspicuous effect, such as the observation of deformities or deaths. In these cases, the challenge is to characterize more fully the spectrum of effects that may be occurring. For example, the investigation of recurring fish kills may begin by gathering information on the sexes, life stages, and species affected, the timing and frequency of the kill, and the symptoms of the dead and dying fish. The causal assessment may encompass all of the aspects of the kill or may focus on a particular aspect of the effect, like its timing, frequency, or a particular symptom.

BOX 7.1 MULTIMETRIC INDEXES AND CAUSAL ASSESSMENT

Multimetric indexes are designed to respond to many different stressors (Davis and Simon, 1995). However, a low index value observed in a particular case is often driven by large changes in just one or a few of the constituent metrics. Identification of those metrics as the effect to be analyzed increases the likelihood of identifying a single dominant cause. For example, in the Long Creek case (see Chapter 22), three of the 30 invertebrate metrics in Maine's index were responsible for most of the index deviations relative to the comparison site. In the case of the Willimantic River, the adverse effect was originally defined in terms of an overall low index score, but the causal assessment used just one metric, reduced EPT taxa richness, as it was responsible for most of the decline in the index. The analysis might have been even clearer if data were available for individual families, genera, or species, because some caddisflies (Trichoptera) are more tolerant to metals, which were high in the unpermitted effluent that was eventually identified as the probable cause. A subsequent perusal of the data showed that 97% of the samples were made up of cheumatopsychid caddisflies, which prefer fine particular matter and have been shown to tolerate high metal concentrations in other studies (e.g., Pollard and Yuan, 2006).

7.4 Defining the Spatial and Temporal Extents of the Assessment

7.4.1 Identifying Where Effects Occur

When defining effects, we also recommend defining the location and the spatial and temporal extents of the assessment as narrowly and precisely as feasible. A natural place to start is the spatial and temporal extents of the observed effects. However, the full spatial and temporal extents may not be known in the early stages of an assessment. Perhaps the effect was observed at a site measured during a probabilistic monitoring survey or noted anecdotally. Initial estimates of extent typically must be refined as the assessment proceeds.

It may seem advantageous to broaden the geographical or temporal dimensions of an assessment because sources and human activities often occur over a larger area than an individual site. However, even if environmental factors and sources operate at coarser scales, they may impact individual sites to different degrees (see Box 7.2). A broader geographical or temporal scope will increase the number of observations that can be analyzed. However, a broader scope also increases the likelihood of encountering different causal processes resulting from different sources, human activities, or natural factors. In general, narrow geographic and temporal scopes benefit causal assessments by isolating and restricting contributing factors.

There are four strategies for dividing a geographic area into smaller units for analysis.

Subdivide by source. Subdividing by source may make sense if different sources are affecting different parts of the system. For example, a river system can be partitioned by segments between outfalls or between tributary confluences. The Northern Rivers case (see Chapter 24) focused on stream reaches below several different outfalls from municipal waste treatment plants and pulp and paper mills.

Subdivide by pattern of effects. In a causal assessment of the Little Scioto River, Ohio, the study reach was divided into three subreaches based on the observation that the biological effects qualitatively changed in different subreaches (Norton et al., 2002a). Effects in the first subreach were eventually attributed to stream channelization, effects in the second to channelization plus PAHs from a creosote site, and effects in the third to channelization, metals, and nutrients (Cormier et al., 2002).

Subdivide based on environmental characteristics. During the investigation of acid precipitation in the Adirondacks, lakes receiving similar amounts of acid precipitation had very different responses. Some became acidified, while others did not. Explaining the differences became a major prong of the investigation. Differences in response were eventually traced to differences

BOX 7.2   CAUSAL ASSESSMENT AND WATERSHEDS

It has become a truism that watersheds are the proper spatial unit for aquatic ecosystem management, but watersheds are not always the appropriate unit for causal assessment in aquatic ecosystems. Watersheds are often appropriate units for organizing collaborative management programs involving stakeholders, government agencies, and the public. In addition, watersheds are appropriate analytical scales if the issue of concern is the export of nutrients, sediment, or other pollutants to a downstream resource such as the Chesapeake Bay or Gulf of Mexico.

However, if one is concerned with effects within the watershed, the appropriate unit for analysis of adverse biological effects and their causes is typically the individual tributary or reach. This is because of the importance of local sources, such as tilled fields, cattle access areas, storm drains, mine drainage and waste dumps, and local differences in slope, substrate, or other geological features. Even if all important sources and geological features are uniform across the watershed, small-scale analyses are important because of the differential sensitivities of biotic communities and ecosystem processes at different locations within a stream and across streams of different sizes. If a watershed-scale cause is operating, it is relatively easy to combine the results of multiple reach-scale or tributary-scale causal analyses or to extrapolate the local-scale results to the rest of the watershed. However, if a causal analysis is performed at the whole-watershed scale, important local causes cannot be identified after the fact and will be missed.

in watershed characteristics, such as soil depth and the pathways by which shallow groundwater and surface runoff reached the lakes (Jenkins et al., 2007).

Subdivide based on natural history attributes. Some species may have extensive ranges and encounter different causes at different points. For example, anadromous salmon encounter one set of causal processes in their spawning stream and a different set during their time at sea.

If larger scale processes are expected to contribute to the effect it may be worthwhile to expand the geographic or temporal scope. Beware of two pitfalls when using a broader geographical definition of the case, such as a watershed. First, it may be tempting to attribute all effects to one cause, especially if the effects seem to be similar. Ecological causal assessments frequently suffer from overdetermination, that is, more than one cause can produce a given effect. Second, because multiple causal processes are likely operating within a broader spatial scale, it may be tempting to conclude that multiple stressors are interacting to cause all events within that area. Instead, we recommend considering the alternative that different stressors may be the dominant factors at different locations.

In some circumstances, it will not be possible to define the spatial extent of the causal assessment in a way that isolates different causes. For example, multiple outfalls and nonpoint sources may influence the same reach of a river. Experimental approaches (see Chapters 15 and 16) may be needed to reach conclusions about cause in these cases. It may also be possible to isolate events in time, for example, the kit fox case study focused on the part of the temporal record that coincided with greatest declines (see Chapter 25). Another example comes from the study of acidification in Adirondack lakes. Although the investigation initially examined effects throughout the year, it eventually became clear that the processes going on in the spring were sufficiently distinct that they represented a potentially different set of causal processes. By focusing more narrowly on spring conditions, the investigators identified that nitrates were playing a more important role than sulfates at that time of year (Jenkins et al., 2007).

7.4.2 Identifying Where Effects Do Not Occur

The discussion above has focused on defining where effects occur. Equally important for causal assessment is to identify locations where the effects do not occur or occur in a different way. Conditions at these locations substitute for the conditions of actual interest, that is, the conditions that would have been observed at the site of interest if the effect had not occurred. These conditions cannot be observed—measurements can only reflect what actually happened. Conditions at locations where the effect does not occur are practical surrogates for those that would have been observed in the unexposed, unmeasured (i.e., counterfactual) no-effect scenario.

Contrasting conditions where effects do and do not occur is one of the fundamental ways that evidence is developed for evaluating cause. This strategy, discussed further in Chapter 10, dates back to Mill's Method of Differences (Mill, 1843; Lipton, 2004, see Section 3.1). Confidence that a difference in response was caused by a candidate cause is increased when other environmental conditions in the two situations are similar. Evidence that a candidate cause differs between a situation where the effect occurs and an otherwise similar situation where it does not occur supports the argument that the stressor caused the effect. When only one candidate cause differs between the two situations, then it is the cause.

The first response to the question of where effects are not occurring is typically "reference sites." Identifying reference sites has become a standard part of monitoring surveys. Unfortunately, contrasting conditions at a site where an effect occurs with those at the highest quality regional reference sites rarely points to only one factor, because many attributes and stressors typically differ. For this reason, another strategy is to identify locations where effects are observed but to a lesser degree or in a different way. We call these "comparison sites" to differentiate them from high-quality reference sites. The most useful comparison sites share as much causal history and as many natural factors as possible with the sites under investigation. For example, comparison sites in causal assessments involving streams ideally would be located within the same watershed or even adjacent to the site under investigation, share the same soils and climate, and be located in a similarly sized stream. In the investigation of the effects of acid rain in the Adirondacks, lakes adjacent to each other responded very differently to acidification despite receiving the same amount of acid rain, allowing researchers to focus on other characteristics that differed (Jenkins et al., 2007). In the Little Floyd River case study (Haake et al., 2010b), a less affected stream reach at the confluence of a tributary was key for identifying not only the causes but also pointing to mitigation measures that would improve the rest of the river.

Demarcating where or when effects began sometimes leads to the cause. For example, the timing of a decline of unionid mussels coincided with the invasion of zebra mussels (Martel et al., 2001). Similarly, identifying where the effects begin occurring may lead to a point source that can be remediated. This process is a familiar strategy for investigating spills. For example, pesticide concentrations were followed upstream to the location of a pesticide spill responsible for a fish kill (U.S. EPA, 2013a). In the Willimantic case study, the investigators began by contrasting water quality at the site under investigation with several higher quality sites within the watershed. They used this information to target additional sampling to demarcate where effects began. Doing so enabled them to identify the point source discharge just upstream of where a large decline in the EPT taxa richness was observed (Bellucci et al., 2010).

7.5 Summary

Defining the case is one of the major parts of formulating a causal assessment problem.

The case description identifies:

- The effects that will be investigated and their spatial and temporal extents. We recommend defining both the effects and their extent as specifically as possible. In general, narrow scopes benefit causal assessments by isolating and restricting contributing factors, allowing specific causal relationships to be defined and better understood.

- A series of comparison sites (e.g., regional reference sites or local comparison sites) where the effect is not occurring or is occurring in a different way. Comparison sites are most useful when they share most of the causal history and natural environmental factors with the locations under investigation. Time can also be used to define a situation where effects did not occur, although this strategy is less common.

8

Listing Candidate Causes

Susan B. Norton, Kate Schofield, Glenn W. Suter II,
and Susan M. Cormier

This chapter describes the process of developing the list of candidate causes. Strategies are described for refining the list and dealing with multiple causes. The construction and effective design of conceptual models are discussed not only for the purpose of describing and organizing candidate causes, but also to guide the derivation of evidence and communication of results.

CONTENTS

The second major part of problem formulation for causal assessment identifies the group of the candidate causes that will be considered. Deliberately defining a set of alternatives for evaluation is a simple step, yet it is one of the most powerful strategies for countering hypothesis tenacity, the common cognitive error in which the first or most memorable explanation for an effect is favored and evidence of alternatives is ignored.

Identifying which causes will be investigated requires balancing the objectives of efficiently managing resources for the assessment while ensuring that the most important candidate causes are considered. Including more candidates increases the chances that the assessment will evaluate the real cause as well as causes of interest to different audiences. However, each candidate cause requires resources for data collection and analysis. It is important for assessors to carefully consider available information for all relevant alternatives and document the process of considering and selecting the candidate causes that will be evaluated.

This chapter is organized by the process of developing the list of candidate causes. A preliminary list is generated (see Section 8.1) and is explored and organized further using conceptual models (see Section 8.2). The final list is honed by splitting some causes, combining others, and deferring still others (see Section 8.3). An example of a final list used in the Long Creek case study is shown in Box 8.1.

8.1 Initiating the List of Candidate Causes

Causes may be described as agents, processes, or events (see Chapter 2), but typically the list is begun with agents. Candidate causes are agents that the investigators have some reason to believe could have produced the effect being studied. Candidate causes are commonly referred to as stressors, even

---

**BOX 8.1   CANDIDATE CAUSES OF THE DECLINE
IN MACROINVERTEBRATE ASSEMBLAGES IN
LONG CREEK, MAINE, USA (SEE CHAPTER 22)**

- Decreased dissolved oxygen
- Altered flow regime
- Decreased large woody debris
- Increased fine sediment
- Increased in-stream organic matter production
- Increased temperature
- Increased toxic substances

though the term, which has a negative connotation, is an imperfect fit for agents that are required for life (e.g., dissolved oxygen).

We recommend that the description of a candidate cause focus on the proximate cause, that is, the agent that actually contacts or co-occurs with the susceptible entity, producing the observed effect. Minimizing the steps between the candidate cause and the earliest change in the entity increases the chance of making a definitive link between cause and effect. As discussed below, the description of a candidate cause can include additional information to distinguish among alternatives, for example, by specifying a source or providing details on the form of the agent or the process by which an event could result in exposure.

The first major choice in developing the list of candidate causes is whether to limit the investigation to only one candidate cause or to consider all alternatives. This choice is often strongly influenced by the decision context. For example, an assessment may evaluate whether toxic substances in the effluent from a wastewater treatment plant are responsible for observed downstream declines of macroinvertebrate diversity. Constraining the causal assessment greatly simplifies data collection and analysis. One need generate data only for one candidate cause, and if the evidence indicates that this cause is unlikely or highly likely, the investigation may end.

Limiting the investigation to only one candidate cause may lead to missing additional or even the most influential causes. For example, establishing that the wastewater treatment plant effluent is not toxic ignores other ways an effluent might change downstream biota, for example, by altering temperature or increasing nutrients. In addition, concluding that a cause is not responsible for effects is more convincing when the likely cause is identified. For example, an initial causal assessment of kit fox declines on the Elk Hills Naval Petroleum Reserve addressed only contaminants from oil production and concluded that they were not responsible (Suter et al., 1992). That conclusion was made more convincing in a subsequent causal assessment that addressed many candidate causes and identified coyote predation as the predominant cause (see Chapter 25).

The process of developing the list of candidate causes provides an opportunity to ponder which ecological and physiological mechanisms might be operating to produce the effect. Evidence for a candidate cause may seem adequate until other causes are considered that may have stronger evidence. Assessment of other candidate causes may also reveal weaknesses in the evidence for the cause of concern that would not otherwise be apparent.

Reasons for including a candidate cause on the list typically draw upon a combination of experiences from other locations, subject area knowledge from the literature, observations available from the site, local history, and stakeholder concerns. Strategies include the following:

- *Review common causes of the observed effect in your state or region.* Do not forget the usual suspects. Common causes may be identified by asking local experts or analyzing regional data. For example, low dissolved oxygen levels and diseases are listed as common causes of fish kills in Virginia ponds (Helfrich and Smith, 2009). Analyses of regional associations may suggest stressors or sources that should be considered. In the mid-Atlantic region of the United States, a high likelihood of a poor macroinvertebrate index score was associated with poor quality sediment, acid deposition, and mine drainage (van Sickle et al., 2006). The scientific literature can be queried to identify causes that have been associated with the effect in other circumstances. Published lists of causes of some effects are available (e.g., Table 8.1) and can be used to prompt ideas.

- *Consider which organisms are affected and their natural histories.* Different organisms respond to different stressors in different ways. When the observed effect is a decline of an individual species, its life cycle and habitat requirements suggests candidate causes. For example, an investigation of the decline of a unionid mussel population would need to consider effects on the specific fish that host their larval life stage (Bogan, 1993).

- *Visit the site.* Visits to the affected and comparison sites are essential sources of insights and clues. Observations of existing sources suggest candidate causes. The occurrence of some causes can be directly observed, for example, fine sediments embedding cobble habitat in a stream. Some causes produce distinctive and observable symptoms, such as whirling disease in trout and distinctive lesions in corals (see Chapter 17). The CADDIS website has lists of observations that indicate that a stressor might be present (U.S. EPA, 2012b).

- *Think about uncommon or unique aspects of the observed effect or situation.* Observations that are unusual for a particular area may suggest a new or previously undocumented stressor or source. For example, a new and conspicuous increase in nonnative zebra mussels was investigated as the cause of native mussel declines in the Rideau River in Canada (Martel et al., 2001).

- *Consider unknown sources or stressors.* Including a placeholder for an unknown source or stressor can serve as a reminder throughout the investigation that the true cause may not have been on any of the initial lists, like the contaminant mixture from the broken industrial effluent pipe discovered during the Willimantic River case study (see Chapter 1).

- *Consider legacy sources or stressors.* Causal assessments may also need to consider influences of past land uses, in addition to current activities. For example, fine sediments in streams have been linked to historic logging practices (Harding et al., 1998), and land-use prior to urban development can be an important predictor of stream community response to urbanization (Brown et al., 2009).

TABLE 8.1

Candidate Causes That Can Potentially Affect Freshwater Biota, Including Algae, Macroinvertebrates, and Fish

DO regime
Hydrologic regime (includes flow or depth conditions, timing, duration, frequency, connectivity, etc.)
Nutrient regime
Organic-matter regime
pH regime
Salinity regime
Bed sediment load changes, including siltation
Suspended solids or turbidity
Water temperature regime
Habitat destruction
Habitat fragmentation (e.g., barriers to movement, exclusion from habitat)
Physical crushing and trampling
Toxic substances
Herbicides and fungicides
Halogens and halides (e.g., chloride, trihalomethanes)
Fish-killing agents (e.g., rotenone)
Insecticides
Lampricides
Metals
Molluscicides
Organic solvents (e.g., benzene, phenol)
Other hydrocarbons (e.g., dioxins, PCBs)
Endocrine disrupting chemicals
Mixed, cumulative effect
Interspecies competition
Complications due to small populations (e.g., inbreeding, stochastic fluctuation, etc.)
Genetic alteration (e.g., hybridization)
Overharvesting or legal, intentional collecting or killing
Parasitism
Predation
Poaching, vandalism, harassment, or indiscriminate killing
Unintentional capture or killing
Vertebrate animal damage control
Radiation exposure increase (e.g., increased UV radiation)

Source: Adapted from Richter, B. D. et al., 1997. *Conserv Biol* 11 (5):1081–1093.

- *Engage the broader community.* Processes for involving stakeholders in assessments vary greatly depending on the decision and assessment context and may range from informal conversations to consultations required by law. The involvement of the broader community can benefit the initial phases of the assessment. Those familiar with the area know local history and current and legacy sources. Subject area experts can describe influential environmental and ecological processes. Managers and interested parties can provide insights into factors they think should be investigated.

8.2 Developing Conceptual Models

The initial process of developing the list of candidate cause typically generates a wide variety of factors that might contribute to an effect, including agents (e.g., stressors such as fine sediments, metals, ammonia), sources (e.g., agricultural fields, mining sites), or human activities (e.g., urban development). Conceptual models help organize these factors into a framework for analysis and communication.

Although we recommend developing conceptual models during the problem formulation phase of a causal assessment, conceptual models have many benefits throughout the assessment. At the outset, they clarify thinking and provide a structure for communicating current understanding. They can be used to refine the list by identifying which causes might be evaluated separately, be operating jointly to produce effects, or be evaluated as a step in a causal pathway rather than a proximate cause (see Section 8.3). They can be used to identify where data collection or analysis efforts might be used to help distinguish among causes. They provide a structure for quantitative analysis of associations and model development. Finally, they provide a visual aid for presenting the assessment's conclusions (see Chapter 20). For example, in the kit fox case study (see Chapter 25), the initial conceptual model diagram was pared down and reorganized to present the final causal explanation.

8.2.1 Using Diagrams to Explore and Present Conceptual Models

Although conceptual models can be described in text or pictorial format, their development typically begins with a diagram. Conceptual model diagrams (also called graphical organizers, concept maps, or node-link displays) organize information and relationships between concepts. They have many uses relevant to causal assessments, including eliciting expert knowledge, communicating relevant scientific concepts to interested parties, and supporting the development of quantitative models (Bostrom et al., 1992; Suter, 1999; Tergan, 2005; Dennison et al., 2007; Allan et al., 2012).

Diagrams that use a box-and-arrow format are well suited for depicting cause–effect relationships (e.g., Hyerle, 2000; Figures 8.1 and 8.2). In these diagrams, each set of two shapes and a connecting line represents a cause–effect linkage hypothesized to be occurring in the system. The shapes

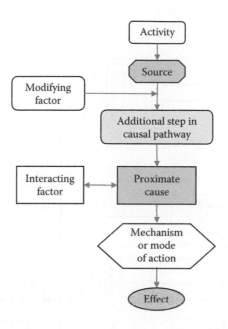

FIGURE 8.1
A template for conceptual model diagrams with different shapes used for different elements, for example, source in an octagon and effect in an oval.

represent natural or anthropogenic variables that have been or could be measured in the environment, either directly or indirectly by using surrogate indicators. The lines reflect causal processes. In the most straightforward cases, a line implies a direct causal influence of one variable on another. In other cases, lines may reflect associations expected from indirect (i.e., mediated through other variables) or unknown causal mechanisms.

Conceptual models for many candidate causes associated with the decline of fish and macroinvertebrate assemblage are available on the CADDIS website (U.S. EPA, 2012b). These models (including low dissolved oxygen, metals, nutrients, ammonia, sediments, pesticides, dissolved minerals, altered flow, toxic chemicals, physical habitat, and high temperature) provide a starting point for case-specific diagrams.

When creating case-specific diagrams, the typical starting point is the effect, which appears at the bottom of the diagram template shown in Figure 8.1. Sources or activities, at the top, are identified next. Most conceptual models trace the pathways from land uses or human activities to identify possible targets for management intervention. In the Long Creek conceptual model diagram, land-use activities and other sources such as urbanization, industries, impervious surfaces, and in-stream impoundments were traced to proximate causes. Assessments conducted under other frameworks may extend the scope to include economic and social drivers (e.g., DPSIR, Figure 8.3; Box 8.2; NAS, 2011).

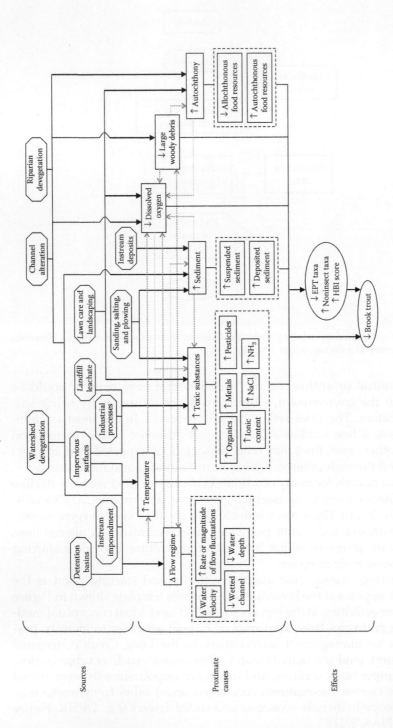

FIGURE 8.2

A conceptual model diagram from the Long Creek case study (see Chapter 22). Candidate causes (also listed in Box 8.1) are shown in rectangles. Dark lines depict pathways potentially linking sources, proximate causes, and effects. Light gray lines depict pathways by which candidate causes may interact. Boxes within dotted lines provide more specific descriptions of candidate causes.

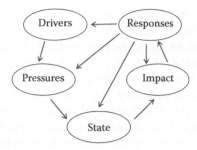

FIGURE 8.3
The DPSIR framework. (Adapted from Kristensen, P. 2004. The DPSIR Framework. Paper read at Comprehensive/Detailed Assessment of the Vulnerability of Water Resources to Environmental Change in Africa Using River Basin Approach, September 27–29, 2004, Nairobi, Kenya.)

The diagram should identify the proximate causes, that is, the stressors that contact or co-occur with the biota. Focusing analyses on the relationship between proximate causes and effects can improve the associations developed as evidence. For example, regional studies of the effects of acid deposition on wood thrush occurrence showed little relationship with soil pH or calcium content as causal variables, but a strong association with the abundance of calcium-rich invertebrate prey, the apparent proximate cause (Hames et al., 2006).

BOX 8.2   THE DPSIR FRAMEWORK

The DPSIR framework (Smeets and Weterings, 1999; Kristensen, 2004) (see Figure 8.3), used frequently by members of the European Union, was developed to organize environmental indicators in a sequence of events:

- **Driving forces**, that include economic policies, societal needs, and wants lead to
- **Pressures**, that include emissions and waste products from sources, lead to
- Physical, chemical, and biological **States** that produce
- **Impacts** on ecosystems and human health, lead to
- Management and political **Responses** that can be directed toward any step of the sequence.

Although the naming conventions are different, the concepts reflected by the terms driving forces, pressures, states, and impacts are analogous to the terms activities, sources, proximate causes, and effects shown in Figure 8.1.

More details of the steps in the causal pathway show antecedents of the proximate cause; for instance, increased nutrients lead to increased algal biomass, resulting in low dissolved oxygen after the algae die which negatively affects invertebrates. In practice, measurements of the proximate cause (e.g., dissolved oxygen) may not be available, and a measurement earlier in the causal pathway (e.g., nutrients) might be used instead as a substitute. Using measurements of an antecedent of several proximate causes may be the best option until more knowledge is obtained. Using nutrients as an example again, effects on invertebrates are produced through pathways other than algae including increasing bacteria levels or decomposition rates (Lemly, 2000; Yuan, 2010a,b). Fecal coliform was used as an indicator of organic enrichment in the Clear Fork case (see Chapter 23). Intermediate steps in the causal pathway may also provide the best target for management action, for example, reducing nutrient inputs rather than increasing dissolved oxygen levels by installing aerators.

Interacting or modifying factors are environmental attributes or stressors that can alter the proximate cause, the susceptible entity, or the relationship between two steps in the diagram. Identifying variables that are better treated as an interacting or modifying factor rather than a proximate cause shortens the list of candidate causes (discussed further in Section 8.3.2).

Defining the biological effect as several more specific changes can suggests ways of distinguishing among candidate causes. For example, one of the effects in the Long Creek conceptual model diagram is a decline in EPT taxa. But which taxa declined? Some are sensitive to deposited sediment and some require sandy bottoms. Identifying specific changes can suggest mechanisms and modes of action (see Box 8.3) that have led to the effect and ways to distinguish among alternatives. For example, it may be useful to understand whether fine sediments bury salmon eggs or abrade the gills of adult fish, even though both mechanisms can lead to population declines.

8.2.2 Strategies for Creating Effective Diagrams for Causal Assessments

Diagrams are useful for visualizing conceptual models because they increase the amount of material that can be mentally processed at one time. Relationships between boxes are easier for readers to perceive and use than text descriptions of the same information (Okebukola, 1990; Robinson and Kiewra, 1995). Diagrams accomplish these feats of efficiency by communicating information through both their individual elements and the way those elements are arranged in space. Searching for relevant information is easier because related concepts are given similar shapes or colors and are grouped closely together. Lines linking boxes provide cues that guide readers to the next piece of relevant information (e.g., the next step in a proposed causal pathway) (Larkin and Simon, 1987).

Conceptual diagrams have been shown to increase knowledge retention and transfer to problem-solving activities across a broad range of educational

BOX 8.3 MODE AND MECHANISM OF ACTION

The terms mode of action and mechanism of action are similar and often confused or used interchangeably in the literature. In toxicology, mechanism of action is used to describe changes at the molecular level and mode of action is the functional or anatomical change that produces the effect of interest. For the purpose of this book, we extended these definitions for application to ecological effects that may be above the organism level of biological organization (e.g., population declines and degraded species assemblages). We use mechanism of action when describing the processes by which effects are produced, especially at a finer level of detail or with more specificity (e.g., at a lower level of biological organization). We use mode of action when emphasizing the functional outcomes of those mechanistic processes. For example, acute lethality is an organism-level mode of action that has the same implications at the population level no matter what mechanism brings it about. As an illustration, acute lethality to kit foxes occurs by predation, road kills, and other mechanisms, but they all have the same influence on the outcome of the population model because of their common mode of action (see Chapter 25).

Causal analyses often benefit by grouping stressors by common mechanisms or modes of action. Mechanism of action is frequently used to categorize the ways toxic chemicals produce their effects, for example, cholinesterase inhibition, narcosis, and reactive oxygen generation (Russom et al., 1997; Escher and Hermens, 2002; de Zwart and Posthuma, 2005; McCarty and Borgert, 2006; Suter, 2007). Modes of action can be used to combine causes that share the same functional outcome, as in the kit fox population model described above. Candidate causes can be grouped by mode of action even when the details of the underlying mechanisms are not known.

levels and applications (reviews by O'Donnell et al., 2002; Vekiri, 2002; Mayer and Moreno, 2003; Nesbit and Adesope, 2006). They are particularly effective for communicating information to audiences that have less prior subject area knowledge, like the general public (O'Donnell et al., 2002; Mayer and Moreno, 2003).

There is no doubt that diagrams created for causal assessments can get complicated. The feeling of being overwhelmed by the scale and complexity of an image, called cognitive overload or map shock (Blankenship and Dansereau, 2000; Mayer and Moreno, 2003), is sufficiently common that different strategies for combating it have been investigated. Many of these stem from the principles of similarity, continuity, and proximity used in graphic design (see Table 8.2).

TABLE 8.2
Recommendations for Creating Effective Conceptual Model Diagrams

Recommendation	Rationale	Application for Causal Diagrams
Group related concepts	Space, shape, and color are three ways related concepts can be grouped. Maps that consistently used these characteristics improved users' ability to recall information (Wiegmann et al., 1992; Wallace et al., 1998).	Causal diagrams often use shapes and colors to distinguish effects, proximate causes, sources, human activities, and natural factors (e.g., rainfall) that might influence exposure. The shape or color isn't as important as consistent usage.
Use a hierarchical structure	Students using concept maps organized hierarchically retained more information than those using maps having a web-like configuration, with nodes surrounding a central concept (Wiegmann et al., 1992). In another study, students using concept maps organized in a top-down orientation outperformed those using a concept map with a left to right orientation (O'Donnell, 1994).	A top-down, hierarchical structure is a natural fit for many causal diagrams that begin with human activities and end with biological effects.
Use simple lines to suggest that concepts are related	Lines direct users to related concepts. Some concept mapping programs require that links be labeled. However, embellished links can introduce redundant or confusing information that may make the diagram more difficult to process (Wiegmann et al., 1992; Mayer and Moreno, 2003; Wallace et al., 1998).	Cause-and-effect diagrams are simpler than other types of concept maps, because they tend to use lines to communicate only two concepts: (1) that boxes have an association reflecting a causal process or (2) that several boxes are subset(s) of a more general box. It is relatively easy to visually distinguish between these two linking concepts by using lines with arrows for the first concept, and brackets for the second. Simple, straight lines are effective for this task. Minimize the use of broken or curved lines. Use restraint when embellishing lines or using different types to communicate the importance or type of a connection.

Break up more complex diagrams	Breaking up information is an effective method for managing complexity (see Chapter 5). However, breaking up a complex diagram into a series (called stacking) can interfere with an audience's ability to subdivide the information in ways that make most sense to them. In studies with students, the stacking strategy was most effective with high-spatial-ability students, but decreased information retention in low-spatial-ability students (Wiegmann et al., 1992).	Consider presenting both an overall diagram and more detailed diagrams for each candidate cause, source, process, or activity (e.g., Suter, 1999).
Write it out	Highly verbal audiences may find a sequential text easier to follow than a diagram. Not surprisingly, placing text descriptions near the diagrams makes it easier for readers to scan between the two (Hegarty and Just, 1993).	We recommend including both text and graphical representation. The written narratives provide a way to review that the graphical relationships make sense.
Narrate or discuss it	According to dual-coding theory, receiving information by both verbal and visual modes increases our ability to retrieve it from long-term memory (Vekiri, 2002). Narrating animated diagrams of scientific explanations improved students' performance on problem-solving tests compared with using on-screen text to accompany the animations (Mayer and Moreno, 2003). Discussing conceptual diagrams in teams enhanced learning over individual study (Patterson et al., 1992).	Seek opportunities to present the diagram to colleagues and the broader community.
Animate it	Allowing self-pacing improved students' ability to recall information from a complex diagram and apply the information in a problem-solving task (Blankenship and Dansereau, 2000; Mayer and Moreno, 2003).	Animation features in software programs introduce pieces of the diagram one at a time, providing another way to reveal the model incrementally.

8.3 Refining the List

After generating the initial list of candidate causes and organizing it using conceptual models, the list is revisited and refined to produce the final list of candidate causes that will be evaluated further. The objective is to refine the list in a way that supports the analysis of evidence to explain the effect and also anticipates alternative management actions. Three strategies for refining are: (1) splitting broadly defined candidate causes when they encompass multiple modes or mechanisms of action (see Section 8.3.1; Box 8.3); (2) combining candidate causes when they share a common mechanism of action or source (see Section 8.3.2); and (3) identifying candidate causes that may be deferred, thereby winnowing the list down to what Woodward (2003) calls the "serious possibilities" (see Section 8.3.3).

8.3.1 Splitting Candidate Causes

The initial list of candidate causes may have included some that are broad categories. Sometimes there are reasons that groups of agents are best analyzed in combined fashion (see Section 8.3.2). However, most of the time both the purposes of analysis and management are better served by disaggregating them by causal pathway and mechanism of action.

8.3.1.1 Disaggregate by Causal Pathway

Broad categories like land use can be disaggregated by using the conceptual modeling approach described above to identify the different sources and stressors and effects that they produce. For example, agricultural land use and suburban development are commonly described as causes of undesirable effects. Such broadly defined causes cannot be analyzed with any precision or remediated. However, management actions can reduce inputs of sediment, nutrients, pesticides, or other agents commonly associated with those land uses if they are determined to be causal.

8.3.1.2 Disaggregate by Considering Mechanisms

Multimetric indices for habitat combine measurements for individual proximate causes such as suspended sediment, substrate texture, woody debris, flow velocity, and channel depth. Disaggregating indices like habitat indices into their constituent metrics can help focus analyses on the variables that are most mechanistically related to the effect. For example, decreased pool depth is often an important variable explaining alteration of fish assemblages, because deeper pools can accommodate more and larger fish, but substrate embeddedness is more important for explaining degradation of macroinvertebrate assemblages, because it reflects the loss of interstitial habitat and because invertebrates require little water for immersion.

Sometimes agents are combined without knowledge of how to analyze them in combination. For example, low dissolved oxygen and altered food sources were combined in the Bogue Homo case study, because most of the evidence (nutrient concentrations and chemical oxygen demand) was relevant to both, and it seemed reasonable to the authors that those agents could have combined effects on the macroinvertebrate assemblage (Hicks et al., 2010). However, no mechanism for combined action was identified and it was not possible to combine those variables in a way that allowed the development of an exposure–response model. In the end, the evidence pointed to low dissolved oxygen and little could be said about food resources. Hence, combining candidate causes without a way to test or model their combined effects is likely to be inconclusive.

8.3.1.3 Disaggregate by Considering Management Actions

Some candidate causes have a source in common. For example, "flashy flow" is actually two proximate causes: greater high flows and lesser low flows. Although these causes are produced by the same source (impervious surfaces) they may require different remedial actions. For example, stormwater storage structures can reduce peak flows without enhancing low flows.

8.3.2 Combining Candidate Causes

The same principles of considering causal pathways, mechanisms of action, and management approaches can be used to combine candidate causes. Combining causes can reduce the number of candidate causes to be assessed and compared but it must be done in a way that is consistent with the requirements of the assessment and the management decision.

The information available prior to analysis may be inadequate for deciding whether candidate causes should be combined. For example, causes should not be combined unless there is a reason to believe that they occur together. The same approaches discussed below may also be used after evidence has been derived and each individual cause evaluated. The decision as to whether to combine agents at this stage or later must be made in each case, depending on how much is known about pathways and mechanisms by which effects are produced, and whether the management actions might address individual causes together or separately. In general, we recommend keeping causes separate unless there is a clear rationale for combining them.

8.3.2.1 Combine Proximate Causes with Antecedents into a Single Candidate Cause

Sometimes candidate causes on the initial list are not proximate causes but instead are part of the causal pathway leading to a proximate cause. For example, phosphorus, organic matter, and DO may all be proposed as candidate causes, but in reality they are often part of one causal pathway, leading to low DO as the proximate cause. Hence, they may be listed as low DO due to decomposition of added organic matter or low DO due to phosphorus increasing algal production, respiration, and decay.

Similarly, natural features of a region, such as climate or soil pH, are best included as contributors to the effects produced by a proximate cause. For example, naturally acidic soil pH may influence bioavailability and exposure of biota to metals. Metals would be treated as a candidate cause and low soil pH as a factor contributing to the effects of the metals. It is important to retain naturally occurring candidate causes when they are exacerbated by human activities, that is, they are not present at background levels. For example, phosphorus concentrations and pH are naturally high in many rivers, but fertilizer application can increase phosphorus (and nitrogen) levels leading to increased photosynthesis and large swings in pH over the day.

8.3.2.2 Combine Causes Produced by the Same Source

In some cases, multiple agents stemming from the same source are best listed as a single cause. For example, individual constituents of an effluent can each be listed as a candidate cause, but it may be more appropriate to list the effluent as a single candidate cause, perform whole effluent toxicity tests, model or measure dilution and transport of the effluent, and analyze the evidence that the effluent is the most probable cause (see Chapter 15). This strategy works well when the exposure and effects of the constituents are measured together as in an effluent toxicity test, thereby avoiding the need to generate an exposure–response model from measurements of the constituents and tests of their individual effects. The candidate cause is the effluent and the expression of exposure is the proportional dilution of the effluent. Similarly, the mixture of toxic chemicals in ambient water, sediment, or soil may be treated as a candidate cause and characterized by toxicity testing. For example, soil toxicity, defined by a seedling growth test, may be listed as a candidate cause and compared to soil compaction and infertility as alternative causes of low plant production.

In other situations, an important goal of the assessment is to distinguish or quantify the relative contributions of different constituents of an effluent. For example, an objective of the Athabasca River case study (see Chapter 24) was to clarify the different roles of effluent constituents. In addition, not all causes with a common source can be effectively analyzed in combination. For example, stormwater flow from impervious surfaces is a source of many agents, including toxic chemicals; high flows that remove organisms, damage physical habitat, and remove woody debris; elevated temperatures; and low flows between events. These different agents have different modes of action and types of effects and may be addressed by different management actions.

8.3.2.3 Combine Causes by Mechanism of Action

The effects of causes that share a mode or mechanism and have combined exposures should be treated as a single cause when they co-occur. When modes or mechanisms are known, models can be an effective way to combine causes.

Most of the methods for modeling multiple stressors were developed for analyzing the risks from multiple chemicals. Exposure additivity models (i.e., concentration addition or dose addition) are used for chemicals with the same mechanisms of action such as organophosphate pesticides causing cholinesterase inhibition or neutral hydrocarbons causing baseline narcosis. This may be done by converting individual concentrations into a common toxicity-normalized concentration (e.g., toxic unit (TU)) and then combining them into a measure of combined exposure such as the sum of toxic units (ΣTU). For example, individually measured PAHs have been combined by adding their toxicity-normalized concentrations to estimate their combined toxic effects (Di Toro and McGrath, 2000). Alternatively, the exposure levels for a set of chemicals may be normalized to that of a single chemical with well-characterized toxicity using toxicity equivalency factors (van den Berg et.al., 1998). For nonspecific toxicity (e.g., baseline narcosis) one may assume equal potency on a molar basis (Escher and Hermens, 2002).

Nonchemical causes with the same mechanism of action may be similarly combined. For example, rocks and large woody debris may be combined as hard substrates. For this approach, the exposure to the combined candidate cause is most frequently expressed as the summed amount of the similarly acting agents (e.g., total fines, total habitat structure, total suspended solids). However, not all factors that appear to be the same have the same mechanism of action. For example, deposited sand, silt, and clay are often combined as "fines." Combining these three sediment size categories may be appropriate for gravel-spawning fish, but benthic invertebrates may perceive sand as a different substrate from silt and clay. Similarly, suspended mineral particles (sediment) may be combined with suspended algae and organic particles when the mechanism of action is reduced light for submerged aquatic vegetation or inhibition of visual predation, but not if it is gill abrasion or interference with filter feeding.

8.3.2.4 Identify Causes that may Induce the Effect Jointly

Sets of agents that cause a common effect through independent mechanisms should not, in general, be combined into a single candidate cause. Still it is worth identifying causes that might be working jointly at an early stage so that analyses might be designed to evaluate that possibility. At a later stage in the assessment, responses could be combined when their effect is a component of a modeled, higher-level effect. An example is combining causes of lethality to kit foxes in a demographic model (see Chapter 25). The total effect expected from multiple agents acting independently is generally estimated by summing the responses expected from each individual cause (i.e., response addition).

8.3.2.5 Identify Causes that May Induce the Effect Interactively

Some causes interact with each other to induce effects (i.e., are complex causes) (see Table 8.3). For example, low dissolved oxygen levels and low flow velocities interact to produce asphyxiation of some aquatic insects that rely on the flow of water to transport oxygen to their gill surfaces. If a model is available to quantify the interaction, then the combination can be analyzed as a candidate cause. Interactive models for some pairs of chemicals in laboratory tests can be found in the literature (e.g., atrazine and organophosphate insecticides; Belden and Lydy, 2000). Unfortunately, most interactions are simply identified in experimental results and no general model of the interaction is generated. Pairs of chemicals can usually be adequately represented by summing the responses expected from each (i.e., using concentration or response addition models; U.S. EPA, 2000b). However, a chemical and a natural environmental factor such as temperature or pH are likely to be interactive, requiring a more complex model (Laskowski et al., 2010).

Even when agents are appropriately combined into a single candidate cause, it is important to consider whether any of the agents can cause an effect independently. For example, elevated temperature inevitably contributes to the effects of low dissolved oxygen by reducing oxygen solubility and increasing oxygen consumption by biota, but elevated temperature may also be sufficient to cause effects even when dissolved oxygen is high.

8.3.3 Deferring Candidate Causes

At this point, the list of candidate causes should include only alternatives that someone has argued are worth evaluating further. For this reason, eliminating causes prior to analysis would fail to address a legitimate candidate cause. When the list of candidate causes is long, it may be tempting to shorten the list to those that are considered most likely. However, in most cases, data must be available and analyzed to defend eliminating a cause from consideration, and so assessing it may require no more effort than justifying its elimination from the list. By including them in the assessment, these candidate causes can be compared to others, making the assessment more complete and transparent and decreasing the likelihood of overlooking a true cause or of alienating a stakeholder.

TABLE 8.3

Some Complex Causes that Produce Freshwater Biological Effects

Multiple Agents	Nature of the Combined Effect
Ammonia and DO	Ammonia decreases the oxygen-carrying capacity of fish blood (Smart, 1978).
Ammonia and pH	pH is the primary determinate of the proportion of un-ionized ammonia (NH_3—the more toxic form) versus ammonium (NH_4^+) and affects the toxicity of both forms. U.S. Water Quality Criteria are adjusted for pH (U.S. EPA, 2013b).
Ammonia and temperature	Increasing temperature increases the proportion of NH_3. Invertebrates are more sensitive to NH_3 at higher temperatures, but fish toxicity is not significantly or consistently influenced by temperature. Invertebrate data are adjusted for temperature in U.S. Water Quality Criteria (U.S. EPA, 2013b).
DO and flow	Because most aquatic invertebrates do not actively ventilate their respiratory surfaces, they withstand lower dissolved oxygen levels if flow rates are high (Jaag and Ambühl, 1964).
DO and metals	Low dissolved oxygen increases the toxicity of metals in most studies (Holmstrup et al., 2010).
DO and various chemicals	Low dissolved oxygen increases the toxicity of most chemicals in most studies (Holmstrup et al., 2010).
DO and temperature	Increasing temperature decreases the solubility of oxygen while also increasing respiration in many organisms, thus depleting oxygen and increasing demand in air and water (Materna, 2001).
Freezing and metals	Freezing temperatures increase the toxicity of metals, apparently due to membrane damage (Holmstrup et al., 2010).
Metals and pH	Increasing acidity increases the proportion of metals in the form of free ions, the most toxic aqueous form. It also influences the binding capacity of organic matter and competes for biotic ligands (discussed in Chapter 18).
Metals and temperature	In general, the toxicity of metals increases with increasing temperature, but the effect is variable (Heugens et al., 2001; Gordon, 2005; Holmstrup et al., 2010).
Divalent metals and calcium or magnesium	The toxicity of other divalent metals (e.g., Ag, Al, Cd, Co, Cu, Ni, and Zn) is decreased by calcium and magnesium (Paquin et al., 2002).
Pathogens and temperature	Fish diseases are more common and more severe at higher temperatures (Materna, 2001).
Pathogens and various chemicals	In most studies, the virulence of pathogens or parasites was increased by chemicals, but results were mixed, and in some cases the pathogens and parasites became less virulent (Holmstrup et al., 2010).
Pesticides and pesticides	Pesticide mixtures are typically a little less than concentration additive, and in US agricultural settings, one will dominate the toxicity of a sample and only two or three will significantly contribute (Belden et al., 2007a,b).
Temperature and pesticides	Increased temperature typically increases the toxicity of pesticides (Holmstrup et al., 2010). An exception is the class of pyrethroid pesticides, which show increased toxicity as temperatures decrease (Harwood et al., 2009).

continued

TABLE 8.3 (continued)

Some Complex Causes that Produce Freshwater Biological Effects

Multiple Agents	Nature of the Combined Effect
pH and temperature	The combined effects of high pH (>9) and elevated temperature are independently additive in fish, but low pH and temperature have more than additive effects (Materna, 2001).
Temperature and various chemicals	In general, toxic chemicals decrease the critical thermal maximum (Heugens et al., 2001; Gordon, 2005).

Note: Entries are ordered alphabetically by the first word of the entry.

It might be argued that a candidate cause could be omitted if there is no plausible mode of action linking it to the effect being investigated. For example, overharvesting could be omitted from a causal assessment of liver cancer in fish. Even in such cases, caution is warranted because it could be that the mechanism of action just has not been demonstrated. Even more problematic is excluding a cause based on comparing site concentrations with effect benchmarks, such as water quality criteria. Criteria and other effect benchmarks are intended to protect most species most of the time, but they may not be applicable to a particular effect, species, or site, and sampling may miss periods of high concentrations.

If prioritization is needed to manage a long list of candidate causes, we recommend deferring the least plausible candidates for later analysis. These second-tier candidate causes may be revisited if the results of the causal analysis are weak or ambiguous and an iteration of the process is needed. Documenting the rationale for deferral increases transparency and helps ensure that the deferred candidates are not forgotten. Furthermore, it may be appropriate to defer the consideration of candidate causes that are at an inappropriate spatial or temporal scale for the current decision. For example, a causal analysis of a localized problem may defer consideration of region-wide increases in temperature. The presence of regionally or globally distributed causes should not prevent the identification of local causes that can be remedied.

Some have suggested narrowing the list to candidate causes that have potential management options (Gentile et al., 1999). This is efficient in terms of supporting the decision-makers, but it runs the risk of eliminating an important cause and exaggerating the importance of minor but readily remediated contributors to undesirable conditions. It also precludes the possibility that creative options might be found for remediating causes that are not part of the a priori set of options.

Finally, if a candidate cause lacks data or when available data are untrustworthy, analysis may best be deferred until data are obtained.

The existence of a list of deferred items does not mean that the causal assessment is incomplete. Ideally, analysis of deferred candidate causes will not be needed, because a probable cause is identified, remediated, and the biological condition improved. But if an additional assessment iteration is needed, the deferred list is ready for use.

8.4 Summary

Together with the case description (see Chapter 7), the list of candidate causes defines the problem that will be investigated. Developing the list requires a balance between inclusiveness and restraint. If the true cause is not on the list for consideration, the assessment will either be inconclusive or identify a false or less influential cause. On the other hand, some restraint is needed because each candidate cause requires resources for evaluation. Striking an effective balance requires professional judgment and often diplomacy.

Conceptual models capture the alternative causes and provide a useful framework for the analysis and communication tasks. The models also depict knowledge of causal pathways, a benefit because management actions are often targeted at the sources or human activities that produced the stressors.

With the definition of the case and a list of candidate causes completed, the assessment process proceeds to analyzing data to develop evidence, the subject of the next chapters.

Part 2B

Deriving Evidence

Evidence is information used to evaluate whether an apparent relationship is causal. It is derived from the analysis of data using summary statistics, quantitative models, or logical arguments.

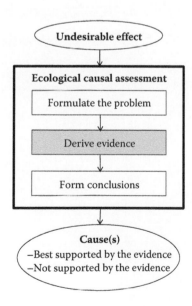

The chapters about deriving evidence are organized based on the source of information:

- Near-site data (see Chapters 9 and 10)
- Regional data (see Chapters 11–13)
- Experimental systems (see Chapters 14–16)
- Symptoms and simulation models (see Chapters 17 and 18)

9

Case-Specific Observations: Assembling and Exploring Data

Susan B. Norton and Michael G. McManus

The chapter describes the acquisition and exploration of data from the affected site and nearby comparison sites. Organizing data by candidate cause and placing them in spatial and temporal context are steps that later support the analysis of associations.

CONTENTS

This chapter is the first of several that discuss data and analyses that are used to derive evidence of causation. Data from the case are the most directly relevant to the causal assessment and provide the best chance of isolating or even directly observing the causal processes that have led to the effect. Although larger data sets support more robust estimates of variability and sophisticated statistical analyses, most large data sets come with a price. They broaden the geographic scope and with it the probability that the data will reflect the influence of many different causal processes which may or may not be relevant to the case. Furthermore, evidence derived using those data sets must be related to case-specific observations. Focusing first on information from the case begins the process of understanding what data are available for deriving evidence directly from the case or in conjunction with models and knowledge from other similar situations.

9.1 Identifying and Acquiring Relevant Information and Data

Because case-specific observations of a biological effect are what prompted the investigation, at least some observations are likely to be available or readily obtainable even in the early stages of an assessment. Chapter 14 describes designs and methods useful for planning studies specifically for causal assessment. Identifying and acquiring available case-specific data and documenting their relevance and quality are not trivial tasks. However, the source and quality documentation systems begun early in the analytical process provide the foundation of a credible assessment.

9.1.1 Sources of Information

Online sources of information, such as Google Earth and Google Maps, provide some of the first pieces of information available to assessors. Spatially referenced data (i.e., data layers) such as surficial geology, stream networks, and watershed boundaries are used to place site data into geographic context. In the United States, stream network information and watershed boundaries are available from the National Hydrography Dataset (FGDC, 2014). Other potentially useful data layers include land use and land cover information and boundaries for ecoregions, which are areas that are similar in vegetation, climate, soils, and geological substrate (U.S. EPA, 2014a). A growing number of databases make information on the location of hazardous wastes sites and toxic releases available to the public (e.g., U.S.EPA's Envirofacts database; U.S. EPA, 2014b). Soil, water, and other data layers can be accessed from a compilation of over 200 resources for spatial data and analysis from U.S. EPA's Geospatial Toolbox (Hellyer et al., 2011).

Site visits provide an on-the-ground reality check that cannot be duplicated by remote data sources. Some observations may have been made during the problem formulation process (see Chapters 7 and 8). These can be reviewed or supplemented with additional visits as analytical tasks begin. Documenting observations in site notes, annotated maps, and photographs are useful memory aids, and in some circumstances can be used directly to develop evidence (see Chapter 10).

Additional data potentially relevant to the assessment come from a variety of sources. Data sources can be identified by contacting local government agencies and universities. Businesses, industries, and community monitoring groups may also have data they are able to share. For example, in the United States, much information on water quality is collected by states. Every state collects basic water quality parameters (e.g., DO, pH, conductivity) through field meters or spot samples, and many also collect samples of algae, macroinvertebrates, fish, and bacteria and document habitat information. Tissue and sediments may have been collected for contaminant analysis. Many US states also perform toxicity testing or have requirements in permits for others to perform toxicity tests and submit reports. Regional water boards or sewer districts may have water quality data, county health departments may have bacterial data, county (or regional) soil and water conservation districts likely have water quality data, and water utilities with surface intakes have water quality data. Outside the United States, many countries have similar programs to collect and store environmental information, especially if the area is suspected to have chemical contamination or poor biological condition.

9.1.2 Assessing Relevance and Quality

Most government agencies and private entities require a plan and systems for documenting data sources and the data's relevance and quality for scientific studies. Even if not required, such systems are good practice for causal assessments. Relevance and quality are two of the factors used to weigh evidence when forming final conclusions. Being able to provide the origins of the information used to form conclusions increases the assessment's credibility when results are communicated. Data may be available in a variety of forms, including hand-written records, spreadsheets, relational databases, and maps. In addition to the data themselves, descriptors associated with the data (i.e., metadata) such as sampling and measurement methods, location, times, and quality assurance codes should be compiled. Tables useful for tracking data include ones that document the origin of each data set and the variables that it includes, and ones that list the measurements that are relevant to each candidate cause (see Box 9.1). Each data set should be traceable back to documents or other records that describe sampling designs, methods, and quality assurance procedures that are later described and referenced in the assessment report.

Outside of site visits, assessors are often faced with limited opportunity to collect new data, especially in early stages of the assessment. For this reason,

---

BOX 9.1   MEASURED VARIABLES RELEVANT TO EVALUATING DECREASED DISSOLVED OXYGEN IN THE LONG CREEK CASE STUDY (EXCERPT FROM TABLE 5 IN ZIEGLER ET AL., 2007A)

- Canopy shade
- Chlorophyll *a*
- Water chemistry, 2000 and 2001 storm flows: total phosphorous, ortho-phosphorous, total Kjeldahl nitrogen, nitrate, and nitrite.
- Water chemistry, 2000 base flows: total phosphorous, ortho-phosphorous, total Kjeldahl nitrogen, nitrate, and nitrite water quality, dissolved oxygen.

much, if not all, of the information available is likely to have been collected for purposes other than causal assessment. Understanding the reason the data were collected can help determine their relevance to and utility for the causal assessment. Status and trend studies may have measured many different parameters, but at a limited number of locations and times. Monitoring programs conducted to meet regulatory requirements (e.g., permits) may repeatedly sample relevant locations over time, but measure few parameters. Targeted studies may focus on one proximate cause or set of causes by controlling for others. For example, a study of the effects of water chemistry may deliberately avoid sites with poor habitat. Even a well-designed study designed to evaluate causation may have investigated one, but not all of the candidate causes.

In addition to relevance, the quality of data determines whether it is appropriate for the purpose of the assessment. The level of quality and documentation needed will depend on the type of causal assessment being conducted. Investigators conducting causal assessments for legal actions will need to carefully document sampling, processing and handling procedures (e.g., chain of custody). On the other hand, preliminary causal assessments, for example, those used to identify the types of measurements and locations for additional sampling, have lower requirements for data quality and documentation. For example, the available data for the Groundhouse River, MN, was used in a preliminary causal assessment which identified excess deposited sediment as the dominant cause (U.S. EPA and MPCA, 2004). These results were used by Minnesota Pollution Control Agency (MPCA) to guide the collection of more data in 2005, which were subsequently used to confirm the initial assessment and identify some other less influential causes (MPCA, 2009). Additional quality issues are discussed in Chapter 11 in the context of analyzing larger regional data sets.

9.2 Organizing Data Using Conceptual Model Diagrams

The conceptual model diagrams described in Chapter 8 provide frameworks for identifying and organizing potentially useful data. The search usually begins with identifying data relevant to quantifying the biological effect and proximate causes (the dark oval and rectangle, respectively, shown in Figure 9.1). However, data relevant to any of the shapes is potentially useful, for example, data on activities or sources such as pesticide application rates or locations of hazardous waste sites.

Measures of the effect characterize the biological responses of primary interest (the dark oval at the bottom of Figure 9.1). The data sets that are used to identify the effects of concern may also include more detailed information on specific responses. If responses are very specific, they may be diagnostic

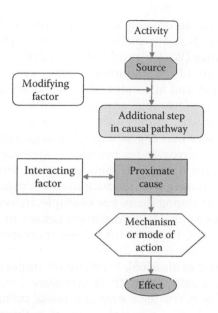

FIGURE 9.1
A conceptual model diagram (also described in Chapter 8) provides a useful structure for identifying and organizing information relevant to a causal assessment.

of the cause or they may eliminate a candidate cause that cannot induce that effect (see Chapter 17).

Measures of proximate causes in the environment include stressor measurements, such as degree of siltation, dissolved oxygen concentrations, or chemical concentrations. These are used to establish whether stressors occur at elevated levels when compared to local comparison sites, regional reference sites, or some other standard. In some cases, the candidate cause is the lack of a required resource, such as nesting habitat. In cases of the absence of a resource, measurements establish that the resource is indeed missing at the place or time it would have been required by the affected organisms.

Measures of sources are useful for quantifying antecedents to candidate causes, for evaluating whether sites differ in the sources that are present, and, after the causal analysis, for identifying actions that can be taken to improve conditions. Source measurements can be difficult to use directly in a site-specific causal assessment because sources often are spatially extensive (e.g., impervious surfaces in an urban area), distribute stressors over large areas (e.g., sulfur oxides from coal-fired power plants), or may contribute multiple stressors (e.g., pesticides, sediments and nutrients from agricultural fields). However, understanding the location and dispersion characteristics of sources can focus sampling efforts and sometimes can be used to eliminate candidate causes. For example, in a Middle Eastern study, an air pollution plume exhibited a continuous concentration gradient through an area of damaged orchards. The smooth spatial gradient provided evidence that the air emission source was an unlikely source of stressors because the degree of decline in the orchards was spatially random rather than decreasing with distance from the source (Wickwire and Menzie, 2010).

When measurements of the proximate causes are not available, information on the location and attributes of possible sources are sometimes used as surrogates. For example, source information can be useful for intermittent stressors (e.g., impervious surfaces as a surrogate for high flow events) or stressors that degrade quickly (e.g., agriculture fields and application rates as a surrogate for pesticide exposures). Fate and transport models use source data to estimate exposure levels at the affected site. Information on sources that produce many proximate causes cannot be used to distinguish among them. For example, increases in impervious surface area have been linked to proximate causes in streams including increased flow extremes, temperature spikes, increased toxic substances, increased salinity, and decreased dissolved oxygen (e.g., Paul and Meyer, 2001; CWP, 2003; Walsh et al., 2005). Estimates of impervious surface alone would not help distinguish among these stressors.

Measures representing intermediate steps in a causal pathway provide opportunities to evaluate whether a complex causal pathway is complete. For example, one pathway by which excess nutrients affect stream biota is by stimulating periphyton growth, which reduces dissolved oxygen through respiration and decay. To evaluate this pathway, data on dissolved oxygen concentrations may be supplemented with data on those steps in the causal pathway such as nutrient concentrations and periphyton biomass.

Interacting or modifying factors are environmental attributes or stressors that can alter the proximate cause, the susceptibility of biota, or the relationship between the two. For example, low pH increases the toxicity of metals (Holmstrup et al., 2010). Lower atmospheric pressure at higher elevations reduces the solubility of oxygen in water (Hem, 1985).

Finally, evidence of relevant *mechanisms or modes of action* may be used to verify that a biologically relevant interaction with the proximate cause has occurred. Measurements might include biomarkers of exposure, tissue residues, or abundances of organisms representing different functional feeding groups (e.g., increase in filter-feeding insects).

9.3 Exploring the Data Using Maps and Timelines

Maps and timelines are essential tools for placing sampling events into spatial and temporal context. Exploring the data in time and space reveals patterns that should be considered when evaluating the association between stressors and responses (the subject of Chapters 10 and 12). Maps and timelines help join data sets that come from different sources. They also are used to identify ways of grouping and comparing sampling events to minimize the influence of natural factors. For example, samples taken within a stream reach constrained by a geological fault would not be comparable to those in an unconstrained reach. Alternatively, it may be advantageous to identify the locations of samples that straddle an influential tributary.

Maps in schematic form (see Figure 9.2) are simplified to emphasize features thought to be important. Plotting sampling locations on aerial images gives a more realistic bird's-eye view (e.g., Figure 9.3; see also Figure 22.1 from the Long Creek case study and Figure 23.2 from the Clear Fork case study). Using GIS can be a valuable way to assemble, organize, and visualize the data for the case (see Box 9.2). Online base maps of topography and hydrography can be combined with sampling locations obtained from local sources. However, additional effort is typically required to obtain spatial coordinates of sampling locations. For environmental assessments, permits issued by government agencies are often available as geospatial data and have been used to investigate cumulative impacts (Lindberg et al., 2011). While National Land Cover Data are available in the United States, users are cautioned that such data are not designed for local applications, such as at the county level (Homer et al., 2007).

A timeline of sampling events is another strategy for organizing and exploring data and placing them in context of natural and anthropogenic gradients. Temperature and precipitation records provide insights into temporal trends and episodic events like storm flows. Stream flow is often an important variable for aquatic assessments. For example, placing the sample timing in context of the hydrograph (see Figure 9.4) helped determine which sampling events to use to characterize exposure

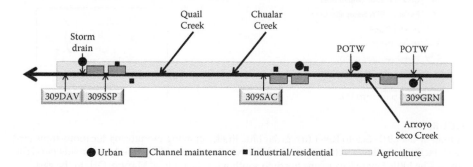

FIGURE 9.2
Schematic map of a portion of the Salina River, California, showing sampling points (in boxes), major tributaries, and potential sources. The Salinas River flows from east to west, and so the direction of stream flow is depicted from right to left (Adapted from Hagerthey, S. E. et al. 2013. In *Causal Assessment Evaluation and Guidance for California*, edited by K. Schiff, D. Gillette, A. Rehn, and M. Paul. Long Beach, CA: Southern California Coastal Water Research Project.)

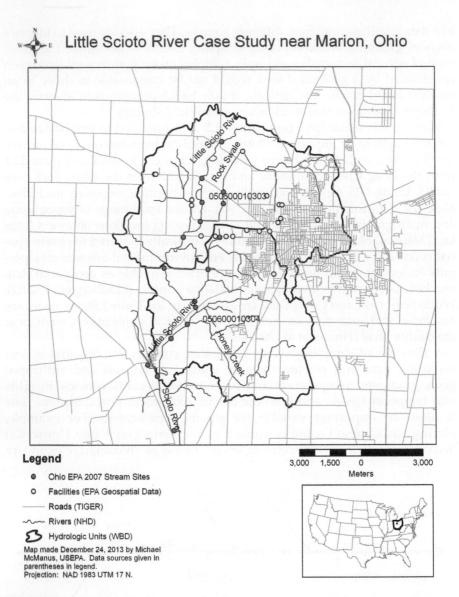

FIGURE 9.3
A map of the Little Scioto River (irregular blue lines), showing monitoring locations (blue circles), facilities (yellow circles), watershed boundaries (heavy black lines), and roads (straight lines). The Little Scioto River runs north to south with the town of Marion, OH, to the east.

of macroinvertebrates to nitrogen in the Salinas River, CA, USA. In the United States, the Geological Survey website StreamStats provides stream-flow statistics (U.S. Geological Survey, 2014a,b). Although small streams are frequently ungauged, flow data from nearby gauges can be interpolated to estimate stream flow at the site using Version 2 of NHDPlus (Horizons

---

BOX 9.2   GIS AND CAUSAL ASSESSMENT

GIS is a technology designed to acquire, store, manage, analyze, and visualize georeferenced data (Goodchild et al., 1993). Software and data sets for use in GIS are rapidly evolving. Current examples of GIS software include commercial products, such as ESRI® ArcGIS, as well as open-source software, such as QGIS and various packages in R (Bivand, 2014). These examples are typically run on a desktop computer. Services for using a GIS online are available including the Geospatial Platform from the partner agencies of the Federal Geographic Data Committee in the United States (FGDC, 2014).

A GIS provides a platform for combining data from many different sources into an integrated map (Waller and Gotway, 2004). Data sets in a GIS format support spatial queries, overlays of different data sets, and calculation of measures of proximity between monitoring sites and potential sources of stressors. More details on performing such spatial queries using GIS tools, functions, and operations, can be found in de Smith et al. (2013). Further advice for applying statistical descriptive techniques and models to spatially-referenced data is provided in Waller and Gotway (2004) and Bivand et al. (2008).

Systems Corporation, 2012). For stressors with daily and seasonal cycles (e.g., dissolved oxygen and temperature), timelines may need to be developed on several temporal scales. Diurnal cycles in aquatic concentrations of stressors such as metals (Nimick et al., 2003) and nutrients (Scholefield et al., 2005) may also be present.

9.4 Pairing Observations in Time and Space

The analyses of associations discussed in Chapters 10 and 12 require that the two variables of interest (e.g., stressor measurements representing the proximate cause and response measurements representing the effect) are paired in time and space. For example, at each location, sediment samples should be taken at the same time and place as the sample of the fish or benthic assemblage. In addition, the time periods must be consistent across all locations used in the analyses. For example, temperature and biota measurements taken in spring at the affected site typically should not be paired with temperature and biological samples taken in summer from the comparison site, because the seasonal shifts may obscure human influences.

When the data come from different sources, sampling locations and times may not exactly coincide, and observations are paired based on professional

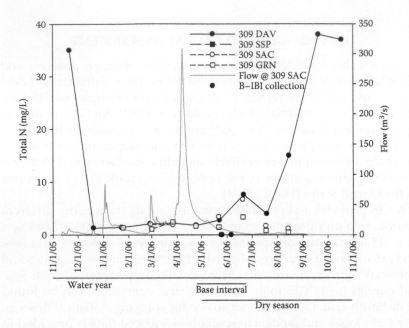

FIGURE 9.4
Biological sampling (i.e., the B-IBI) in the Salina River occurs after scouring flows in spring but before nitrogen concentrations substantially increase in the dry season. Nitrogen concentrations measured in the time period between scouring flows and biological sampling were considered to be most relevant to the assessment. Sampling points 309DAV, 309SAC, and 309GRN are shown in Figure 9.2. (Adapted from Hagerthey, S. E. et al. 2013. In *Causal Assessment Evaluation and Guidance for California*, edited by K. Schiff, D. Gillette, A. Rehn, and M. Paul. Long Beach, CA: Southern California Coastal Water Research Project.)

judgment. A starting point is to pair the sampling events that are the closest to each other in time and space. The temporal stability of measurements should be considered when pairing observations. For example, in the absence of other disturbances, the measurements of large woody debris are fairly constant whereas total suspended solids vary greatly over time and under different flow conditions. Similarly, land cover data taken from national land cover databases need not be matched as closely in time to stressor or biological data as, for example, noise levels at a site.

When pairing stressor and response measurements, we recommend considering how and when the most biologically relevant exposure occurs and to consider alternative ways to pair data other than by the exact time and place. For example, "grab samples" of instantaneous stream temperature collected at the same time as a biological sample may be less relevant than the seasonal average or maximum stream temperature. Dissolved oxygen is best measured when it reaches its diurnal extremes to determine whether critical concentrations occur. The potential for time lags between exposure and effects also should be considered. For example, when a stressor, such as a diversion of water flow, prevents salmon from reaching the sea on their outmigration, the effect (i.e., destruction of the salmon run) will not be observed until that year class returns to spawn years later. For terrestrial systems, spatial variability in both stressor levels and habitat usage are often considered when estimating biologically relevant exposures. When in doubt, analyses can evaluate multiple pairing options, for example, by examining average concentrations along with frequencies of extreme values or by analyzing different time lags. The investigators of the kit fox case study (see Chapter 25) analyzed the relationship between kit fox and prey abundances in the same year and in the previous year, because of the importance of the vixen's nutritional state to reproductive success.

9.5 Summary

Identifying, assembling, and organizing observations from the case are the first steps toward analyzing data for causal assessment. Good systems for documenting the origin, relevance, and quality of data early in the assessment will reap benefits throughout the process, particularly when conclusions are formed and communicated. Conceptual model diagrams help identify data relevant to evaluating different candidate causes as well as important gaps in information. Maps and timelines help place sampling events in the context of spatial and temporal patterns and trends.

Case-specific observations provide the most relevant evidence to the investigation. However, the following issues should be considered before proceeding to analysis.

- Data often come from different sources, may have been collected at different dates and times, and were likely collected for purposes other than supporting the causal assessment. Understanding the reason the data were collected helps determine their strengths and limitations for using them in the causal assessment.

- Data must be paired in time and space in order to support the analysis of associations. Maps and timelines help organize and identify data collected at similar times and places. Subject area knowledge may suggest alternative pairing approaches that are more biologically relevant.

- Exploring the data using both maps and timelines suggests overall patterns and correlations between environmental variables that must be considered when analyzing the data for associations between stressors and responses, which is the subject of the next chapter.

10

Case-Specific Observations: Deriving Evidence

Susan B. Norton, David Farrar, and Michael Griffith

This chapter discusses approaches to use, analyze, and interpret observations from the site where the effect has been observed and nearby comparison site(s) where the effect has not been observed or has been observed in a different way. Evidence from these observations is often the first to be derived in a causal assessment.

CONTENTS

Case-specific observations are often the first data to become available for analysis in a causal assessment. Evidence derived from case-specific observations is valued because it is indisputably relevant to the specific causal event of interest. Most frequently, case-specific observations are used to provide evidence that a proximate cause co-occurred or covaried with the effect. They are also used to link sources or human activities to the occurrence of a proximate cause. Less frequently, they are used to associate the effect with measurements reflecting exposure or a mechanism (e.g., biomarkers). When time-series data are available, they also provide evidence that exposure to the cause preceded the effect in time.

A thorough analysis of case-specific observations also provides the foundation for combining these data with data or information from other sources, discussed in later chapters. For example, Chapters 12 and 13 discuss combining case observations with additional data from larger regional data sets. Evidence also can be derived by comparing case observations to the results of laboratory test results, discussed in Chapter 15.

This chapter begins by discussing sensory evidence from the case, such as might be documented during site visits (see Section 10.1). Although qualitative in nature, this evidence can be just as useful and compelling as that derived using quantitative methods. The remainder of the chapter (see Section 10.2) describes the types of analyses that can be applied to quantitative measurements potentially available at the early stages of an assessment. As described in Chapter 9, these data were likely collected for purposes other than causal assessment. Chapter 14 describes designs and methods useful for planning studies specifically for causal assessment.

10.1 Observing the Presence of Sources, Proximate Causes, and the Steps In-between

Site observations most often are used to document steps in the pathway from human activities to the proximate cause. For example, the location of a large parking lot and stormwater outfall just upstream of the affected site can be documented as a potential source of proximate causes such as salt and oil. The smell of untreated sewage is unmistakable, indicating the presence of a source of organic carbon and bacteria. Bank erosion upstream of the affected site can provide a source of fine sediments. Observations also can be used as evidence that a causal pathway is incomplete. For example, in the Willimantic River case study, the presence of abundant riffles to aerate the water weakened the argument for low dissolved oxygen.

For some candidate causes, site observations can provide visible evidence that proximate causes have co-occurred with organisms. For example, precipitates of "yellow boy" were observed coating the stream bed in Stonecoal Branch (see Chapter 23) providing supporting evidence that acid mine drainage caused the declines in macroinvertebrate assemblage condition. In Buffalo Creek, a tributary to Clear Fork, the observation of heavy deposits of coal fines during monthly reconnaissance visits was considered to be more reliable evidence than sedimentation measurements taken on the day of biological sampling (Gerritsen et al., 2010). Exposure of organisms at the affected sites can also be directly observed. Iron and manganese precipitates have been observed directly on caddisflies in streams below mountaintop mines and valley fills (Pond, 2004). Bacteria have been observed coating mayflies in waters with high nutrients (Lemly, 2000).

Although direct observations make for vivid evidence, as with all sources of evidence, the observations still must be evaluated for quality and relevance to the effect being investigated. For example, it is better to document that a particular source or exposure observed at an affected site does not occur at the comparison site, rather than assuming that it is absent. In addition, the timing or location of an observation may be irrelevant to the effect being investigated. For example, a new housing development adjacent to a stream may have started construction after the time period of interest.

Typically, only a subset of the variables of interest (if any) in a causal assessment can be directly observed and evaluated in terms of presence and absence. The analysis of variables that are always present in some amounts is discussed in the next section.

10.2 Analyzing Associations between Variables

The next two sections describe analytical techniques that can be used to evaluate the strength of association between two variables. The analysis of associations is a useful approach for deriving evidence because many proximate causes are always present in some amount. For example, water bodies have at least some dissolved oxygen. The approaches described in this section extend the concept of "co-occurrence" to the tendency for changes in the level of a stressor to be associated with changes in biological quality.* Section 10.2.1 describes methods that can be used to quantify differences between two locations or times. Section 10.2.2 discusses the methods used for quantifying covariation when paired measurements are available from multiple locations or times.

For simplicity, the discussion will focus on associations between measurements of the proximate cause and the effect. However, the analytical techniques can be applied to any two shapes in the conceptual model diagram described in Chapter 9. In particular, it is good practice to explore the degree of association between different proximate causes. These results are used when interpreting associations between each proximate cause and the effect, discussed in Section 10.2.3.

The analysis of associations described below emphasize visualization techniques such as dot plots and scatter plots, and simple statistics that describe differences in magnitude and degree of association. The use of confidence intervals is discussed, with the caveat that they will be imperfect reflections of variation. Applying conventional statistical methods to site observations can be problematic. Many statistical approaches are based on the

---

* A "greater" level of a stressor may correspond to a lower absolute concentration. For example, lower concentrations of dissolved oxygen in water would be considered a greater degree of degradation—a greater stressor level—for aerobic organisms like fish.

assumption that data are normally distributed, which is rarely the case with environmental data. Environmental data often have a lower bound of zero and a higher frequency of extreme upper values than a normal distribution. Outliers may indicate extreme events that may be more important in terms of biotic response than the central tendency (e.g., mean or median values) of the data. In addition, many statistical methods assume that observations are independent of each other. Environmental data are frequently autocorrelated, that is, observations taken closely together in space and time may tend to be more similar than those taken farther apart. Finally, sample sizes are frequently small (e.g., fewer than 10 samples), precluding the use of many modeling approaches. A larger array of statistical methods can be used with larger data sets and are discussed in Chapters 12 and 13.

10.2.1 Quantifying Differences

The first approach for analyzing data from the case contrasts conditions between locations where the effect is and is not observed. Although differences are typically evaluated in space, they can also be evaluated over time. For example, in the kit fox case study (see Chapter 25), coyote predation was compared between two time periods to associate changes in predation with declines in the kit fox population.

As discussed in Chapter 7, the more similar the environmental conditions are at the affected and comparison sites (or times), exclusive of the candidate cause being evaluated, the more confident the conclusion that the difference in the response was caused by that stressor and not something else. Picture, for example, a case on a stream reach with historical data from before the onset of effects. If ammonia concentrations were the only factor that changed between the times when effects were not observed and then observed, then there is strong evidence that ammonia played a role in the onset of the effects. This simplified example is powerful for two reasons. First, because the location is the same, we have some confidence that many attributes of the environment (e.g., stream size, elevation) are also the same between the two observations. Second, only one candidate cause changed.

Unfortunately, historical data that used the same measurement and sampling procedures are rarely available. Instead, data from the affected site are compared with data from sites where the effect has not been observed. Data from reference sites (i.e., high-quality sites minimally exposed to stressors) may be available as part of status and trends studies (e.g., Bailey et al., 2004 and Box 11.1). However, reference sites will likely differ from the affected sites in many ways, including natural differences. Local comparison sites differ in fewer ways.

Differences can be visualized using maps or timelines. For example, a river-mile diagram of the Athabasca River (see Chapter 24) showed conspicuous increases in chlorophyll *a* near municipal waste and pulp and paper mill outfalls.

The magnitude of the difference is used to evaluate whether stressor levels are greater where effects are observed. Differences can be quantified using many different approaches. Calculations that evaluate the degree to which an observation is considered to be unusual or surprising are useful when few samples from the affected site are available (see Section 10.2.1.1). When more data are available, the magnitude of the difference between the affected and comparison sites can be estimated (see Section 10.2.1.2).

10.2.1.1 Improbability of an Observation

Calculating the degree to which a stressor observation is unexpected or unusual is one way of quantifying the difference between the affected site and a comparison site. Some simple calculations can be used even when very few measurements are available (see Table 10.1).*

One approach sets expectations by quantifying the degree to which a small data set is capable of identifying high values as improbable. The probability of observing a high value relative to a set of observations from a comparison site is placed in context by noting that the comparison site observations define a range of possibilities. That is, N random observations from the comparison site divide the range of possible values into $N + 1$ segments. Therefore, the probability that a subsequent observation is higher than the highest comparison site value is $1/(N + 1)$. In the example described in Figure 10.1, the probability of observing any observation greater than the highest value (15.7) is $1/(5 + 1)$, or 17%. The value of 17% reflects the limited ability to identify a high value as very unusual from small data sets.

If more samples from the comparison sites are available, prediction limits can be used to quantify the degree to which affected site observations would be considered unusual (see Table 10.1). Prediction intervals require the selection of a level of confidence (such as 95% or 90%). For example, observations outside a prediction interval calculated using a confidence level of 95% would be expected to occur only 5% of the time (see Figure 10.2). Multiple prediction intervals can be calculated to evaluate different probabilities of occurrence.

10.2.1.2 Magnitude of Difference

If multiple stressor measurements are available from both the affected and comparison sites, then the magnitude of difference can be evaluated. The magnitude of difference (also called the measure of effect) estimates the degree of change in a response variable associated with a specific change in the stressor variable.

The magnitude of difference can be visualized using quantile–quantile (Q–Q) plots, mean-difference (M–D) plots, or dot plots (see Figure 10.3)

---

* Although a sample size of less than 10 is used to define very small sample sizes in this chapter, there are no rules for the number of observations required for these calculations.

TABLE 10.1

Some Statistics That Estimate the Degree That an Observation is Unexpected

Statistic	Formula	Notes
Very few (<10) observations from comparison locations, 1 from affected location		
Probability of exceeding maximum (or minimum) value (PE)	$$PE = \dfrac{1}{n+1}$$	PE is the probability that any new value from the same population as the n data points would be more extreme than the most extreme value observed in n observations
More than 10 observations from comparison locations, 1 from affected location		
Nonparametric one-sided prediction bounds	$PI_l: x < X_{[\alpha \cdot (n+1)]}$ or $PI_u: x > X_{[(1-\alpha) \cdot (n+1)]}$	The prediction interval side (i.e., upper or lower) must be selected a priori based on biological knowledge. $X_{[1]}, \ldots, X_{[n]}$ are sample values ranked from smallest to largest, e.g., $X_{[1]}$ is the smallest observed value, and so on ("order statistics")
Parametric one-sided prediction bounds	$PI_l: x < \bar{X} - t_{(\alpha, n-1)} \cdot s \cdot \sqrt{1 + \left(\dfrac{1}{n}\right)}$ or $PI_u: x > \bar{X} + t_{(\alpha, n-1)} \cdot s \cdot \sqrt{1 + \left(\dfrac{1}{n}\right)}$	The equations used for data that are or can be transformed to a normal distribution. The prediction interval side must be selected a priori based on biological knowledge

Equations adapted from Helsel and Hirsch (1992).
α is the level of confidence.
n is the number of observations.
$t_{(\alpha, n-1)}$ is the value from the Student's t statistical table corresponding to confidence level α and $n-1$ observations.
s is the standard deviation.

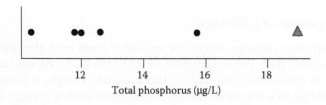

FIGURE 10.1
Example of comparing a single affected site observation of total phosphorus (triangle) with five samples collected at comparison sites (dots). The triangle is outside the range of phosphorus levels that occurs at comparison sites. However, based on this small sample, the probability of observing any value above the maximum is 17%.

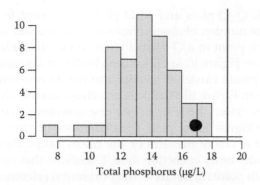

FIGURE 10.2
The observation (line) would be considered to be unusual, because it is above the 95th prediction interval based on comparison site concentrations (the dot).

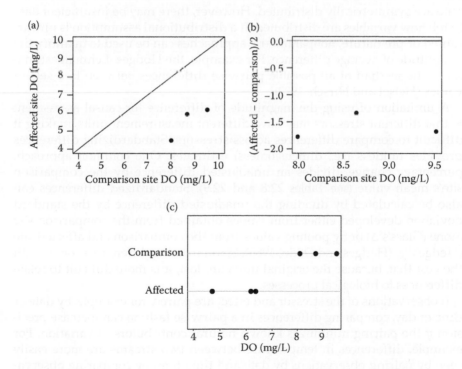

FIGURE 10.3
Plots for visualizing differences in variable values between two locations. (a) Q–Q plot, (b) mean-difference (M–D) plot, and (c) dot plot. The three plots show different ways of showing that DO concentrations are lower at the affected site than at the comparison site. In the Q–Q plot, all affected sites are below the 1:1 line; in the M–D plot, affected sites are displaced from the 0 comparison line; in the dot plot, affected and comparison sites are clearly separated.

(Cleveland, 1993). Q–Q plots and M-D plots are easiest to generate when there are the same number of observations for the affected site and the comparison site.* Each point in a Q–Q plot corresponds to rank-ordered values of each data set (see Figure 10.3a). Locations with similar concentrations produce plots with points clustered around the one-to-one reference line (i.e., the diagonal line in Figure 10.3a). Locations that consistently differ produce plots with points offset above or below the reference line. Each point in an M-D plot (see Figure 10.3b) plots the difference between pairs of points against the value at one of the sites or the mean value. Data can be paired either by rank order or by collection date. Locations that consistently differ produce plots with points offset from the horizontal reference line. Dot plots (see Figure 10.3c) present categorical data with respect to the categories (i.e., comparison or affected site) and the variable of concern (DO).

The question of how much the value of a variable differs between two locations often begins by comparing differences on average over space or time (see Table 10.2). The difference between mean values is most useful when data are symmetrically distributed. However, there may be insufficient data to tell how variables are distributed. If a distributional assumption is unwarranted or premature, nonparametric approaches can be used to quantify the magnitude of average difference. For example, the Hodges–Lehman estimator is the median of all possible pairwise differences between two sets of values (Helsel and Hirsch, 1992).

A limitation of using the magnitude of difference for causal assessment is that different stressors may have different measurement units, making it difficult to compare differences across stressors. Standardizing differences provides unitless (i.e., dimensionless) estimates. One common approach, percentage change, divides an unadjusted difference by the comparison site's mean value (see Tables 22.8 and 22.9). Standardized differences can also be calculated by dividing the unadjusted difference by the standard deviation developed either from values obtained from the comparison site alone (Glass's Δ) or by pooling values from the comparison and affected site (Hedges' g) (Hedges and Olkin, 1985). Standardizing differences comes with the cost that, because the original units are lost, it is more difficult to relate differences to biological processes.

If observations of the stressor and effect are paired, for example, by date or time of day, comparing differences in a pairwise fashion can increase precision *if* the pairing minimizes known natural contributors to variation. For example, differences in temperature between two streams are more easily seen by pairing observations by date and time than by comparing observations that have been averaged over a year.

Uncertainty in the magnitude of a difference can be captured at least partially by calculating confidence intervals. A confidence interval reflects

---

* When the number of observations is unequal, Q–Q plots can be constructed based on interpolated quantiles.

TABLE 10.2

Some Statistics That Estimate the Magnitude of Difference between Sites

Statistic	Formula	Notes
Multiple measurements from two locations (x and y). Measurements are not paired (e.g., by date).		
Nonparametric: Hodges–Lehmann (H–L) estimator[a]	$\hat{\Delta} = \text{median}[x_i - y_j]$ For all combinations of $i = 1, 2, ..., n$ and $j = 1, 2, ..., m$	The H–L estimator is the median of all possible pairwise differences between each x (from a sample of size n) and y value (from a sample of size m). There will be $n \times m$ possible difference
Parametric: mean difference (unpaired)[a]	$\bar{D} = \bar{x} - \bar{y}$	The mean difference is the difference (D) between the means of observations from location 1 (x) and location 2 (y)
Relative difference: percent change (PC)	$PC = \dfrac{\bar{y} - \bar{x}}{|\bar{x}|}$	PC is the unitless difference expressed relative to the mean value observed at the comparison site (x)
Relative difference: Glass's delta (Δ)[b]	$\Delta = \dfrac{\bar{y} - \bar{x}}{s_x}$	Δ is the unitless difference expressed relative to the standard deviation (s) observed at the comparison site (x)
Multiple measurements from two locations (x and y). Measurements are paired (e.g., by date)		
Nonparametric: median difference (paired)[a]	$\bar{D} = \text{median}[x_i - y_i]$ For $I = 1, 2, ..., n$	$\bar{D}$ is the median difference (D) between multiple paired observations
Parametric: mean difference (paired)[a]	$\bar{D} = \text{mean}[x_i - y_i]$ For $I = 1, 2, ..., n$	$\bar{D}$ is the mean difference (D) between multiple paired observations.

[a] Equations adapted from Helsel and Hirsch (1992), which also includes equations and examples for confidence intervals.
[b] Equations adapted from Glass (1976).

uncertainty associated with use of limited, variable data to estimate the true value of a parameter like a median. Confidence intervals reflect the value that a mean or median estimate would likely take if the same sampling program was repeated numerous times. The uncertainty quantified by a confidence interval is only one source of uncertainty in an estimate. Other sources such as biased measurement methods, unmeasured confounders, or uncertainties in conceptualization may also be important.

Nonparametric confidence interval approaches are available, in addition to the more familiar parametric approaches (Snedecor and Cochran, 1989; Helsel and Hirsch, 1992; Hahn and Meeker, 1991). Confidence intervals can also be calculated for standardized estimates of differences, for example, the percentage difference between means or a ratio of response probabilities. These calculations tend to be more involved.

All of these confidence interval calculations are based on the assumption that observations are (drawn randomly from the target population) for example, all samples of benthic organisms in a stream reach during a given year. The calculations also are based on the assumption that each value for each sample is independent of values for another sample. This assumption may be violated if samples are correlated in space or time (i.e., they are auto-correlated) or if the effect of interest is contagious, for example, the incidence of a communicable disease. Confidence intervals can still be calculated if the randomness and independence assumptions do not hold. However, they will reflect sampling and measurement variation only and thus may not fully represent the uncertainties that apply in a given situation.

Confidence interval calculations are closely related to statistical hypothesis tests. However, statistical hypothesis tests are a poor fit for analyzing observational data for site-specific causal assessments. Statistical hypothesis tests were designed for analyzing data from randomized experiments (discussed further in Chapters 15 and 16), where the study is designed to answer a specific question, with adequate power to detect important differences. Ideally, treatments (e.g., exposure to a chemical) are randomly assigned to experimental units (e.g., animals) that are isolated, so that one unit cannot influence the treatment or response of others. The random assignment of treatments will tend to neutralize the influence of any confounding factors, so that a significant difference can be confidently attributed to the effect of the treatment. In observational studies, exposures to stressors are not randomly assigned. For example, the amount of runoff from agricultural fields is not randomly assigned to different streams. For this reason, a significant difference in response cannot be confidently interpreted as an effect of a candidate cause.

In addition, for the types of processes investigated in causal assessment, there is no reason that the usual null hypothesis of zero difference is a reasonable expectation. For example, two locations would not be expected to have identical mean values for most variables. Statistical significance depends not only on the magnitude of biological effect, but also on the amount and variability of the data (see also Box 3.2). With enough data, a result might be "statistically significant" when the magnitude of effect (e.g., percent difference in means) is not large enough to be biologically important. Conversely, a biologically important effect might not be found statistically significant, if assessed based on a small data set. Alternative approaches explicitly formulate the null hypothesis in terms of meaningful differences selected by the investigator based on biological significance or on the distribution of values at the comparison site (e.g., Kilgour et al., 1998). The small sample sizes usually available at this point in the analysis will limit the ability to detect differences, no matter how they are specified.

The size of a confidence interval constructed using observational data is also highly dependent on sample size, the degree of autocorrelation, and the influence of confounding variables. Small sample sizes will result in large confidence intervals. A high degree of positive autocorrelation (e.g., observations closer together in space are more similar than those farther apart) makes confidence intervals artificially narrow. For these reasons, the degree to which confidence intervals overlap does not have a direct causal interpretation; it is just another piece of information that places the magnitude of difference in perspective. An analyst can avoid boiling the analysis down to one often inscrutable number (e.g., a *p*-value) while simultaneously providing more information useful for judging how much stressor values differ. We recommend reporting both the magnitude of difference and confidence intervals, thus providing the information necessary to evaluate each aspect of the results.

10.2.2 Quantifying Covariation

Covariation is the degree to which two variables move together, either both increasing in tandem or in opposite directions (i.e., one increasing as the other decreases). It is most useful for quantifying the association between effects and stressors that influence all of the sites in the analysis, but to different degrees.

The attributes of the association of greatest interest are the direction and the strength of the covariation, that is, the degree to which the level of the stressor variable accounts for the level of the response variable. A strong association, in a direction that is consistent with biological theory, increases confidence (1) that the observed pattern was produced by a direct or indirect causal relationship, (2) that the association is strong enough to be observed over measurement error and natural variation, and (3) that the association (which may have been hypothesized from other studies) is being manifested in the system under investigation.

Scatter plots are a familiar means of visualizing how two continuous variables covary (see Figure 10.4). If one of the variables is expressed as a categorical variable (e.g., the presence or absence of an organism or a habitat feature like large woody debris), dot plots can be used with the categorical variable as the classification variable (analogous to site designations in Figure 10.3c). Using different symbols for the observations from the affected site and comparison sites provides a qualitative check that the overall pattern of covariation is relevant to the case.

Correlations (see Table 10.3) provide a dimensionless expression of covariation. Results range from −1 to 1 with values of +1 indicating a perfect positive relationship (i.e., both variables increase or decrease in tandem), values of −1 indicating a perfect negative relationship (i.e., the two variables increase or decrease in opposite directions), and 0 indicating no association.

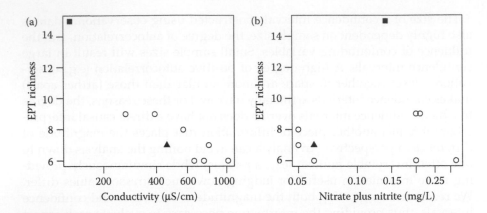

FIGURE 10.4
Example scatter plots. The affected site is the solid triangle; the comparison site is the solid square. Additional nearby observations are shown as open circles. In the Long Creek case study (see Chapter 22), these plots provided evidence that salts (measured as conductivity) were associated with EPT richness, but weakened the case for nitrate plus nitrite.

Contingency tables are also used to quantify the degree of covariation (see Figure 10.5). Contingency tables enumerate observations in different categories and so require that continuous variables of both the stressor and response be classified (e.g., above or below a benchmark value). Using the values of the stressor and response measurements from the affected site as the benchmarks for the classification makes it easier to relate the results to the case.

There are many ways to calculate statistics based on contingency tables (e.g., see review by Fielding and Bell, 1997). Table 10.3 includes calculations for relative odds ratios and relative risk, which are frequently used by human health epidemiologists. Odds ratios and relative risk calculations are unitless, which facilitates comparison across stressors, but can make mechanistic interpretations more difficult.

As with the calculations used to quantify differences discussed in Section 10.2, confidence intervals are preferable to statistical significance testing. Confidence interval calculations for correlations are more involved than those for the magnitude of effect, but fortunately are included as part of most standard statistical packages. Because of the small number of samples that are usually available, exact methods are preferred (e.g., Agresti, 2002).

The statistical calculations shown in Table 10.3 express the strength of the association in relative terms, making it easier to compare results across stressors. However, most calculations assume a linear (e.g., Pearson's correlation) or monotonic relationship (e.g., Spearman's correlation, Kendall's Tau). Some stressor-response relationships would be expected to show a different form. For example, algal productivity increases as phosphorus concentrations increase, but then decreases at higher concentrations. Correlations and reliability statistics can still be useful when observations are from parts of

TABLE 10.3

Some Statistics That Estimate the Degree of Covariation for Multiple Sites with Both Stressor (x) and Response (y) Measurements

Statistic	Formula	Notes
Stressor (x) and response (y) are continuous variables		
Parametric correlation: Pearson's r	$$r = \frac{1}{n-1}\sum_{i=1}^{n}\left(\frac{x_i - \bar{x}}{s_x}\right)*\left(\frac{y_i - \bar{y}}{s_y}\right)$$	Pearson's r estimates the linear dependence of y on x and is dimensionless. If squared (i.e., r^2), it estimates the amount of variation in y that is explained by x
Stressor (x) and response (y) are ordinal (i.e., rank ordered) variables		
Nonparametric correlation: Spearman's ρ	$$\rho = \frac{\sum_{i=1}^{n}(Rx_i * Ry_i) - n*((n+1)/2)^2}{n(n^2-1)/12}$$ where R = the observation rank ignoring group	Spearman's ρ estimates the monotonic dependence of y on x. It can be computed using the equation for Pearson's r on the ranks of observations (Snedecor and Cochran, 1989)
Stressor (x) and response (y) are categorical variables		
Nonparametric correlation: Kendall's Tau (τ)	$$\tau = \frac{S}{n(n-1)/2}$$ where $S = P - M$ and P = the number of $y_i < y_j$ for all $i < j$ M = the number of $y_i > y_j$ for $i < j$ For all $i = 1, ..., (n-1)$ and $j = (i+1), ..., n$ For pairs ordered by x value	Kendall's τ estimates the monotonic dependence of y on x. τ can also be calculated for continuous variables, but will yield lower (absolute) values than Spearman's ρ for the same data
Relative risk (RR)	$$RR = \frac{a/g}{c/h}$$	RR is the ratio of frequency of effects observed at exposed sites, over frequency of effects at unexposed sites
Relative odds ratio (ROR)	$$ROR = \frac{a \cdot d}{b \cdot c}$$	ROR is the ratio of the odds of observing responses at exposed sites over unexposed sites

Correlation equations adapted from Helsel and Hirsch (1992). *RR* and *ROR* adapted from Rothman et al. (2008).

A confidence interval for both Pearson's r and Spearman's ρ can be based on the z transformation approach (Sokal and Rohlff, 1995).

n is the number of observations; s is the standard deviation; a, b, c, d, g, and h are defined in Figure 10.5.

		Undesirable effect		
		Effect observed	Effect not observed	Totals
Exposure	Exposed	a	b	g = (a + b)
	Not exposed	c	d	h = (c + d)
	Totals	e = (a + c)	f = (b + d)	n = (a + b + c + d)

FIGURE 10.5

A contingency table. Variables in the cells are used to calculate odds ratios and relative risk, among other statistics.

the relationship that only increase or decrease. They should not be used if a unimodal relationship is anticipated based on subject area knowledge or the pattern observed in a scatter plot.

10.2.3 Interpreting Associations

The results obtained from analyzing case-specific observations are typically the first to be scrutinized. As additional evidence is accumulated, the interpretation will likely evolve. At first, results are taken at face value. A finding of no association provides evidence that the two variables are not causally related. Conversely, a strong association in the expected direction is evidence that the two are more likely to be causally related. Confidence in face-value interpretations is increased when the variables and their association are clearly linked to the conceptual diagram created at the beginning of the process. If subject-matter knowledge indicates which variables are capable of producing others, that knowledge can and should be used to interpret results.

Associations have greater weight if they are large in magnitude and based on high-quality data (see Chapter 19). Overall, however, these results usually provide weak positive evidence, because stressors frequently covary, resulting in many associations. For this reason, it is good practice to explore the degree of covariation between stressors. Conducting the analyses for all candidate causes is another way to make it clear that multiple stressors occur jointly with the effect. For this reason, the causal analysis rarely ends here. Instead, the results are brought forward and combined with other evidence (e.g., from regional studies or experiments).

It could be that the difference or association may not have been produced by a direct causal relationship between the two variables. Instead, the causal relationships may be more complex. For example, increased runoff from impervious surfaces increases peak flows and decreases base flows of receiving streams. Both peak and base-flow data will likely be correlated with macroinvertebrate richness (see Figure 10.6), even when only decreased base flow is the true cause.

The pattern of associations can give insights into the underlying causal structure that might have produced it and provides the basis for structural

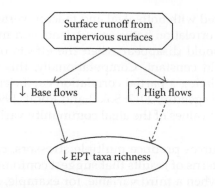

FIGURE 10.6
Decreased base flows and increased high flows will be correlated if they are both produced from surface runoff from impervious surfaces, complicating the analysis of the cause of decreased EPT taxa richness.

equation and path analysis (Shipley, 2000). These analytical approaches can require more observations than are usually available from a case, but the concepts can be applied when interpreting results from smaller data sets. There are only a few reasons, besides direct and indirect causation, that two variables may be associated.

- Noncausal associations between two variables can occur when they are produced by the same antecedent (e.g., a human activity or source). In the example shown in Figure 10.6, decreased base flows and increased peak flows will be correlated because they are both produced by increased surface flow from impervious surfaces.

- Noncausal associations can be produced if both variables have a temporal or spatial trend. For example, *Amelanchier* shrubs (aka shadbush) bloom when the shad run in New England rivers. Or, a correlation could result from mixing data from two regions that have different average values for both variables.

- Noncausal associations may be produced as an artifact of the sampling process. If two variables cause a third (either independently or jointly), they will be correlated in samples selected on the basis of the third variable (aka "collider" bias) (Greenland et al., 1999). For example, if declines in stream stonefly abundances are caused by a combination of temperature and siltation, the subsample of sites with low stonefly abundances will show a correlation between temperature and siltation, even when these two variables are independent across the entire population of sites.

- An association may reflect an indirect causal relationship. For example, nutrients may cause changes in macroinvertebrate abundances by first altering the algal community. If this pattern is true, nutrients will be correlated with both algal and macroinvertebrate endpoints. However, the correlation between nutrients and macroinvertebrate abundances should disappear when the effects of the algal community are held constant. Computationally, this is accomplished either by calculating a partial correlation coefficient (Shipley, 2000; Legendre and Legendre, 2012; Sokal and Rohlf, 1995) or by stratifying on different values of the algal community variable.

Because many sources produce multiple stressors, causal investigators often encounter patterns of results that suggest confounding. A confounding pattern occurs when a third variable, for example, another stressor or a natural spatial or temporal gradient, is correlated both with the stressor and the response. The pattern raises a warning flag because the influence of the third variable can bias estimates of the strength of association between the first two. Strategies for minimizing or statistically adjusting for confounding require larger amounts of data and are discussed in Chapter 13.

In some investigations, there may be a reason to suspect that an association or difference should have been detected, but was not. Some issues that can be mitigated with additional sampling effort or with a different sampling or measurement strategy include the following:

- The stressor variable or response variable has a high degree of measurement error.
- The measurement methods were not sensitive enough to distinguish differences.
- The stressor or response variable are highly variable in time or space (e.g., stressors associated with episodic storm flows).
- The stressor and response variables are not paired appropriately in time or space.
- Different sites are impacted by different stressors which obscures the association.

In rare cases, associations or differences may not be detected because of the influence of other causal processes. Another cause may be influencing the effect in the opposite direction so that the association is not observed. For example, turbid water can shade algae to the extent that they do not respond to increases in nutrients. A stressor may be so common or severe that it obscures the effects of another. In both of these cases, we hope that our process will lead to the identification of a first, most conspicuous group of stressors for management action. Additional associations may become apparent only after the management actions that reduce the first group of stressors have taken place.

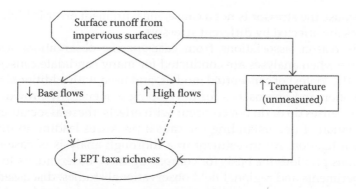

FIGURE 10.7
An unmeasured cause (high temperature in this example) could be the true reason that EPT taxa declined.

Finally, there is always a chance that an unmeasured variable is the dominant cause (see Figure 10.7). Although the issue of unmeasured causes should be kept in mind throughout the causal assessment, in our experience, it is particularly relevant when, at the end of an assessment iteration, none of the stressors or their combinations emerges as a good explanation for the observed effects. Revisiting sources and human activities to ask what additional stressors may be occurring is one way to identify additional causal hypotheses for follow-on monitoring efforts and investigation in a subsequent iteration.

10.3 Summary

Case-specific observations frequently provide the data to derive the first pieces of evidence in a causal investigation. Direct qualitative observations provide useful evidence of presence and absence of sources, proximate causes, and the steps in between the two. For stressors that always occur but in different amounts, associations between stressor and effect measurements can be evaluated using visualization methods and calculations such as standardized differences, correlations, and statistics based on contingency tables.

Simple associations between two variables should be approached cautiously. Many effects can be caused by many different agents and environmental factors and those agents and factors often co-occur or covary. An association between two variables could be produced by a direct causal relationship, by a variable that is part of the causal chain of events (e.g., a source or human activity), or by a confounding factor. Weak relationships can be found because the stressor is not a cause, measurement error is high, or different sites are affected by different stressors.

For this reason, associations from case-specific observations are most informative when analyses are conducted for many candidate causes, interpreted in the context of conceptual models, and used with additional sources of information and approaches for developing evidence. By providing evidence that causes do or do not co-occur with effects, these associations form the foundation of understanding the causal processes leading to the effect under investigation. An investment in a thorough analysis of case-specific observations provides the basis for comparing these observations to results from experiments and regional field observational studies, discussed in the following chapters.

11

Regional Observational Studies: Assembling and Exploring Data

Jeroen Gerritsen, Lester L. Yuan, Patricia Shaw-Allen, and David Farrar

This chapter expands the discussion of data acquisition and exploratory data analysis to observational studies beyond the specific study site. These initial activities are important for identifying the strengths and limitations of the data for deriving evidence (discussed in the next chapter).

CONTENTS

In Chapters 11 through 13, we describe how larger regional data sets can be used with case-specific observations (see Chapters 9 and 10) to develop evidence for site-specific causal assessments. In this chapter, we review types of observational studies that are conducted at places other than the sites under investigation. We describe some of the considerations and potential pitfalls in assembling and exploring these data. In Chapter 12, we describe some methods we have found useful for deriving and interpreting evidence. Chapter 13 describes approaches for identifying and mitigating the influence of confounding variables.

Many of the same considerations relevant to organizing and analyzing data from the case (see Chapter 9) apply to the analysis of broader-scale observational studies, especially selecting biologically relevant measurements and statistics and pairing observations in time and space.

11.1 Studies and Data Sets

The topic of this chapter is the assembly of observational data in preparation for deriving evidence. By regional observational data, we mean measurements that are not associated with a direct manipulation of environmental conditions (e.g., measurements collected during monitoring) and that may or may not have been collected from a probability-based survey (e.g., Cochran, 1965).* Observational data also include measurements collected by sensors on satellites or aircraft. Observational studies include published reports and scientific papers describing insights derived from observational data as well as uninterpreted observational data collected in databases. Information from published articles and reports are subject to the same scrutiny as applied to analyzing new data.

Although the larger number of observations in regional data sets is an opportunity to conduct different and potentially more informative statistical analyses, an upfront warning on analysis of observational studies is in order:

> In our experience, data preparation (assembly, cleanup, and quality control) is the single most time consuming part of using outside observational data. If the data are not already "yours," it may consume half of your resources for analysis.

Much of the data used to generate evidence comes from routine monitoring programs conducted by government agencies at all levels. At the local level, drinking water utilities often monitor water quality at their intakes, and dischargers are often required to report the constituents in their effluents and to monitor some sites downstream from the effluent. In the United States and in many other countries, national agencies monitor weather and climate, hydrology, water quality, air quality, aquatic biology, forest condition, wildlife populations, fisheries, coastal zones, and more. There are also some long-term academic studies in single places, such as the red deer study in the United Kingdom (e.g., McLoughlin et al., 2008) and the U.S. Long-Term Ecological Research (LTER) program (e.g., Hobbie et al., 2003). Some websites provide lists of potential data sets (e.g., Hellyer et al., 2011).

---

* Strictly speaking, data from randomized surveys are not considered "observational" by statisticians, but many state-monitoring databases include mixtures of designed surveys and nondesigned observational data. We categorize both under the term "observational data" in this chapter.

The objectives and design of a monitoring program determine its relevance to the case-specific investigation. Its technical foundation determines its quality. Both relevance and quality influence the weight that is appropriate to assign to the evidence when integrated with other evidence to form conclusions (see Chapter 19).

11.1.1 Assessing Relevance

At the most basic level, the data must contain information about both candidate causes and the biological effects. It is usually not possible to obtain data for every candidate cause, but observational studies that are relevant to even a subset of causes are still useful to obtain and analyze. The following steps can help identify relevant observational data:

- Look for data that, alone or in combination with other sources, link two or more variables shown in the conceptual model diagram (see discussions in Section 9.2). The variables may be relevant to sources, intermediate steps, proximate causes, or the effect.

- Determine whether and where the information exists; obtain metadata to examine methods and the time period that data were collected. Document data gaps. There may be unknown or unmeasured intermediate stressors or factors; these do not disqualify the data, but the greater the complexity and the greater the number of unknown or unmeasured factors, the more difficult the analysis and more ambiguous the results.

- Determine whether there are observations where the stressor and response measurements can be considered to coincide spatially and temporally. For example, it is usually not appropriate to pair stressors that were measured during 1990–2000, with responses measured only after 2005. This does not mean that all observations must be taken at the same time and location; for some measurements, annual averages or periodic observations are acceptable. We will discuss more specifics on pairing observations in time and space below (see also Section 9.4).

- Consider combining data sets from multiple agencies and from multiple studies within agencies to obtain the desired stressor information. As an example, the U.S. Geological Survey (USGS) measures streamflow from many stream gauges throughout the United States. Examining effects of stressors such as flow and flow alteration may require USGS streamflow data in addition to another agency's biological monitoring data.

- Ascertain whether the locations and ecosystem types of the observational studies are relevant to the case. For example, observational studies from mountain streams of West Virginia, United States, are clearly relevant to similar streams and similar organisms from mountain streams of the nearby states of Maryland, Pennsylvania, Virginia, and Kentucky (U.S. EPA, 2011a), but they may or may not be relevant to a region with different geology and topography, such as the coastal plains of Florida, Georgia, and South Carolina.

- Determine whether the stressor gradient in the data set is relevant to the case. Stressor-response models require measurements of an effect (e.g., biomass, abundance, or overt anomalies) in conjunction with varying levels of the stressor. The ideal is to have response data observed at stressor levels ranging from very low to high values. This means the stressor levels in the wider, observational data should bracket the values found at the affected site. The least stressed sites of the region (reference sites; see Section 11.1.4) provide the low end of the relationship, but reference sites are not necessary to develop a stressor-response model.

11.1.2 Assessing Database Consistency and Quality

Monitoring programs vary widely in methodology, design, and scientific rigor (e.g., Yoder and Barbour, 2009). We recommend obtaining data from the original source if possible (agencies that collected the data), rather than from data warehouses (e.g., U.S. EPA STORET/WQX). Also, obtain whole files rather than querying through the warehouse's interface, because important options may not exist in interfaces. Some older data warehouses (e.g., U.S. EPA STORET legacy) lack consistent quality assurance (QA), and may lack metadata. For example, study objectives and sampling design may be missing or very difficult to find in the data warehouse, rendering usability of the data questionable at best. In another example, discharger-submitted data in the National Pollutant Discharge Elimination System (NPDES) database may in some cases consist of permit limits rather than actual observations.

Evaluating the quality of biological assemblage data presents some unique challenges. The data typically consist of taxonomic names of species (or higher taxa) and estimates of abundance, such as counts of the number of individuals, biomass, or percent cover. Considerations for evaluating the quality of a biological assemblage database are described below:

Consistency in sampling methods—Sampling methods should be consistent or made consistent in the initial data analysis, especially for taxonomic information (taxa and counts). Relatively minor differences in sampling effort and level of taxonomic identification among monitoring programs can be reconciled by using the "least common denominators." That is, subsample data to approximate equal sampling effort (i.e., randomly subsampling large intensive samples to the level of effort of the study with a smaller effort) and aggregate taxonomic information to the lowest (finest) common identification (i.e., aggregating species data to genus, or genus to family, as necessary, to attain a consistent level of identification in all the data). Major differences in objectives and sampling design typically cannot be reconciled, and affected data sets should be kept separate in analysis. It is especially important to examine methods when data from multiple sources are to be pooled. Elements of sampling methodology that should be checked include the following:

- Biological response information—The biological response data can range from composition and abundance from one or more biological assemblages, to rate measurements (e.g., photosynthesis, respiration, other gas exchange), to individual measurements on target species or groups (e.g., frequency of diseases or anomalies, stress proteins).

- Sampling frequency—Many aquatic biomonitoring programs sample only during fixed seasonal index periods (e.g., July–October), which is a compromise to try to maximize information while controlling costs, logistics, and safety. Index periods are selected based on known ecology to reduce natural variability, optimize gear efficiency, and maximize the information about the assemblage (Barbour et al., 1999), or to sample at times when stresses are likely to be highest, such as initial stream loadings after dry spells or during base flow when pollutants are least diluted. Large organisms (e.g., standing vegetation, corals, fish, macroinvertebrates) are rarely sampled more than annually to characterize a single site. Periphyton, phytoplankton, chlorophyll *a*, and water chemistry are more likely to be sampled several times a year to characterize a water body because the short-term variability of these measures is very high (e.g., Knowlton and Jones, 2006; Barbour and Gerritsen, 2006).

- Sample collection and processing—Reported field methods should be consistent and well documented. Ideally, the objective of the sampling methods is to obtain consistent samples that are representative of the target biological assemblage or target response and the relevant environmental attributes.

- Taxonomic resolution and consistency—The "lowest practical" identification, to genus or species when possible, is favored because it yields more detail, especially when considering traits of the species (e.g., Lenat and Resh, 2001). Nevertheless, useful information can be derived from less resolved taxonomic identifications such as families (e.g., Gerritsen et al., 2000a). Birds, mammals, corals, plants, and fish are typically identified to species, whereas macroinvertebrates are most often identified to genus. In analysis, mixing levels of identification creates ambiguity (e.g., identifications to family only in a genus-level data set are ambiguous, because the genus is unidentified and unknown). Cuffney et al. (2007) examined consequences of different handling methods for ambiguous taxa. They determined that methods that preserve the largest numbers of taxa in resolving ambiguities were most effective for retaining useful information. Again, the taxonomic resolution rules must be applied consistently across all sites and dates, so the data are comparable. Developing and applying these rules for "operational taxonomic units" are time-consuming but is generally necessary for analysis of the taxonomic data (see, e.g., Cao and Hawkins, 2005).

General QA screening—Every data set is likely to have at least some internal errors that can be hunted down and fixed such as multiple spellings, capitalization and punctuations of class variables, variation in units and surprising or suspicious consistencies (e.g., abundances are multiples of 3, or repeated patterns of numbers). Software may efficiently detect errors such as false categories due to capitalization or spelling differences. Notifying database owners of found problems is good stewardship.

Many recent research and monitoring data sets have an associated, formal sampling and analysis plan or quality assurance plan. Data sets with such plans are generally preferred, and the plans should be reviewed to help assess applicability to the questions being addressed. Nevertheless, a quality assurance plan does not guarantee good quality or relevant data, nor does absence of a plan indicate poor quality. In either case, one needs to assess the sampling design and methods for quality and applicability. Rejection of data for minor misdemeanors can lead to decisional paralysis.

11.1.3 Pairing Observations in Time and Space

Examination of associations between stressor and responses requires adequate co-location of stressor and response measurements in time and space. This does not mean that all observations must be collected at the same time and place, but that they are *representative of the same time and place*. For example, most assemblage information (species composition and abundance) is considered to represent a generation time or more of the assemblage, from several weeks to multiple years. Similarly, integrative estimates of stressors can be developed from single-point-in-time measures. A data set of paired stressor and response observations may need to be "built up" from separate observational studies. In the past, many U.S. state water quality monitoring programs were split into separate biological and chemical programs, with separate sampling designs, locations, and schedules. These uncoordinated programs produced a great deal of unusable data, having numerous locations with chemical water quality data but no biological information, and vice versa.

It may be possible to salvage at least parts of disconnected databases, as well as pairing sites by mapping the sampling stations. We have used the NHDPlus (Horizons Systems Corporation, 2012) as the basis for mapping and pairing biological with chemical and hydrologic stations. Some considerations for assembling a data set from multiple sources, with NHD and flow data as examples include the following:

- Associate each sampling station with an NHD reach. Some monitoring programs have already done this, but in many cases it will be necessary to use the position information [latitude–longitude, Universal Transverse Mercator (UTM) grid] as well as the station description. Determine an acceptable distance between the station location and the stream reach beyond which the data point is rejected as a mismatch.

- Identify biological and water quality stations, and gauging stations on the same NHD reach. Examine these reaches for permitted discharges, as well as nonpoint sources, between the sampling locations. This may require accessing discharge (e.g., NPDES) databases, as well as inspection of aerial images (e.g., Google Earth). If there are no intervening potential sources, the NHD reach can be considered to be the "sampling location," and all samples on the reach refer to that location.

- Determine a time period for characterizing more frequent observations (chemistry, flow, chlorophyll) to associate with the less frequent observations (fish, benthic macroinvertebrate composition). The time period should be biologically meaningful. Depending on the situation, measures of central tendency, or maximum or minimum values, of chemical, flow, and chlorophyll measurements for either 1 year or one growing season prior to each biological sampling event could be considered a single observation space.

In addition to assisting in database development, mapping sampling sites is also useful for identifying potential sources and spatial relationships.

11.1.4 Reference Sites and Conditions

Reference sites are often defined and identified in monitoring programs conducted by government entities such as states, tribes, and provinces (see Box 11.1). The inclusion of reference sites in a data set has the advantage of ensuring that the best (i.e., least disturbed) sites in a region have been sought and included. Such sites would not have been deliberately included in a probability sample or sampling for enforcement. Hence, reference-site data allows for development of a more complete stressor gradient for generating quantitative stressor–response relationships and identifying confounders (see Chapters 12 and 13). However, reference sites are not always relevant comparison sites to determine co-occurrence. For example, some candidate causes may not have been considered when defining the reference or their levels may be too high to provide a no-exposure or even a low-exposure condition. In such cases, comparison sites must be identified ad hoc.

BOX 11.1   REFERENCE CONDITION

We summarize briefly two considerations in using regional reference condition and reference sites, but for further analysis and development, we refer the reader to the rather extensive literature on bioassessment and biocriteria (e.g., Barbour et al., 1999; Bailey et al., 2004; Stoddard et al., 2006). In general, analysts in causal assessment will use reference sites when they are available, but causal assessment does not define or develop reference condition.

Developing Reference Condition

Ideally, regional reference sites are "minimally disturbed"—nearly pristine with minimal or no detectable biological effects of human activity (Bailey et al., 2004; Stoddard et al., 2006). Realistically, "minimally disturbed" is seldom achieved, and the selection of reference criteria is often a mixture of data analysis and professional judgment. Reference systems represent the least disturbed conditions typical for a region and minimally disturbed when possible. Geologically or morphologically atypical sites should be excluded from consideration, because the goal is to define the average and typical regional condition in the absence of human disturbance (Stoddard et al., 2006). The development of reference site selection criteria is a consensus process and draws upon the experience and knowledge of local professionals, many of whom have sampled biological communities across large regions. Land use/land cover data, extent of point-source inputs, habitat surveys, presence of impoundments, human population density, and road density are often used in identifying regional reference sites (Stoddard et al., 2006).

Natural Classification of Biological Data

Many natural regional and habitat characteristics (such as stream size, slope, dominant natural substrate, etc.) also affect the species composition of undisturbed water bodies. Accordingly, a critical step in using data from reference sites is to account for natural sources of variability in biological indicators through discrete classifications or continuous models (e.g., Barbour et al., 1999; Hawkins et al., 2000). Failure to properly account for natural variability can lead to confounding of responses by natural factors. In most cases, biological monitoring programs of widely distributed resources have either identified natural classes (e.g., biota in streams, forests, grasslands) or developed appropriate models.

BOX 11.1   (continued) REFERENCE CONDITION

Accounting for natural variability in reference sites requires examination of biological gradients or assemblage types and associating these biological gradients with natural variables. Potential analyses include nonmetric, multidimensional scaling (NMS), indicator species analysis, correlations, cluster analysis, metric distribution plots, and regression analysis (e.g., Jongman et al., 1987; Wright, 2000; Hawkins et al., 2000).

11.2 Exploratory Data Analysis

Rather than going straight to quantitative analysis, becoming familiar with the data is important for identifying strengths, limitations, ways to improve subsequent analyses, and potential pitfalls. For example, correlations among stressors may limit our ability to infer causes (e.g., Zuur et al., 2010). We become familiar with the data and its quirks through exploratory analyses, which range from very simple descriptive statistics and graphics, to more complex multivariate analyses.

The most important exploratory activity is becoming familiar with the data sets by examining them with tables and graphs, which can include scatter plots, correlation matrices, and box plots. Correlation matrices show stressors that covary and may confound causal assessment. Scatter plots, especially, show the extent of relationships between pairs of variables. Some of these topics were introduced in Chapter 8, where, we discuss additional considerations for analysis of observational data.

Maps are a graphical way to explore and present data that complement scatter plots and correlation coefficients. Scatter plots and correlation coefficients are nongeographic summaries of the variables, whereas maps display the spatial patterns of the variables. Data sets that have the same correlation coefficient and scatter plot pattern may exhibit different spatial trends (see examples in Monmonier, 1993). Concordance among graphs, correlation coefficients, and maps may suggest an important underlying factor to consider for a causal analysis.

11.2.1 Autocorrelation and Independence

Statistical models and tests typically include an assumption that data observations are independent, but observations close together in space and time may not be independent. Consider a measurement of bottom-water dissolved oxygen (DO) at 4:00 am in a lake. Would the DO at 5:00 am be expected to differ? Lake-bottom-water DO goes through predictable fluctuations determined by photosynthesis, respiration, water stratification, and wind-driven circulation, and unless a very unusual event has occurred, DO measurements 1 hour apart will be very similar. Such measurements are not independent and are termed autocorrelated in time. Similarly, water chemistry measurements separated by several hundred meters on a stream reach are spatially autocorrelated, if no intervening tributaries or discharges are present. Land forms and geologic formations extend spatial autocorrelation over longer distances, for example, the expectation is that most streams in a mountainous area will have relatively steep slopes.

The simplest way to deal with autocorrelation is to define sample units (e.g., individual lakes, stream reaches, land areas) in such a way that autocorrelation is minimized, for example, individual lakes are considered independent sample units for most studies, as are stream reaches that are neither tributaries nor their receiving streams. Multiple measures within a lake or stream reach are not independent and usually a single estimate of central tendency (mean, median) should be used as the observation representing that unit.

Autocorrelated observations can be used to develop seasonal or annual estimates of the variable for a relevant sample unit, for example, seasonal averages of nutrients and chlorophyll concentrations for a lake basin (e.g., Knowlton and Jones, 2000). The single seasonal estimates are then independent of similar observations from other places (lakes) or years, and do not artificially inflate sample size. There may be situations where autocorrelation of biotic measures may have been caused by stressors (e.g., responses to multiple wastewater discharges to a stream); in such cases, the apparent autocorrelation is part of the response and is not likely to be inherent to the biotic measurements. These observations can be retained as individual observations.

Widespread spatial autocorrelations, such as those from mountains or geologic formations, may be less of a concern because slope, elevation, and alkalinity are often the primary classification variables for both terrestrial and aquatic systems (e.g., Hawkins et al., 2000). If the biota is sensitive to these factors, a classification prior to examining stressor–response associations and poststratifying the analysis according to the classification will take into account the natural variables (see example in Section 11.2.2).

Spatial or temporal autocorrelation can be examined with an autocorrelogram. This requires data that are evenly spaced in space or time, either set distances or set intervals (lags). Autocorrelation coefficients are calculated for lags $(1, 2, 3, \dots, N - 1)$ and plotted by lag. For methods, see a textbook that covers time-series analysis (e.g., Chatfield, 2004; Legendre and Legendre, 2012). In practice, few environmental monitoring data sets are sampled at consistent spatial or time intervals, so decisions on independence of observations most often will require professional judgment.

11.2.2 Ordination Methods

A limitation of scatter plots and correlations (discussed in Section 10.2.2) is that they apply only to relationships between two variables. When several different variables interact, multivariate approaches for exploring data may provide greater insights.

Ordination methods help to identify variables that structure and differentiate habitats and systems from each other. By identifying natural attributes that influence biological community structure and composition, ordination results suggest classifications that reduce variability, making human-induced changes in biota more apparent. Ordination results also identify potential confounders by identifying natural attributes and stressor variables that covary with each other.

Ordination refers to a group of commonly used statistical techniques that reduce the complexity of many variables (e.g., the abundance of 200 species from 50 sites) into a smaller number of synthetic variables, such that the sites and species are arranged ("ordered") on the new variables. Ordination reduces the complexity of data so that it can be depicted graphically, and relationships among objects can be examined. Samples (sites) that are similar to each other display close together on the ordination graph. Data are typically depicted and expressed as relationships in two or three dimensions.

Three families of ordination procedures that have been used successfully with ecological data include principal components analysis (PCA) and related methods; correspondence analysis (CA) and related methods; and NMS, a distribution-free method. Computationally, all ordination methods use a distance or a similarity matrix among sites or among variables and calculate eigenvalues of the distance matrix to define the principal axes, or use a numerical approximation technique (NMS). There are numerous distance and similarity coefficients (Legendre and Legendre, 2012). For a more complete explanation of these methods, see any text that covers ordination (e.g., Jongman et al., 1987; Ludwig and Reynolds, 1988; Legendre and Legendre, 2012).

PCA is particularly useful for examining environmental variables and metrics that vary monotonically with one another. It is often used to identify which physical and chemical attributes are strongly associated with each other. PCA can also suggest potential grouping or classification of the sites. A simple PCA of measured water quality variables of Florida lakes was used to confirm a classification of lake types (see Figures 11.1 and 11.2; Gerritsen et al., 2000b). Eight water quality variables (pH, alkalinity, conductivity, chlorophyll-*a*, total nitrogen, total phosphorus, platinum-cobalt color, Secchi transparency) aligned on two major axes corresponding to (1) alkalinity/pH and (2) color/ Secchi transparency. Lakes were classified into two groups on each axis: acidic lakes and alkaline lakes, and clear lakes and colored lakes, yielding four lake classes. The statewide PCA confirmed an earlier classification from a smaller set of lakes (Shannon and Brezonik, 1972). In a subsequent correspondence analysis of littoral benthic invertebrate species composition, species composition was primarily associated with water color (see Figure 11.2).

PCA is not effective for use with species composition data, because species are often distributed unimodally along environmental gradients. Also,

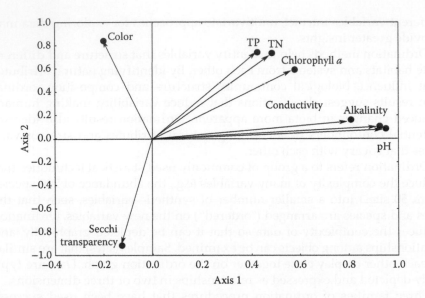

FIGURE 11.1
Principal components ordination of water quality in 570 Florida lakes. Dots and arrows show the projection of each variable onto the first two principal components, which explained 78% of the variation of the data set. Conductivity, alkalinity, and pH are close together and parallel to the first axis and Color and Secchi transparency each point in opposite directions (increased water color results in reduced transparency) along the second axis. (Data from Florida Lakewatch; Gerritsen, J. et al. 2000b. *Development of Lake Condition Indexes (LCI) for Florida.* Tallahassee: Florida Department of Environmental Protection.)

species-by-site matrices often have large numbers of empty cells, representing sites where a given species was not found. Absence falsely contributes to similarity in PCA because the analysis uses correlation as the measure of similarity. For species composition data, we recommend NMS or correspondence analysis.

11.3 Summary

Observational data from outside a particular case provide important support for causal assessments. Although substantial effort is required to develop data sets, it is worthwhile because, once assembled and organized, the data can be used for many investigations.

The process of obtaining data, determining quality, filtering, and reduction, is time- and resource-intensive. Database construction cost is typically

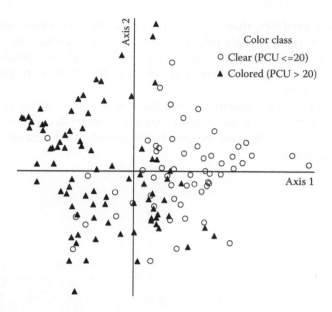

FIGURE 11.2

Ordination (detrended correspondence analysis) of littoral benthic macroinvertebrate assemblages of Florida lakes, showing color classes identified from water quality PCA (open circles; closed triangles). The plot shows that invertebrate species composition differs among lakes according to water color, supporting the use of color as a classification variable. (Modified from Gerritsen, J. et al. 2000b. *Development of Lake Condition Indexes (LCI) for Florida*. Tallahassee: Florida Department of Environmental Protection.)

grossly underestimated in scoping a causal assessment. People wrongly assume that "because data have been collected" in a relevant region that

- The data will be easy to obtain from the original sources;
- The data will effortlessly fall into a relational database with working queries;
- The data are relevant to the questions at hand;
- The data contain all relevant parameters measured everywhere; and
- The data are error-free.

The reality is that obtaining the data, developing a database for the project, and identifying and correcting errors are enormously tedious and time-consuming, and further, after the database is complete, the relevant sample size is substantially smaller than originally estimated. We caution planners of causal assessments to be realistic, in order to allocate at least 50% of data analysis resources for obtaining and assembling data and QA. This still represents a significant cost savings compared to sampling in the field.

Initial data analyses that explore autocorrelation of observations and covariation among variables help improve subsequent analyses by identifying promising classification variables to reduce natural variability and potential confounding factors that may obscure the stressor–response relationship of interest.

Once the data sets are assembled and their potential for analysis has been explored, it is time to reap the benefits and use them to develop associations that can be related to the specific case being investigated. The next chapter begins that process.

12

Regional Observational Studies: Deriving Evidence

Jeroen Gerritsen, Lester L. Yuan, Patricia Shaw-Allen, and Susan M. Cormier

This chapter reviews several approaches for deriving evidence from regional observational studies. Observational studies have the advantage of reflecting realistic exposure conditions, but analyses may be hampered by high natural variability and the influence of confounding factors.

CONTENTS

Regional observational studies provide larger data sets to support the development of empirical models or distributions. Evidence relevant to a specific investigation is derived by comparing observations from the case with the results from the larger studies.

In this chapter, we share experiences analyzing observational studies for causal assessments and point the reader to resources with methods, formulas, and software that have been employed successfully by us and others. Libraries of books have been written on statistical methods. We do not provide detailed instructions on statistics because our goal is to show that there are a variety of ways to develop evidence and to interpret it. Our examples here are all drawn from aquatic causal assessments, but many of the same techniques and considerations apply to terrestrial assessments.

This chapter focuses on the use of observational studies to provide evidence that the proximate cause and the biological effect co-occur or that the level of exposure to the stressor is sufficient to induce the effect. However, the same approaches can be applied to evaluate causal events that lead up to the proximate cause. For example, observational studies have been used to evaluate whether nutrient enrichment has caused periphyton growth in streams, thus increasing diurnal variability in DO (higher highs and lower lows). In this example, low DO is a proximate cause of the degraded benthic invertebrate community (e.g., Miltner, 2010).

12.1 Comparing Stressor Levels and Comparing Effects

One of the simplest ways to generate evidence from observational data is to compare stressor levels at the biologically affected sites of the case to stressor levels at comparison sites that are unimpaired or less degraded. For example, comparisons of levels of a stressor between the site and regional reference sites define whether the levels of stressors at the site under investigation differ from those at sites with high biological quality. Data showing that a stressor co-occurs with the effect at levels outside the range associated with high-quality biological conditions is evidence that supports that candidate cause. Data showing that levels of the stressor at the affected site are within the range associated with high-quality biological conditions weakens the case for that candidate cause (see Chapter 23, Clear Fork case study). This approach is most applicable to investigations prompted by effects observed as part of a biological monitoring or assessment program. It uses the definition and description of the reference condition defined as part of that program, typically conducted by a state, province, or other government entity (see Box 11.1). For the most part, regional reference sites and regional reference conditions have already been developed by the state or other agencies prior to assessing biological condition in the case itself. Because reference sites are identified for purposes other than causal assessment, the criteria for their selection should be reviewed before using them.

Comparisons of stressor levels to those at reference sites can be complemented by comparing stressor levels to groups of sites where similar effects have been observed. Data showing that, at the affected site, a stressor occurs at levels at or above the range associated with similar effects provides evidence that supports a candidate cause. Data showing that the stressor was below the ranges associated with similar effects weakens the case for that candidate cause.

We describe two broad approaches for performing these comparisons: simple categorical comparisons and models to account for natural gradients that may be seen with continuous data. Simple comparisons include graphical comparisons of stressor levels or responses; contingency tables to determine association; and use of contingency tables to estimate relative risk, which calculates the risk of observing effects based on given stressor levels. Natural gradients that may confound the simple comparisons could include temperature, habitat or catchment area, slope, elevation, and others, and can be modeled with regression models.

12.1.1 Graphical Comparison

Stressor levels associated with biologic effects in the case can be compared with site groups categorized by different stressor levels or different effect levels. Categorical comparisons can be made graphically (e.g., using box plots) or in tabular form (e.g., using a contingency table of frequencies of observations in different categories).

As an example, consider data on mayflies in streams and the potential effect of dissolved aluminum on mayfly abundance (see Figure 12.1). The data are

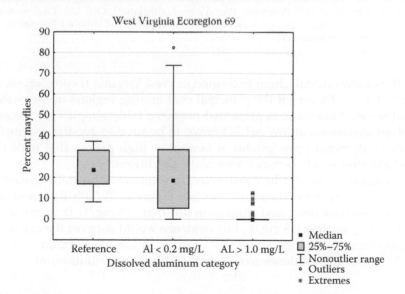

FIGURE 12.1
Box plot showing the percent of mayflies in benthic invertebrate samples from streams in the Central Appalachians of West Virginia (Ecoregion 69) with three categories of dissolved aluminum (Al): reference sites (Al often not measured), Al < 0.2 mg/L, and Al > 1.0 mg/L. (Data from West Virginia DEP, also used in Gerritsen, J. et al. 2010. *Inferring Causes of Biological Impairment in the Clear Fork Watershed, West Virginia.* Cincinnati, OH: U.S. Environmental Protection Agency, Office of Research and Development, National Center for Environmental Assessment. EPA/600/R-08/146.)

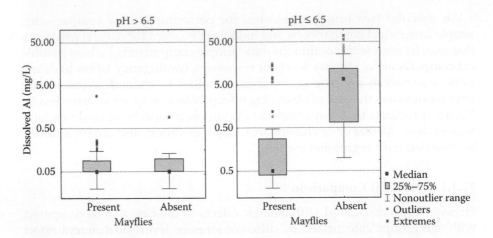

FIGURE 12.2

Box plots of dissolved aluminum concentrations in sites with and without mayflies, in neutral or alkaline pH (left), and in acidic conditions (right) in the Central Appalachian Ecoregion (69) of West Virginia. Mayflies are rarely present at sites with dissolved aluminum >0.5 mg/L and pH < 6.5. Dissolved aluminum >0.5 mg/L almost always occurs only in sites with low pH. (Data from West Virginia DEP, also used in Gerritsen, J. et al. 2010. *Inferring Causes of Biological Impairment in the Clear Fork Watershed, West Virginia.* Cincinnati, OH: U.S. Environmental Protection Agency, Office of Research and Development, National Center for Environmental Assessment. EPA/600/R-08/146.)

from the Central Appalachian Ecoregion of West Virginia (Ecoregion 69; see Woods et al., 1999), one of the principal coal mining regions in the eastern United States. Aluminum is presented here as a fairly simple case, because dissolved aluminum at low pH is known to be toxic to aquatic organisms. Dissolved aluminum precipitates at neutral or high pH, so that toxic dissolved aluminum only occurs under acidic conditions.

Site categories can also be constructed to compare aluminum concentrations at sites with and without mayflies (see Figure 12.2). Note that most sites with mayflies have dissolved aluminum less than 0.5 mg/L. If aluminum at the site is higher than 0.5 mg/L, this evidence would support the argument that aluminum caused the absence of mayflies. If the concentration is lower than 0.5 mg/L, this evidence would weaken the case for aluminum.

12.1.2 Categorical Data: Contingency Tables and Relative Risk

The examples above examine either a categorical stressor (aluminum high or low; see Figure 12.1) or categorical response (mayflies present or absent; see Figure 12.2). These same data can also be analyzed in contingency tables to estimate the probability of the relationships, but more appropriately for causal analysis, to estimate relative risk of finding an adverse effect given certain conditions. Relative risk is frequently cited in public health news, for example, the risks of behaviors or condition (e.g., smoking, drinking, obesity) contributing to health outcomes (e.g., cancer, heart disease, diabetes).

Contingency tables are easiest to relate to the case being investigated if at least one category for the table spans the range of exposure conditions at the affected site. For example, if the affected site has a pH of 5.5 and dissolved Al of 0.75 mg/L, the category thresholds for Table 12.1 would be appropriate.

The familiar chi-square statistic calculated from contingency tables tells us that the results of the contingency table are extremely unlikely to be due to chance, that is, a strong relationship exists between dissolved aluminum concentration and absence of mayflies. We cannot analyze the effect of dissolved aluminum at pH greater than 6.5, because dissolved aluminum does not occur at those pH values.

TABLE 12.1

Frequency of Mayfly Occurrence in Benthic Samples of Ecoregion 69 in West Virginia, for Two Conditions of Dissolved Aluminum

Mayflies	Low Al (<0.5 mg/L)	High Al (≥ 0.5 mg/L)	Row Totals	Chi-Square
A. All sites, Ecoregion 69				
Present	397	10	407	
	(83.2%)	(2.1%)	(83.3%)	
Absent	31	39	70	
	(6.5%)	(8.2%)	(14.7%)	
Column totals	428	49	477	183.8
	(89.7%)	(10.3%)		$p = 0.0000$
B. Ecoregion 69, pH > 6.5[a]				
Present	345	1	346	
Absent	22	1	23	
Column totals	367	2	369	
C. Ecoregion 69, pH ≤ 6.5				
Present	52	9	61	
	(48.1%)	(8.3%)	(56.5%)	
Absent	9	38	47	
	(8.3%)	(35.2%)	(43.5%)	
Column totals	61	47	108	47.18
	(56.5%)	(43.5%)		$p = 0.0000$

Note: A. All Sites; B. Sites with pH > 6.5 (Corresponds to Left-Hand Side of Figure 12.2); C. Sites with pH ≤ 6.5 (Corresponds to Right-Hand Side of Figure 12.2.)

[a] Chi-square value and percentages are not shown because there were too few high aluminum observations.

The chi-square told us that we could be confident that there was an association between dissolved aluminum and absence of mayflies, but it told us nothing about the magnitude of the association, only that it is "more likely than expected by chance." Relative risk gives us magnitude.

Relative risk (RR) (see equation in Table 10.3) is defined as the probability of an adverse event (say, lung cancer) given exposure (smoking), divided by the probability of the adverse event, given nonexposure. For the aluminum example, RR would be calculated as the probability that mayflies are absent, given that aluminum ≥0.5 mg/L, divided by the probability of mayflies are absent, given that aluminum <0.5 mg/L (U.S. EPA, 2006a; Lachin, 2000).

The estimated relative risk of mayfly absence given high dissolved aluminum is: RR = (39/49)/(31/428) = 10.98. Thus, a stream with dissolved aluminum ≥0.5 mg/L is 11 times as likely to lack mayflies as a stream with dissolved aluminum <0.5 mg/L based on this data set. Aluminum at concentrations higher than 0.5 mg/L at the site would support the argument that aluminum accounts for the absence of mayflies.

Of course, all other things are not equal, and in this case, we also know that high dissolved aluminum only occurs under acidic conditions. Because acidity may also adversely affect mayflies, acidity is a confounding factor for the effects of aluminum. What if acidic sites are examined, where pH ≤6.5? Is the relative risk of high aluminum still greater? Table 12.1C shows mayfly presence under acidic conditions, with high and low dissolved aluminum. The relative risk is now RR = (38/47)/(9/61) = 5.48. A stream with high dissolved aluminum is 5.5 times as likely to lack mayflies, compared to the low aluminum condition, in acidic streams. This is less than the RR of 11 in all streams and indicates that (1) at least some of the risk may be due to acidic conditions and (2) high dissolved aluminum still has RR greater than 5 under acidic conditions.

Table 12.2 shows a contingency table for the association with pH alone, when high aluminum sites are removed (it is the two left columns of Table 12.1B and

TABLE 12.2

Frequency of mayfly occurrence in benthic samples of Ecoregion 69 in West Virginia, for two conditions of pH

| Mayflies | Ecoregion 69. Al <0.5 mg/L | | Row Totals | Chi-Square |
	pH > 6.5	pH ≤ 6.5		
Present	345	52	397	
	(80.6%)	(12.2%)	(92.8%)	
Absent	22	9	31	
	(5.1%)	(2.1%)	(7.2%)	
Column totals	367	61	428	5.97
	(85.7%)	(14.3%)		$p = 0.0145$

Note: Dissolved aluminum (Al) <0.5 mg/L throughout. Data as in Table 12.1.

C). Chi-square has a probability of 0.0145, showing there is also an association between low pH and mayfly extirpation not likely to be due to chance. It is clearly less strong than the association with aluminum, but how much less? The relative risk is now $(9/61)/(22/367) = 2.46$. A tentative conclusion from these tables is that the risk of mayfly loss from dissolved aluminum at low pH is nearly twice the risk of mayfly loss from low pH alone. The conclusion is tentative because we used the same data set (cut and excluded in different ways) for all the relative risk estimates, so the data were not independent.

How is this useful for causal assessment? Evidence is derived by placing site data in the context of the contingency tables, for example, the ones described above that quantify the relationships between acidity, aluminum concentration, and mayfly occurrence. Suppose an affected site lacks mayflies. If the stream is acidic and has high dissolved aluminum, then there is strong evidence that acidity and aluminum are the cause. Aluminum present at relatively low concentrations would weaken the argument for aluminum toxicity. Low aluminum and circum-neutral pH would weaken the argument for either acidity or aluminum causing the lack of mayflies.

12.1.3 Regression Models of Natural Variability

The comparison of stressor levels at the affected site to sites that lack effects can be refined for stressors along natural gradients. For example, temperatures naturally decrease with increasing elevation. Natural gradients can be modeled using reference sites (i.e., sites where effects are not observed), but minimally stressed reference sites are not necessary—only sites with low levels of the candidate stressors to form the less stressed end of the gradient. The stressor values at the affected sites then can be compared with unstressed or low-stress expectations.

Regression analysis methods are one way to describe how stressor values change along natural gradients. Linear regression tools are available in most spreadsheets and statistical programs. Regression analysis develops a quantitative relationship between one or more explanatory (also called independent) variables and a response (also called the dependent) variable. In this application, the explanatory variable is the natural gradient, the dependent variable is the stressor, and the data include samples from only unaffected (e.g., reference) locations.

Typically, the estimated regression line with confidence or prediction intervals is superimposed over the plotted data (see Figure 12.3). Confidence intervals provide an estimate of the range of possible values for the estimated mean response for any given values of explanatory variables. Prediction intervals provide an estimate of the range of possible values of the response of an individual sample and are usually the most appropriate interval to use when placing individual site observations into context. Selection of a prediction interval, say 90 or 95%, depends on one's confidence in the data and what they represent, the purpose of the regression in the causal analysis,

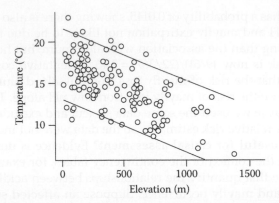

FIGURE 12.3
Stream temperature (°C) vs. elevation (m) in Oregon, USA. Solid black lines are the 95% prediction intervals. The exposure levels outside the boundaries of the lines would support temperature as a candidate cause of biological degradation. (Data collected by Oregon Department of Environmental Quality and used in U.S. EPA. 2006b. Estimation and Application of Macroinvertebrate Tolerance Values. Washington, D.C U.S. Environmental Protection Agency Office of Research and Development. EPA/600/P-04/116F.)

and one's willingness to accept error. For example, high confidence that the data reflect "minimally stressed" reference sites (Stoddard et al., 2006) would suggest a 95% prediction interval or higher.

How is this information useful for causal assessment? Stressor levels that fall outside the prediction interval would be considered to be beyond the range observed at reference sites. That is, the site is no longer similar to reference for the particular stressor. This would support the case for the candidate cause. For example, in Figure 12.3, observing a stream at 1000 m elevation, with a degraded biological community, and having temperature of 20°C would support the argument that temperature was the cause, because temperatures were higher than those expected at that elevation. The more stressor levels depart from the unstressed condition, the stronger the evidence for that stressor.

Regression analysis can be used to fit relationships between any variables with little consideration of the underlying assumptions. However, when the estimated relationships are used to predict likely values of *y* at new values of the explanatory variables, or when the estimated relationships are interpreted with respect to whether they accurately represent the underlying physical or biological relationships, the theoretical assumptions must be considered more carefully. More specifically, one must assess whether the assumed functional form (straight line in Figure 12.3) is sufficiently representative of the actual relationship, whether the sampling variability in *y* is distributed as assumed, whether the magnitude of the sampling variability in *y* changes across the range of predictions, whether the samples used to fit the model are independent, and whether errors in the measured values of the explanatory variables are small enough to be ignored. The assumptions, and how to examine them, are discussed in almost any statistics or regression textbook, as well as on the CADDIS website (Suter et al. 2010b).

12.2 Developing and Using Stressor–Response Models

A powerful use of external data is for building empirical models of biological responses to stressors, often with ranges of stressor values beyond those found in the case itself. Comparing the levels of stressor and responses observed at the case to models with continuous variables developed from data observed elsewhere can provide convincing evidence that a candidate cause both occurs at the site and is sufficient to produce effects.

A typical stressor–response model quantifies a change in a biological variable with changing (increasing or decreasing) exposure to a single stressor or a set of stressors that consistently and strongly covary. The biological variable is usually some attribute of an organism, population, or assemblage. Examples of response variables include abundance, biomass or occurrence of a sensitive species or taxa groups, occurrence of anomalies, or levels of biomarkers in sampled individuals.

The use of field observational data to estimate stressor–response relationships has strengths because many pollutants and effects do not lend themselves to laboratory testing. Migration, spawning, predation, and other behaviors are seldom included; tests of large species are logistically prohibitive. Endangered species are protected from routine testing. Complex exposure pathways and bioaccumulative chemicals are not readily tested. Susceptible species and sensitive life stages may be difficult to maintain and test in the laboratory. Effects that involve interactions among species are not included. Long-term effects due to short-term exposure (e.g., reproductive effects resulting from exposure during a critical stage of development) are rarely measured. In addition, the relative sensitivity of most species is not known a priori, and it is impractical to test even a substantial fraction of the species inhabiting an ecosystem. Also, some exposures are impractical to replicate, such as highly variable concentrations and interactions within mixtures and with the environment.

Limitations of using field observational studies include natural variability, which may reduce the ability to detect the signal, and the potential influence of confounding factors. In this section, we discuss building regression models in circumstances where confounders are assumed to be negligible, unimportant, or mitigated. Then we discuss quantile regression, where it is recognized that confounders are present, but assumptions are made about responses that allow the confounders to be ignored. Finally, we discuss the development of species-sensitivity models to obtain estimates of thresholds of stressors that may be important for environmental management. The identification and mitigation of confounding variables are discussed in Chapter 13. Several analytical approaches can help reduce the confounding effects of multiple, associated stressors. However, it is generally not possible to completely eliminate confounding stressors through analysis of observational data.

12.2.1 Models of Stressor–Response Relationships

Regression models are frequently used to quantify the relationship between stressors and biological responses, for example, macroinvertebrate species richness as a function of fine sediments. The application of regression techniques for stressor–response modeling is similar in concept to estimating how a stressor varies along natural gradients discussed in Section 12.1.3. If a scatter plot looks like it can have a line drawn though it, then regression can define the line. An ideal stressor–response model possesses a clearly defined functional relationship, for example, a straight line (see Figure 12.4), unimodal (see Figure 12.5), or S-shaped (see Figure 12.7, Ephemerella).

How is this information useful for causal assessment? Observations from the affected site that are consistent with expectations quantified by a stressor–response model for a candidate cause would support that cause. For example, Figure 12.4 shows a regression model of the number of EPT taxa as a function of conductivity. (Note that this data set was trimmed to minimize the effects of confounding factors, a technique described in the next chapter.) In the Clear Fork example (see Chapter 23), the explanatory variable is the stressor under investigation (conductivity), the response variable is the biological metric (number of EPT taxa), and data include observations from sites with high and low conductivities. Site values within the prediction intervals are considered to be consistent with the relationship predicted by the model and support the candidate cause. The samples from Clear Fork, indicated by diamonds in Figure 12.4, fall within the prediction intervals of the model, providing one piece of evidence that conductivity caused the decline in EPT taxa, that is, the biological response observed at Clear Fork is consistent with expectations described by the regression model. In contrast, two sites highlighted in the ellipse in Figure 12.4 have much lower numbers of EPT taxa than expected based on the relationship. For the two highlighted sites, the argument for conductivity is greatly weakened by indicating that conductivity alone is insufficient to account for the low numbers of EPT taxa at those sites.

After developing a regression model, most programs provide statistics that describe the characteristics of the estimated fit to the data. These statistics are useful for judging the quality of the model and the strength of the relationship. They include estimated values for the coefficients, the standard errors and p-values for those coefficients, and a measure of the degree the model accounted for observed variability (R^2). Discussion of these statistics and optimal regression methods is beyond the scope of our discussion. Several existing resources provide complete explanations for these different

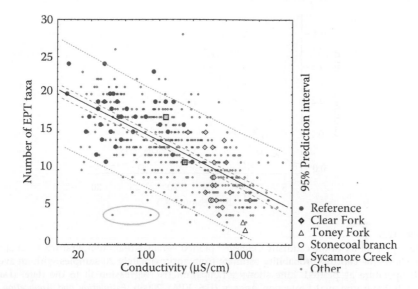

FIGURE 12.4

EPT taxa richness and conductivity in Central Appalachia. Named sites are from the case: Clear Fork and three tributaries. Reference sites and "Other" sites are from outside the case. "Other" sites were selected to remove the influence of some common confounding stressors: acid mine drainage (sites with pH < 6 removed), degraded habitat (sites with habitat score <128 removed), and untreated domestic wastewater (sites with fecal coliform >400 colonies removed). Regression (solid line) and intervals were based on the data subset with confounding stressors removed. Inner dashed lines are 95% confidence interval for the regression; outer dashed lines are the 95% prediction interval. Clear Fork sites (the case) were excluded from the regression analysis, but are plotted to show how they fit on the EPT–conductivity relationship. This evidence shows that the number of EPT genera declines as conductivity increases. Note the two sites in the ellipse. These sites are biologically impaired and well beyond the 95% prediction interval. Conductivity acting alone is highly unlikely to be a cause of the adverse effects at these sites. (See also Chapter 23, Figures 12.1 and 12.2, and Tables 12.1 and 12.2). (Adapted from Gerritsen, J. et al. 2010. *Inferring Causes of Biological Impairment in the Clear Fork Watershed, West Virginia.* Cincinnati, OH: U.S. Environmental Protection Agency, Office of Research and Development, National Center for Environmental Assessment. EPA/600/R-08/146.)

statistics (e.g., Draper and Smith, 1998; Harrell, 2001; Kutner et al., 2004; McDonald, 2008).

Other types of regression models can also be used to estimate stressor–response relationships. For example, nonparametric approaches were used to model the probability of capture of a caddisfly genus (*Calineuria*) as a function of temperature (see Figure 12.5). This regression model would support the argument that a site with temperatures greater than about 20°C would be sufficient to reduce the probability of observing this caddisfly in a sample. Observations of temperatures around 15°C would weaken the argument for temperature reducing the probability of observing the caddisfly. Taxon–stressor relationships have also been used to identify stressor levels

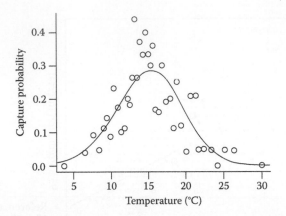

FIGURE 12.5
Capture probability of the caddisfly *Calineuria* plotted versus stream temperature. Each open circle shows the capture probability estimate from approximately 20 samples with an average temperature as plotted. Line shows a nonparametric regression fit to the data. (Data from U.S. Environmental Protection Agency (U.S. EPA). 2006b. *Estimation and Application of Macroinvertebrate Tolerance Values.* Washington, DC: U.S. Environmental Protection Agency, Office of Research and Development. EPA/600/P-04/116F.)

that correspond to a very low capture probability (e.g., extirpation concentrations ($XC_{95}s$)) (U.S. EPA, 2011a). They have also been used to infer the level of stressors, which may be useful when biologically relevant stressor measurements are difficult to obtain (see Box 12.1).

12.2.2 Quantile Regression

Biological responses to candidate stressors frequently show a wedge-shaped plot, as in Figure 12.6. The stressors in Figure 12.6, percentage of sand and fines, and total nitrogen, co-occur with many other stressors in the stream. If the only stressor in the data set were sandy sediment, then

---

BOX 12.1   TAXON–STRESSOR RELATIONSHIPS AND MODELS

If the observational data set is large and includes occurrence and abundance of individual taxa as well as stressor measurements, it may be possible to build taxon–stressor relationship models with some of the same techniques outlined in this section. These models can be used to infer the stressor concentration from the taxa abundances, if the stressors have not been measured in the case data, or to show that the stressor is producing expected biological changes (for more information, see Yuan, 2010a,b).

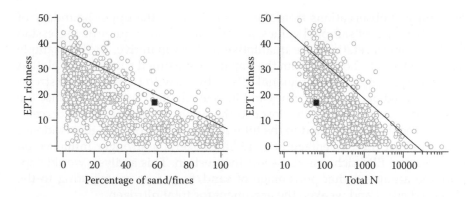

FIGURE 12.6
Quantile regressions depicting the 90th quantile for relationships between EPT richness with percentage of sand/fines (left plot) and log total nitrogen (right plot). The filled squares represent data from a hypothetical affected site. (Data are from the Western United States Stoddard, J., D. Peck, A. Olsen et al. 2005. *Western Streams and Rivers Statistical Summary.* Washington, DC: U.S. Environmental Protection Agency, Environmental Monitoring and Assessment Program (EMAP).)

other stressors (e.g., total N, organic enrichment) would not affect the mayfly richness, and it would be expected that a relationship would look more linear, similar to Figures 12.3 or 12.4. Quantile regression models such a relationship, typically near the top of a wedge, to represent the "best" biological condition in the assumed absence of other stressors. More precisely, quantile regression uses a specified conditional quantile of a dependent (response) variable and one or more independent (explanatory) variables (Cade and Noon, 2003). Modeling the 50th quantile of a response variable produces the median line under which 50% of the observed responses are located, and modeling the 90th quantile produces a line under which 90% of the observed responses are located (see Figure 12.6). Quantile regressions can have more than one explanatory variable, but we limit the following discussion to the univariate case. As with mean regression, the relationship is often assumed to be a straight line.

Quantile regression using an upper quantile (e.g., 75th, 90th) makes the assumption that the stressor being modeled is the only one of importance in the upper quantile. That is, the stressor of interest is limiting the biota, and other stressors are much weaker in their effect. Under this assumption, there is no explicit modeling of the confounding variables, and indeed, no need to measure them or even know what they are.

Tools for quantile regression are available in the statistical software packages R and in newer versions of SAS/Stat. Blossom, the U.S. Geological Survey's freestanding statistical package, also fits quantile regressions (U.S. GS, 2008).

How is this information useful for causal assessment? Interpretation of the results of quantile regressions in causal analysis is based on the proximity of observations from the affected site to the upper boundary of the distribution of sampled values, as estimated from a quantile regression fit. These interpretations are qualitative and comparative. In the example shown in Figure 12.6, data from the impaired site (filled squares) are plotted on scatter plots comparing regional EPT richness with two candidate stressors (increased percent sand/fines and increased total nitrogen). Because the plots show the impaired site closer to the upper boundary of the percentage of sand/fines compared to the total nitrogen relationship, that evidence indicates that the percentage of sand/fines exerts a stronger influence on the observed EPT richness at the site in question. This analysis would support the argument that percentage of sand/fines was contributing to the observed effect and weaken the argument for total nitrogen.

If data from the impaired site are located far above the upper boundary determined from regional data, it may be an indication that the comparison to the regional data is not valid. This situation can arise for a variety reasons. For example, field sampling methods applied at the impaired site may differ significantly from those applied to collect the regional data. In general, large outliers should be inspected carefully to determine whether they can be reasonably compared to regional data.

Although nonlinear quantile regression analyses are available, a simplifying assumption is that the relationships being modeled are linear with respect to the explanatory variables. In Figures 12.6, the relationship between response variables and explanatory variables is assumed to be linear. Many biotic metrics are generally considered to change linearly in relation to stressor gradients, but ecological knowledge of the underlying processes may help one select alternate functional forms. For example, the probability of observing an individual taxon often follows a unimodal function (see Figure 12.5).

12.2.3 Classification and Regression Trees

Research in artificial intelligence and machine learning has yielded new, empirical and distribution-free methods to develop stressor–response associations. A technique called Classification and Regression Tree (CART) analysis consists of successive bifurcations of the data, where each branch is accounted for by a single explanatory variable (classification if the explanatory variable is categorical; regression if it is continuous). At each step, the split or threshold of the explanatory variable is selected to maximize the homogeneity of each of the two resultant groups. Regression trees are an alternative to multiple regression methods, in that multiple explanatory variables are used to predict or explain a single response variable (e.g., De'ath and Fabricius, 2000; Prasad et al. 2006a,b). CART analysis builds a model that is a set of successive, binary decision rules that define homogeneous groups of the response variable. As with any modeling technique with multiple explanatory variables, it may be necessary to reduce the number of explanatory variables to prevent overfitting of the models. Regression trees typically also need to be "pruned" to avoid overfitting, but newer methods based on fitting an ensemble of regression trees (e.g., Random Forests) have improved the robustness of the method.

12.2.4 Species Sensitivity Distributions (SSD), Tolerance Values, and Related Uses of Field Data

Species sensitivity distributions aggregate effect levels for multiple species into a model that estimates the fraction of species that would be affected by a given stressor level (Posthuma et al., 2002). Conventionally, SSDs have been developed from laboratory toxicity test data (discussed further in Chapter 15), but the approach has also been adapted to use with effect endpoints from field observations (Cormier et al., 2008; Cormier and Suter, 2013a,b). SSDs rank affected taxa from lowest to highest and plot the ranks against the explanatory variable. The resulting curve is typically displayed as a cumulative distribution function (see Figure 12.7). Confidence bounds can be generated using bootstrapping techniques (Cormier et al., 2013a).

How is this information useful for causal assessment? To generate evidence for a candidate cause, the SSD is used to estimate the expected reduction in taxa richness associated with the exposure level observed at the study site. The expected reduction in taxa richness is then compared to the actual reduction observed at the study site. The exposure at an impaired site is judged sufficient to cause the effect if the estimated reduction in taxa richness is similar to the observed reduction.

In a case study of Pigeon Roost Creek, Tennessee, USA, an SSD was used to assess whether increased salts measured as specific conductivity was the cause of a reduced number of EPT taxa (Coffey et al., 2014). An SSD (similar to Figure 12.7) of EPT taxa was constructed using extirpation effect thresholds of each genus obtained from U.S. EPA (see Appendix D; U.S. EPA, 2011a). The SSD model was used to predict the proportion of EPT taxa expected to be absent at maximum conductivity levels. That prediction was then compared with the observed reduction of EPT taxa relative to the comparison site. At Pigeon Roost Creek, the EPT taxa were reduced between 69 and 92% of the taxa at the comparison site. However, the observed salt concentration was predicted to reduce EPT taxa by only 40%. Therefore, dissolved salts were concluded to be insufficient alone to account for the reduction in EPT taxa richness observed in affected reaches of Pigeon Roost Creek.

Other approaches have been developed that are conceptually similar to the SSD approach, in that they use responses of individual taxa to stressors, developed from a large field data set. Tolerance values of taxa have been used to infer environmental conditions where the environmental conditions have not been measured, as in paleolimnology of lake beds using diatom frustules. The taxon tolerance values (more properly, taxon optima) are derived from weighted averaging models (e.g., ter Braak and Juggins, 1993;

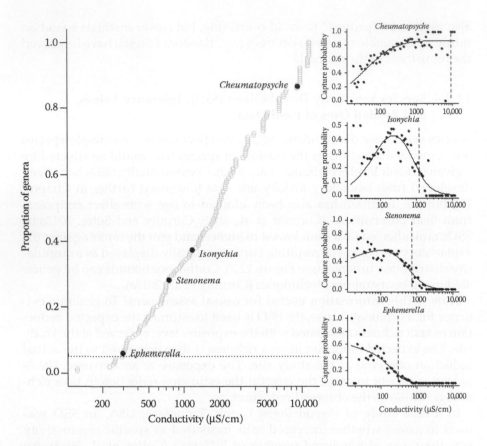

FIGURE 12.7

Example of an SSD depicting the proportion of genera extirpated with increasing salinity measured as specific conductivity in Ecoregion 69. On the left, each point on the SSD plot represents a concentration (XC_{95}) at which a particular genus is captured infrequently, for one of the 163 genera arranged from the most to the least sensitive. Four genera, *Ephemerella*, *Stenonema*, *Isonychia*, and *Cheumatopsyche* are highlighted on the left and their capture probability distributions are shown on the right. The dashed vertical lines indicate the position of XC_{95} values. (Adapted from U.S. Environmental Protection Agency (U.S. EPA). 2011a. *A Field-Based Aquatic Life Benchmark for Conductivity in Central Appalachian Streams.* Cincinnati, OH: Office of Research and Development, National Center for Environmental Assessment. EPA/600/R-10/023F.)

Ponader et al., 2007). These methods have been extended to macroinvertebrate tolerance values (e.g., Carlisle et al., 2007). More recently, Baker and King (2010) described an approach that identifies community thresholds for stressors based on individual species responses, analyzed through multivariate change-point analysis. This methodology would seem to be suited for the same objectives as the SSD approach, but we have not yet seen it applied to developing a threshold for a causal analysis.

12.3 Summary

Regional observational studies can provide supporting evidence that a stressor may have caused an observed effect. Evidence is derived by comparing observations from the case with results from models or data distributions built using large data sets. For example, stressor levels that are associated with effects at other locations may be compared to the case under investigation. Observational studies represent effects from realistic exposure regimes and interspecies interactions, characteristics which make them particularly useful for causal assessment. Approaches that have been successful to support causal analyses include simple box plots, contingency tables, relative risk estimates, and development of more sophisticated stressor–response models from the observational data. Limitations include high natural variability, which may make significant associations difficult to detect, and possible influence from confounding factors (discussed further in Chapter 13). Although regional models of stressor–response relationships require substantial effort to develop, once built they can be used for many investigations.

13

Regional Observational Studies: Addressing Confounding

David Farrar, Laurie C. Alexander, Lester L. Yuan, and Jeroen Gerritsen

Observational studies have the advantage of realistic exposure conditions. However, estimating the effect of an individual stressor in the presence of covarying stressors and natural gradients is challenging because the observed effect can be the result of one or more confounding variables. This chapter discusses methods for identifying confounding variables and mitigating their influence on models used to develop evidence from observational data sets.

CONTENTS

Stressors rarely occur alone. Many human activities are sources of multiple stressors, and multiple sources are likely to occur in any given area. For example, sewage effluent containing endocrine-disrupting compounds may also contain nutrients, metals, ammonia, or organic matter. Any one of these covarying stressors can affect stream organisms; hence, they are said to confound our ability to estimate the true effect of a particular cause.

Confounding, in the most general terms, is an intuitively simple idea: multiple co-occurring stressors and influencing factors may interfere with the ability to quantify the contribution of a specific cause to an observed biological effect. Confounding can occur when multiple stressors are released simultaneously, as in the example of sewage discharge with a mixture of toxic constituents. It may be appropriate to treat the effluent as a single complex cause (see Chapter 8), but if it is necessary to identify a single constituent as a cause, then the other constituents are likely to be confounders.* Confounding can also result when features of a local landscape, such as placement of roads or impoundments, influence the levels of multiple stressors. Natural gradients of temperature, elevation, soil moisture, or other factors can also confound stressor–response relationships that vary on a spatial scale similar to that of the gradient. Confounders are not necessarily stressors or contributing causes. For example, if logging in the riparian zone of a stream has caused biological effects by increasing temperature, the increased woody debris in the stream could be a confounder (it would be correlated with both temperature and the effect and would bias the estimate of temperature effects) even though it provides beneficial habitat structure.

Specific problems caused by confounding in stressor–response models were introduced in Chapter 5 (see Figure 5.1). For example, a simple regression relationship between any single water quality variable and a biological response variable will attribute the combined influences of all correlated stressor variables on the biological effect to the single stressor. If all of the stressor variables influence the biological response in the same direction (e.g., both low base flows and increased stream temperature decrease EPT richness), the relationship between the stressor being modeled (low base flows) and the response variable (EPT richness) will be overestimated. Identifying and accounting for confounding variables improve the accuracy of the stressor–response model and the reliability of any evidence developed from it (see Chapter 19).

This chapter describes strategies for identifying and accounting for confounding variables in the analysis of observational data. Section 13.1 defines confounding and introduces the general methodological strategy of "statistical control." Section 13.2 discusses three strategies for identifying confounding variables that may require statistical control. Finally, Section 13.3 describes statistical methods to control or mitigate the influence of identified confounders in data analysis.

13.1 Concepts of Confounding and Statistical Control

The concepts described in this chapter approach confounding as a form of bias, that is, a systematic tendency for estimates (e.g., of the slope of a

---

* However, if constituents of an effluent interact (e.g., low pH and metals in acid mine drainage), and the interaction is accounted for they should not be treated as confounders.

stressor–response relationship) to be over- or under-estimated (Rothman et al., 2008). A confounder is conventionally defined as a variable that is correlated (or, more generally, associated) with a stressor of interest and has a causal effect on the response of interest.* In a variant of this definition, the confounder is said to be only associated (e.g., correlated) with the response, as in the woody debris example above (see also Pearl, 2009; Section 11.6.4).

In observational data analyses, the term "statistical control" is used to describe techniques for evaluating the effect of a stressor or other causal variable in a data set in which other causal factors are statistically held at relatively constant levels (Sokal and Rohlf, 1995; Shipley, 2000). Statistical control methods are analogous to experimental control methods. In a controlled experiment, the experimenter varies the treatment and attempts to hold constant the other factors that could affect the response variable. With observational data, statistical methods are used to control the influence of such variables after the fact. That is, to evaluate the effects of changing a stressor of interest, other causal factors are analytically held constant to reduce error in the estimate of stressor effect. For example, to accurately quantify the effects of reduced oxygen concentration on aquatic insects, an analysis that quantifies responses in data subsets with similar values for stream temperature, a confounding factor in this example, would reduce or eliminate the confounding effects of temperature. Confounding variables that are included in an analysis for the purpose of statistical control are referred to as control variables. By correcting for bias due to the confounding variable temperature, the estimate of the effect of reduced oxygen on insects in this analysis is said to be "controlled" or "adjusted" for temperature.

With observational data, some form of statistical control is likely to be needed. Note, though, that the terminology of "control" in observational data analysis is not intended to equate the resulting information with that from actual experimental control. Our use of this terminology is only intended to suggest a conceptual basis for methods widely applied to observational data.

13.2 Strategies for Identifying Variables for Statistical Control

How does the data analyst identify a set of variables for statistical control of confounding in a regional observational study? In a data analysis, as in other steps of a causal assessment, decisions about inclusion of variables are based largely on the investigator's knowledge about the ecological processes and causal pathways relating stressors to biological responses of interest

---

* This definition suffices for a three-variable system (stressor, response, confounder) and is helpful generally, but may not be appropriate for more complex systems.

in that region. Statistical or computational methods, including automated variable selection (e.g., stepwise regression), may aid the selection process but are not a substitute for subject area knowledge about relevant causal relationships.

We discuss three general strategies, which are not mutually exclusive, for selecting control variables:

1. The first strategy evaluates variables one by one. Two variations of this strategy are discussed:
   a. Identify variables that plausibly have causal effects on the biological response of interest.
   b. Identify variables that plausibly have causal effects on the response and are also correlated with the stressor of interest.*

2. The second strategy explores all variables simultaneously using a diagram of causal relationships (see Figure 13.1 and Box 13.1). This strategy applies graphical theory to handle relatively complicated causal networks. The set of control variables is identified from a directed acyclic graph (DAG) depiction of plausible causal pathways among all variables, using formal procedures such as those discussed by Shipley (2000).

3. The third strategy is to use an automated scheme such as stepwise regression to select control variables for inclusion as additional explanatory variables in a multiple regression (alongside the stressor of interest).

A qualification for all these methods is that one should not select control variables that are likely to be affected directly or indirectly by the stressor of interest (Cochran and Cox, 1957; Pearl, 2009). For example, one should not statistically control for chlorophyll concentration to examine effects of nutrients on direct measurements of plant growth or biomass because chlorophyll production is a component of plant growth. A possible consequence of violating this requirement is the inadvertent control and elimination of the influence of the stressor of interest—in this example, eliminating the effect of nutrients.

Finally, different audiences for the assessment may have preconceptions about what variables are likely to be confounders. Assessors may need to adjust for such variables or explain why adjustments were not made.

---

* If a variable is correlated with both stressor and response variables, but we do not know how to interpret the correlations, adjustments based on that variable should be approached with caution. For example, as discussed later in this section, if the covariate is caused by the stressor or the response variable, adjusting for it may inadvertently eliminate the effect of the stressor of interest.

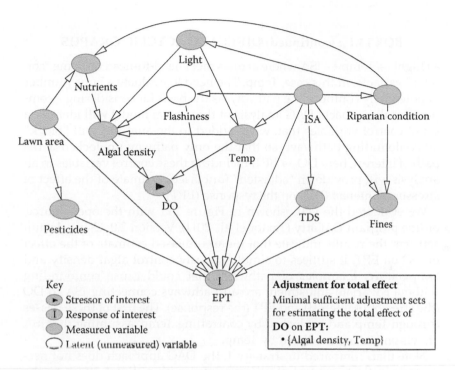

FIGURE 13.1

DAG analysis for effect of dissolved oxygen on EPT richness. This DAG can be used to identify control variables for estimating the effects of DO (the stressor of interest) on EPT richness (the response). The relationships depicted in the DAG represent the plausible causal relationships identified by the analyst. Estimating the effect of DO on EPT in this system is complicated by the presence of numerous confounders and covarying proximate causes. Even in this relatively simple example, the combined effect this network of relationships could have in biasing estimates of the causal effect of DO is difficult to intuit, and a computerized analysis can be helpful (see Box 13.1). The DAG approach works by enumerating all pathways that potentially result in confounding bias. The program output is shown in inset box below the figure: for an unbiased estimate of the effect of DO on EPT, it suffices to simultaneously control algal density and temperature. Abbreviations: Temp, temperature; ISA, impervious surface area; TDS, total dissolved solids.

BOX 13.1   DIRECTED ACYCLIC GRAPHS

In the example depicted in Figure 13.1, the direct causal effect is represented by the arrow from the stressor of interest (DO) to the response of interest (the number of taxa that are EPT). If the directions of the arrows are ignored, DO and EPT are linked by additional pathways (e.g., EPT ← Temp → DO). Of the latter pathways, a subset has the potential for confounding, namely those that include no pattern such

BOX 13.1   (continued) DIRECTED ACYCLIC GRAPHS

as Light → Temp ← ISA where arrows meet head-to-head on some "collider" variable, in this case, Temp. The need to evaluate a large number of potentially confounding pathways is the basis for considering a computational approach. DAG analysis of this causal model will identify a set of control variables that, when added to the analysis, will "block" all confounding pathways, so that the only pathway unblocked is the path of interest, here DO → EPT. Including these control variables in an analysis will provide an "adjusted" (unbiased) estimate of the effect of stressor of interest (DO) on the response (EPT).

We evaluated the DAG shown in Figure 13.1 with the open-source, online program DAGitty (Textor et al., 2011), Version 2.0. Using default settings, the results indicate that for an unbiased estimate of the effect of DO on EPT, it suffices to simultaneously control algal density and temperature. Examples of pathways that could cause confounding without covariate adjustment are all pathways connecting ISA to DO (the stressor of interest) and EPT (the response). The first of these passes through Temp and is blocked by controlling Temp, assuming that ISA effects on DO are mediated by Temp.

Note that, compared to strategy 1, the DAG approach does not necessarily indicate control of all covariates with a direct effect on the response. Including additional variables that have a direct effect, such as TDS or Pesticides, may improve precision in evaluation of causal effects and possibly reduce bias. Approaches 1 and 3 can be combined by starting with an adjustment set generated automatically (e.g., by DAGitty), adding additional variables, if any, that have direct effects on the response (principally to increase precision), then using DAGitty to check that the union of the two variable sets is not associated with bias.

One limitation of the DAG approach is that all confounding pathways are treated as equivalent, whereas in practice there is usually evidence of stronger causal effects for some relationships than others. It may be easier to reason about the relative importance of particular pathways by pursuing strategy 1a or 1b, which focus on smaller sets of variables. The confounding pathways of least concern are arguably those with numerous links of which some have doubtful importance.

For a more complete description, see Glymour and Greenland (2008), Greenland et al. (1999), Morgan and Winship (2007), Pearl (2009, Section 11.1.2), and Shipley (2000).

13.2.1 Strategy #1: Evaluate Variables One by One

The approach to selecting individual variables to control depends on one's definition of a confounder. Strategy 1a is motivated by the concept that confounding results from multiple causes (see Section 10.2.3). In this strategy, we base decisions *only* on the plausibility of causal effects on the response and ignore correlations of variables with the stressor. Because strategy 1a ignores correlation with the stressor of interest which is required in the conventional definition of a confounder (see Section 13.1), including uncorrelated variables may not further reduce *bias* in the estimate of effect. On the other hand, including such variables can increase *precision* of the estimate of effect by accounting for other sources of variation in the response variable. While this chapter is concerned primarily with controlling bias, both forms of error could be important in selecting variables for implementing a method of statistical control.*

Strategy 1b, on the other hand, is a direct application of the definition of a confounder as a correlated variable that causes bias (whether or not it is a direct cause). In this strategy, variables are evaluated individually as with 1a, but now identification of those that should be controlled is based on relationship to the stressor as well as the response. Contingency table methods that address relatively extreme stressor values have been used as an alternative to correlation within a weight-of-evidence approach (Suter and Cormier, 2013a). Both approaches require consideration of an appropriate criterion or cutoff value for the strength of the variable–stressor association. Use of the statistical significance of the relationship of each variable to the stressor of interest is a common practice. It has sometimes been suggested that statistical tests may be used in this context with α greater than the conventional 5% (e.g., Cochran, 1965; Shipley, 2000). However, the use of statistical tests for variable selection to reduce confounding has been rejected by some statisticians (e.g., Stuart, 2010). Alternatively, a cutoff value may be helpful for identification of associations of concern. For example, in the context of a general weight of evidence approach, Suter and Cormier (2013a) used a Spearman correlation of 0.25 for separating "moderate" from "weak" correlations. The implications of different cut-off criteria have not been studied extensively, and deserve further attention.

13.2.2 Strategy #2: Identify Control Variables from a Diagram of Causal Relationships

Simple measures of association, such as standard correlation statistics, are themselves subject to possible confounding. Just as a correlation between a

---

* As a rule, controlling more variables may reduce bias but controlling too many may lead to imprecision, particularly via multicollinearity (see Section 14.2.3). For a more complete discussion of bias and precision in the context of multiple regression, see Myers (1990) and Draper and Smith (1998).

stressor X and response Y is seen to be possibly misleading when considering a system with an additional, confounding variable Z, the X–Z or Z–Y correlations may be misleading when more than three variables are involved. This requires using an approach that can take into account a complex network of plausible causal relationships.

The approach in strategy #2 starts with an organization of information about causal relationships into a diagram of nodes and arrows, where each arrow points from a causal variable to a variable that it affects. Such a diagram is a DAG (e.g., Figure 13.1 and Box 13.1).[*] DAGs can be used to identify sets of confounding variables that can be controlled to ideally eliminate bias in estimates of a stressor-response relationship.[†] In Box 13.1, we give an example of an analysis of the DAG displayed in Figure 13.1. Software programs (e.g., DAGitty; Textor et al., 2011) are available to automatically consider all possible selections of variables and identify one or more sets, called "deconfounding sets" or "adjustment sets," that suffice to eliminate confounding when used as the control variables in an appropriate statistical control method. DAG software can also be used to evaluate whether a given set of control variables—selected by the analyst for other purposes, for example—will be sufficient to eliminate bias due to confounding. The DAG software does not perform statistical data analysis itself: Once a set of control variables has been identified, the data set may be analyzed using one of the statistical control procedures discussed in Section 13.3.

Strategy #2 can be viewed as essentially an extension of strategy 1b, involving the use of more comprehensive networks of causal relationships to provide accurate treatment of more cases (e.g., by inclusion of unmeasured variables such as Flashiness in Figure 13.1). The approach has been featured in some recent epidemiological literature (Rothman et al., 2008). Shipley (2000) provides an introduction to DAGs for biological applications.

13.2.3 Strategy #3: Automated Variable Selection Techniques

The third strategy is to identify statistical control variables using an automated scheme such as stepwise regression. Stepwise variable selection procedures are a common approach for selecting independent variables in regression (despite skepticism long expressed in various disciplines, e.g., Greenland, 2008; Harrell, 2001; James and McCulloch, 1990; Whittingham et al., 2006). Because the approach does not consider correlation with the

---

[*] The diagram is said to be *directional* because directional (i.e., causal) relationships are indicated for some pairs of variables. It is *acyclic* because it has no loops: no variable has a direct or indirect effect on itself.

[†] The word choice of *eliminating* bias follows the technical literature and refers to theoretical results. The actual benefits of the approach would depend, as usual, on how well the underlying theory describes reality. In particular, the effectiveness of a statistical control strategy is contingent on the appropriate, accurate, and precise measurement of the control variables. In the real world, we use such strategies circumspectly, hoping to *minimize* bias.

stressor X as a criterion for selection, the approach is akin to strategy #1a, identifying variables based on their causal effect on the response variable Y.

There are several limitations of using automated methods to select control variables. The most commonly used automated methods use sequences of statistical hypothesis tests for individual variables, and therefore inherit any objections associated with hypothesis tests (see Section 3.4). There is no generally accepted principle that when evaluating the effect of a variable X, additional variables taken into account should be statistically significant (Harrell, 2001).

Automated variable selection methods do not guarantee the most appropriate choice of explanatory variables. Multicollinearity (or simply collinearity) is a problem where alternative choices of the independent variables provide similar or identical quality of fit (e.g., similar R^2) as a result of limited sample size or correlations of explanatory variables. For example, when two variables are closely correlated, then a similar fit will be obtained using one or the other. Basing results on a single selection of independent variables when other selections are just as well supported can introduce arbitrariness and instability into analysis results. When using an automated approach to identify variables, we recommend using more than one approach and considering the sensitivity of results. Backwards variable selection (step-down procedures), which start with all variables in the model and eliminate variables one at a time, are preferred by some authors over forward methods, which add variables one at a time (Harrell, 2001; Greenland, 2008).* Discussions of multicollinearity diagnostics and remedies, with an emphasis on application in ecological data analysis, include Graham (2003), Legendre and Legendre (2012), and Zuur et al. (2010).†

13.2.4 Does Controlling the Confounders Affect Essential Conclusions?

The sections above presented multiple possibilities for identifying variables that may bias or reduce precision of a stressor–response model. Whichever strategy is used, we recommend evaluating whether the effect of statistical control (e.g., including the identified set of variables in a regression model to control for confounding) is large enough to be of practical importance. If not, a simplified model focused on the nonadjusted results may be preferable, at least in concise presentations of the essential findings of an assessment. So long as the set of control variables is valid (e.g., they are not variables affected by the stressor of interest), the degree of bias can be evaluated by comparing

---

* Note that in using a stepwise variable selection approach, the model can be required to include the stressor at each step, for example, using SAS proc REG with an "include" option in the model statement (SAS Inst., Inc., 2008), or comparable options in other statistics packages. (Requesting SAS proc stepwise currently will actually invoke proc reg.)

† The most popular multicollinearity diagnostic in practice, described in each of these sources, is the variance inflation factor.

TABLE 13.1

An Output Table for Two Linear Regression Models

	Parameter Estimate (Slope)	Standard Error
Univariate Model		
Intercept	3.65	0.055
Conductivity	−0.93	0.024
Multivariate Model		
Intercept	3.39	0.11
Conductivity	−0.92	0.029
Habitat score	0.0014	0.0005
Temperature	0.0068	0.0026
Fecal coliform	0.037	0.012

Note: The first is the simple model predicting *Ephemeroptera* taxa richness from conductivity. The second is a multiple regression model with the additional variables habitat score, temperature, and fecal coliform count.

a regression model that includes confounding variables in the set of explanatory variables, with a regression model that includes only the stressor of interest. For example, Table 13.1 shows the parameter estimates for a model predicting *Ephemeroptera* (mayfly) taxa richness in streams based on conductivity, compared with the parameter estimates from a model that includes not only conductivity, but also the confounding factors habitat score, temperature, and coliform count. Including the confounding factors in the model did not substantially change the parameter estimate for conductivity, which suggests loss of about one taxon per unit increase in conductivity based on either calculation.

It may sometimes happen that any subset of explanatory variables, or some subset of models clearly superior to others on ecological and statistical grounds, results in practically the same conclusion. The results of multiple alternative models can be presented, along with indexes of model quality, allowing a determination of whether important results depend on the model, rather than basing findings on a single model (Burnham and Anderson, 2002; Lukacs et al., 2007). Communicating these quality assurance efforts increases confidence in the causal assessment's findings.

13.3 Strategies for Mitigating the Influence of Confounding Factors

In the preceding sections, we described the concept of statistical control and provided some ideas on how to identify variables needing action in order to avoid bias or imprecision. This section describes approaches for executing analyses of stressor–response relationships with confounding variables taken into account. One possible approach (discussed in Section 13.2.3) is multiple regression implemented by including the selected control variables as additional explanatory variables (in addition to the stressor of interest).* Other strategies described below include data restriction and stratification based on the stressor of interest (including propensity score stratification) and bundling of highly correlated stressors.

13.3.1 Restriction

A simple statistical control approach is to analyze a restricted[†] data set, obtained by restricting the range of values for confounding variables to background conditions. This approach removes observations with confounder values extreme enough to pose a significant chance of reduced ecological quality. This strategy was used to develop the regression model to estimate the number of EPT taxa predicted by conductivity levels for samples from the ecoregion of the Central Appalachians (see Figure 12.4 and Table 12.1). Sites with pH < 6 were removed to minimize the effects of both acid mine drainage and acidic deposition; sites with habitat scores <128 were removed to minimize the effects of poor habitat, and sites with fecal counts >400 were removed to minimize the effects of untreated domestic wastewater. This reduced the sample size by half ($N = 515$), improved the confidence interval, and increased the intercept (EPT taxa at low conductivity) by approximately two genera.

13.3.2 Stratification on a Single Variable

In the previous section, we discussed statistical control by analysis of a restricted data set, with data points corresponding to extreme values for potential confounding variables removed. A related statistical control strategy is to divide the data into subsets corresponding to limited ranges of a single confounding variable. An example of using stratification to control for the confounding effect of acidity in a relative risk calculation was discussed in Chapter 12. U.S. EPA (2012a) illustrates stratification using scatter plots (see Figure 13.2 and Table 13.2) to evaluate the effect of total nitrogen (TN) on

---

* Much of the useful literature on applying regression methods in a causal context is found under the topic of "analysis of covariance". Cochran (1957a,b) is still an important source. Also see Huitima (2011) and Snedecor and Cochran (1989). ANCOVA with strata for a covariate is essentially multiple regression with dummy variables used to encode a categorical variable (confounder stratum).

† In this section, "restriction" is the removal of data points based on values of potentially confounding variables, a statistical control strategy. In a causal analysis context, other sorts of restrictions could be helpful. For example, a causal analysis can be restricted to those taxa most likely to be affected by a particular candidate cause (see Section 14.4).

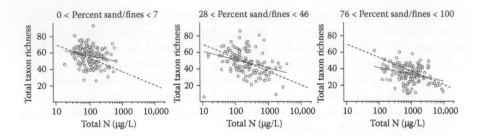

FIGURE 13.2
Example of the use of stratification in regression to control for confounding. The panels in this figure correspond to different strata based on SED values (see Table 13.2). For clarity, only Stratum 1 (0–7%), Stratum 4 (28–46%), and Stratum 6 (76–100%) are shown. Solid lines indicate linear regression fit within each stratum. Dashed lines show linear regression fit using full data set. Note that the regression line using the entire data set is steeper than the lines within the SED strata. This plot shows that decreasing taxon richness is still apparent in each stratum (data subset), even though (see Table 13.2) within-stratum correlation of TN with the confounder SED is low. (Adapted from U.S. Environmental Protection Agency (U.S. EPA). 2012a. *CADDIS: The Causal Analysis/Diagnosis Decision Information System.* Office of Research and Development, National Center for Environmental Assessment. http://www.epa.gov/caddis/ index.html (accessed February 1, 2014).)

macroinvertebrate richness. This analysis stratifies the data set on the percentage of sands and fines (SED), resulting in the six subsets (six strata) indicated in Table 13.2. The selection of SED as a confounder requiring statistical control is supported by its correlation with TN ($r = 0.65$) and the plausibility of its causal effect on taxon richness. As shown in Table 13.2, within-stratum correlations of SED with TN are lower than the overall correlation. Such an effect is expected simply from the fact that the range of SED is lower within strata than in the overall data set. Thus, the potential confounding effects of SED are lowered by conducting analyses within strata because SED and TN are "less confounded" within strata than overall.

TABLE 13.2

Percent Substrate Sand/Fines (SED) in Different Strata

Stratum	SED (%)	r
1	0–7	0.03
2	8–14	0.12
3	15–28	0.08
4	29–46	0.25
5	47–76	0.09
6	77–100	0.15

Note: Column labeled as r shows the correlation coefficient between total nitrogen and SED within each stratum.

Figure 13.2 illustrates the results of regression analysis within each stratum to estimate the TN-total richness relationship. In this example, the slope of the relationship between TN and total richness is similar across strata (solid lines), but noticeably less steep than the slope estimated using the full data set (dashed line). Within each stratum, the strength of correlation between SED and TN is greatly reduced (see Table 13.2), and thus, the potential confounding effects of SED on the estimate effect of TN on total richness is also reduced (U.S. EPA, 2012a).

A stratified analysis might be used in a causal assessment by identifying the stratum most relevant to the site under investigation and using the model from that stratum to develop the evidence. (Results from other strata might be more or less supportive.) There may be no statistical advantage to stratifying a data set on a continuous confounder. Graphical analysis based on stratification provides transparency and increases confidence that the observed effect of stressors is consistent with expectations based on mechanistic understanding. This approach can be used to characterize interactions, for example, when the slope on TN was seen to depend upon SED. Stratification on a single confounder is also a convenient way of introducing a technique for handling multiple confounders, described in the next section.

13.3.3 Advanced Stratification: Use of Propensity Scores

Stratification on a propensity score (Rosenbaum and Rubin, 1983) is a popular approach in some disciplines for handling multiple confounders. The propensity score is a single variable estimated for each observational unit (e.g., a site), which combines all the potential confounders into a single value. For continuous variables, it represents the predicted value of the stressor of interest based on the values of the site's baseline covariates. Because the propensity score represents the effects of many different covariates as one composite variable, it can be used to stratify a data set in the same way as a single variable (see Section 13.3.2). While initially used in epidemiology and social sciences (e.g., Smith, 1997; Joffe and Rosenbaum, 1999; Dehijia and Whahba, 2002), propensity scores have been recently applied to ecological data (Yuan, 2010a,b).

A propensity score is said to be a "balancing score," a combination of the confounding variables such that when the propensity score is held relatively constant, the stressor of interest is approximately independent of the confounding variables. Stratifying on the balancing score variable results in strata in which the stressor is approximately independent of confounding variables that make up the balancing score, so that stratum-specific analyses are not biased by the effects of those confounders. Propensity score analysis has sometimes been viewed as the observational data approach most comparable to the use of randomization in controlled experiments (Austin, 2011; Rubin, 2007). One drawback of the method is that it requires large sample sizes. In addition, as with all methods considered in this chapter, confounding can remain after statistical control methods have been applied, owing to variables not measured or recognized as important.

13.3.4 Bundling of Correlated Stressors

Variables may be so tightly correlated that their effects can only be separated using carefully designed experiments. However, the collective effects of stressors that co-occur can sometimes be evaluated meaningfully by combining or "bundling" blocks of correlated stressors (Harrell, 2001; Van Sickle, 2013). To be as effective as possible for eliminating confounding bias, such blocks should have an ecological interpretation, and the variables included should account for important ecological determinants of the response. As an example, U.S. EPA concluded that excess dissolved ions cause biological degradation in naturally dilute Appalachian streams, but the dominant anions contributing to conductivity (sulfate and bicarbonate in this case) are not physiologically independent (U.S. EPA, 2011a). Measured conductivity combines the effects of all ions and was the single best predictor of biological condition.

Similarly, multiple toxic metals are often discharged from single sources, especially in mining districts, because the metal ores co-occur in geologic formations, and mining and refining operations liberate multiple metals into effluents. A simple approach for examining biological effects of multiple, correlated metals is to use a sum of the estimated toxicity of all toxic metals with a similar mode of action (Clements et al., 2000).

13.4 Summary

The effect of a stressor on a biological response may be underestimated or overestimated if other environmental variables or stressors that also affect the biological response are ignored. In many cases, a simple relationship observed between a measure of biological condition and a single stressor reflects the effects of additional stressors. The additional stressors confound estimates of the relationship between the response and the stressor of interest. Potential confounding variables include all stressors as well as natural variables that may affect a biological measure directly or indirectly. Examples include multiple stressors in water bodies from runoff and from municipal sewage discharges; habitat, elevated temperature, and domestic sewage confounding the effects of leached salts from mining spoil; and sediment confounding the effects of excess nutrients.

Potential confounding variables can be identified from a well-constructed conceptual model diagram. Evaluation of correlations can provide some indication of which additional variables besides the stressor of interest should be taken into account when characterizing a stressor–response relationship. A more comprehensive framework involves the concept of "confounding pathways" in a directed acyclic diagram of causal relationships which can be "blocked" by controlling for an appropriate selection of statistical control variables.

Several methods are used to reduce the influence of confounding variables on stressor–response models. The influence of some strong confounders can be separated from a stressor of interest by stratifying the data according to strength or concentration of the confounder, or by propensity score analysis for multiple confounders. A summary or combined variable can be substituted for tightly correlated variables, for example, using conductivity as a combined substitute for all ions. Different types of statistical control can be combined in a single data analysis. For example, an ecological data set might exclude observations with a high level of urbanization (a restriction), and the data selected might then be analyzed using multiple regression with confounding variables included as additional explanatory variables along with the stressor of interest. Whichever strategy is used, the effect of confounding variables included in the model should be large enough to be of practical importance to the stressor–response analysis. Finally, the web of relationships may be too complicated to characterize using statistical analyses alone. In these cases, carefully designed field studies and laboratory experiments, described in the next chapters, may be required.

The present chapter emphasizes a statistical approach using statistical control methods. However, the objective of the assessment is to determine causes of observed adverse effects using the evidence that is available. Various strategies may be used to eliminate confounding as an alternative explanation of observed effects, by attributing effects to one causal factor rather than another, possibly correlated factor. Many of the simple strategies for minimizing the effects of confounding factors described in Chapter 10 can also be used for larger observational data analyses. For example, a biological response variable may be restricted to taxa sensitive to a smaller set of stressors (a different sort of "restriction" than that covered in Section 13.3), given information on relative taxon sensitivities to different stressors (e.g., abundance of baetid mayflies vs. abundance of all mayflies). Conducting the same analyses across many candidate causes can help make confounding more apparent when the analyses are unable to differentiate among them. In general, confidence in causal findings is increased by determining that specific types of adverse effects are observed when and where they are anticipated based on the action of specific stressors.

14

Assessment-Specific Field Study Designs and Methods

Robert B. Brua, Joseph M. Culp, and Alexa C. Alexander

> This chapter describes the use of field study designs and in situ methods for causal assessment. These approaches help clarify the influence of a cause by reducing variability and the influence of confounding factors.

CONTENTS

Most causal assessments are performed with available data (see Chapters 9–13), but it is sometimes possible to collect assessment-specific data. For such cases, this chapter describes sampling designs (see Section 14.1) and field tests performed with in situ methods (see Section 14.2) (i.e., "in place" experiments conducted in the field; Liber et al., 2007). Field tests allow control of the nature, magnitude, and duration of exposure; control of what is exposed under realistic environmental conditions; and replication of units so that variance can be estimated. Although such tests may be costly, they have advantages over field observational data. Readily available observational data often have been collected for purposes other than causal assessment, such as for monitoring environmental status and trends. Determining the cause of ecological effects through these studies alone is often problematic because confounding factors limit the ability to distinguish among candidate causes (see Chapter 13) (Cash et al., 2003; Culp and Baird, 2006; Baird et al., 2007a,b). Field tests allow the assessor to get data that is targeted to the specific questions left unanswered by routine monitoring data.

These types of field studies have the greatest potential relevance for assessing cause when they are conducted at the affected site. They are often used in later stages of the assessment to confirm a cause (see Chapter 19) or to evaluate the effectiveness of a management action (see Chapter 21). Field studies conducted at other locations can be applied if they are judged to be relevant to the site under investigation. For example, macroinvertebrate diversity in the Androscoggin River, Maine, USA, increased after total suspended solids inputs from a pulp mill were reduced. Those results were used as evidence to identify total suspended solids as a cause of the degraded macroinvertebrate assemblages in a similar river, the Presumpscott River, Maine, USA (U.S. EPA, 2000a).

This chapter discusses study designs (see Section 14.1), in situ methods (see Section 14.2), and the less common approach of whole ecosystem experiments (see Section 14.3). Effective studies are designed to sample a range of different exposures of interest and minimize the influences of confounding factors. In situ methods provide additional opportunities to control exposures and reduce variability in biological responses. When these strategies are used together, they can produce high-quality evidence of cause and effect relationships. Understanding their strengths and limitations informs the design of effective studies for the site under investigation and the interpretation of results from other available studies.

14.1 Study Design Considerations

Components of the study design that influence the quality and relevance of the results include the sampling design (see Section 14.1.1) and the responses and other variables that are measured (see Section 14.1.2). Additional issues associated with specific in situ methods are discussed in Section 14.2.

14.1.1 Sampling Designs

Field study designs that are particularly useful for causal assessment contrast biological responses to physical and chemical characteristics at places where a candidate cause does and does not occur. They can take advantage of a situation where a stressor or its source is deliberately changed, for example, measuring biological responses associated with the removal of a dam (Tuckerman and Zawiski, 2007). They can also take advantage of existing gradients and patterns, for example, measuring biological responses at varying distances downstream from a point source.

Basic sampling designs used for assessment-specific studies include control/impact, before/after, and gradient approaches. This chapter discusses these sampling designs for in situ tests; however, the same designs can be used with biological survey methods, such as those used to sample macroinvertebrate, fish, and algal assemblages. The final sampling design and number of replicates should also consider Type I and II error rates, statistical power, and critical effect size (Osenberg et al., 1994; Underwood and Chapman, 2003; Munkittrick et al., 2009).

14.1.1.1 Control/Impact and Before/After Designs

A variety of approaches have the control/impact design backbone. The simplest of these is the control/impact design, which is also referred to as a reference-exposure design (Green, 1979). The design consists of at least one control and one affected site (e.g., the site under investigation). Results from control/impact designs can demonstrate that an effect occurs where the candidate cause occurs and that the effect does not occur where the candidate cause is absent. Control/impact studies have the opportunity to isolate the contribution of the candidate cause by carefully matching sites for natural factors (e.g., seasonality, flow, elevation, soil type) and other potential stressors (e.g., habitat). If samples are collected from each site close in time, then the influence of temporal issues such as seasonal phenology and storm events can be minimized. Before/after designs apply the same ideas in time by sampling sites before and after a candidate cause (or, more typically, its source) is initiated, manipulated, or otherwise changed. These designs control for many natural site factors but do not control for temporal changes and weather events that occur between the two sampling visits.

Before-After-Control-Impact (BACI) designs combine the two approaches. BACI studies sample at least one potentially affected site and one or more control sites, all of which are sampled before and after the candidate cause or source is manipulated. BACI currently is the preferred study design and has been reviewed extensively by several authors (Stewart-Oaten et al., 1986; Osenberg et al., 1994; Stewart-Oaten and Bence, 2001; Underwood, 1991, 1992, 1994). Stewart-Oaten et al. (1986) outline a modified BACI design for situations where there is more than one affected site, such that similar control and affected sites are paired (Before-After-Control-Impact-Pairs, BACIP). Regardless of the specific design, it is highly recommended that control sites be selected randomly if there is a population of good control sites and that control and affected sites are similarly influenced by both natural factors and other stressors. In aquatic systems, the control site is usually located upstream from the affected site, but it can be located downstream if it is not affected by the source of the impact (Environment Canada, 2002).

BACI designs are seldom used for site-specific causal assessments because causal assessments typically begin after the effect has occurred. However, in some cases a candidate cause is intermittent and the rate of recovery is rapid relative to the interval between causal events. An example might be pesticide applications. In some cases, a candidate cause may be withdrawn for long enough for recovery to occur, as in the temporary shutdown of an industrial source for a process change. BACI designs are more generally useful for manipulations such as adding habitat structure to a stream or liming acidic lakes or for studying the results of a remedial action to determine whether it was effective (see Chapter 21).

14.1.1.2 Gradient Designs

Gradient designs sample responses in a series of sites with decreasing levels of a candidate cause. They are generally useful for observational studies at affected sites. They are also useful for exposing caged organisms or other in situ studies in which units can be placed at intervals along the gradient.

Gradient designs are ideally suited for point source impacts because, when the point source emits the cause, then effects should decline as distance increases from the source of the impact. This design style, along with a reference (upstream)–exposure (downstream) design, was used in the Northern River Basins Study (see Figure 24.2). Control or reference stations in gradient designs are typically located either upstream of or at the location farthest away from the point source.* Ideally, gradient designs consist of a continuum of sampling locations along the gradient. However, if an abrupt change in exposure to the cause occurs, then more sample sites will be needed in the area of discontinuity to characterize the cause–effect relationships (Environment Canada, 2002). As suggested for BACI designs, the exposure gradient should, as much as possible, be independent of any natural or other stressor gradients (Environment Canada, 2002). In addition, the seasonal timing of the study should be selected in the context of the phenology of ecological processes and sensitivity of life stages that vary throughout the course of the year (Clark and Clements, 2006).

14.1.2 Response and Stressor Measurements

Assessment-specific studies can be designed to evaluate many different types of responses, including contaminant uptake, short-term physiological or behavioral changes, and population or community level changes (Liber et al., 2007). The choice of measured responses will influence both the relevance of the test results to the investigation and the way they are used. Responses like increases in body burdens or behavioral changes (e.g., fish coughing) provide evidence that biologically relevant exposure has occurred.

---

* In some gradient studies, there are no completely unexposed sites, just lesser exposed sites.

Responses like changes in growth, reproduction, and mortality can provide evidence that exposures have reached levels to produce changes relevant to population-level effects. A response like changes in species composition provide evidence that exposures have reached levels that can produce assemblage-level changes.

In addition to biological responses, the study design should also identify the variables that measure the occurrence of the stressor(s) under investigation. In consultation with subject-matter experts, designs that include additional water quality variables help determine whether potentially confounding physicochemical variables alter uptake, mode of action, or response. Monitoring programs often have a standard set of measurements that are inexpensive and should always be collected (e.g., temperature, pH, dissolved oxygen, conductivity, hardness). An opportunity to collect new relevant data allows the suite of measurements to be targeted to answer specific questions.

On-site passive sampling methods provide a time-weighted average concentration of a metal, polycyclic aromatic hydrocarbon, pesticide, or pharmaceutical, as opposed to a single spot or single water sample. Samplers include semipermeable membrane devices (SPMD), polar organic chemical integrative samplers (POCIS), or diffusive gradients in thin films (DGT). Passive samplers reduce the risk of nondetected chemical concentrations. Although they do not detect temporal variation in pollutant concentrations, they do not miss high exposure events entirely, because short-term peaks are averaged into the sampling interval (Allan et al., 2006; Alvarez, 2013).

14.2 In Situ Methods for Aquatic Systems

In situ approaches that incorporate sound study design principles can provide high-quality, relevant evidence (Wharfe et al., 2007). Caging techniques (see Section 14.2.1) and colonization substrates (see Section 14.2.2) are two methods used in aquatic systems for controlling exposures and reducing variability. They are useful for demonstrating that the causal agent interacts with organisms at the site and that the exposure is sufficient to induce relevant effects.

14.2.1 Caging Techniques

Caging studies deploy test organisms contained in a holding chamber into a water body. While they allow replication and control the nature, magnitude, and duration of exposure, they generally do not eliminate confounding factors. Test organisms should be selected based on their relevance to the biological effects and sensitivity to the candidate causes under investigation. The use of indigenous, rather than cultured, biota eliminates the

potential for exotic species introduction and decreases the chance that normal temporal or spatial variation in water quality will elicit an effect (Baird
et al., 2007a).

Caging studies are often used to determine causes of adverse effects in
aquatic systems, typically focusing on single species effects (ASTM, 2013)
and occasionally assessing effects on biotic assemblages (Crane et al., 2007).
For example, Allert et al. (2009) evaluated the effects of lead mining activities
on the survival and growth of caged woodland crayfish (*Orconectes hylas*)
after field reconnaissance revealed reduced crayfish densities immediately
below mining sites (Allert et al., 2008). Caged animals exposed to mining
effluent exhibited reduced survival and biomass compared to caged crayfish
at reference sites. These results supported and strengthened earlier laboratory toxicity analyses of other crustaceans that showed increased sensitivity
to chronic exposure to a metal cocktail (Besser et al., 2009) and field biomonitoring results that found metal bioaccumulation in crayfish (Besser et al.,
2007). Other researchers have used cultured organisms of a relevant species
or guild that were then caged and exposed at the site (McWilliam and Baird,
2002; Barata et al., 2007; Alexander et al., 2007). In situ cage testing also lends
itself to evaluating sublethal endpoints, such as post-exposure feeding rate
or body size (Crane and Maltby, 1991; McWilliam and Baird, 2002). A study
of caged mussels downstream from a municipal wastewater plume revealed
that the accumulation and distribution of metals varied among mussel tissues and between exposure types (dissolved or particulate metals; Gagnon
et al., 2006). Caging studies can also provide insights into complex interactions that lead to toxicity. For example, Bowling et al. (1983) deployed caged
juvenile sunfish and showed that a PAH, anthracene, was not toxic except
when irradiated (Bowling et al., 1983; Oris and Giesy, 1985, 1987).

Caging studies are valuable for producing causal evidence because they
provide replication and reduce variance relative to observational studies
while providing a more environmentally realistic exposure than laboratory tests (e.g., ambient temperature regimes). They permit experimental
manipulations to localize stressor exposure within the sediment or water
column, facilitate estimation of bioaccumulation or depuration rates, and can
be used to validate and refine estimates of contaminant transport or fate
parameters (Burton et al., 2005; Palace et al., 2005; Crane et al., 2007). Potential
disadvantages include logistical considerations such as capturing, handling,
transporting, and feeding organisms; locating suitable reference and impact
locations for cage placement; and vandalism (Burton et al., 2005; Crane et al.,
2007). When applying cage methods, care must be taken to avoid fouling that
impedes water flow as this affects contaminant adsorption, sediment and
waste accumulation, and physicochemical conditions (e.g., dissolved oxygen
concentration) (Palace et al., 2005; Liber et al., 2007). Other potential artifacts
that can influence results arise from the size of experimental apparatus (e.g.,
cage size) and stocking densities that may increase stress or competition
(Liber et al., 2007).

14.2.2 Colonization Substrates

Artificial substrates that have been colonized naturally with benthic macroin-vertebrates are frequently used to establish cause-and-effect linkages related to point-source discharges (Clark and Clements, 2006; Roberts et al., 2009). These colonization studies have many of the advantages and disadvantages outlined for caging studies. Notable advantages of this technique include the ability to study how stressor exposure modifies the rate of community succession and the capacity to examine responses of indigenous organisms under a realistic environmental regime. Limitations of the approach include the possible influences of substrate on colonization and sediment deposition during the experiment. Variability in response may be influenced by interspecific interactions and biological processes such as immigration and emigration, but these may also reflect processes that lead to an undesirable effect (e.g., macroinvertebrate drift, food source substitutions) (Clements, 2004; Crane et al., 2007; Liber et al., 2007).

Naturally colonized substrates have been used extensively to establish the ecological effects of metal contaminants in the Rocky Mountain, USA watersheds (Courtney and Clements, 2002; Clements, 2004; Clark and Clements, 2006 and references therein). These researchers allowed trays containing suitable substrates to be colonized by reference stream invertebrates. Then, the trays were transplanted into metal-contaminated streams and to reference stream locations. In the Arkansas River, the metal-contaminated site had significantly lower benthic macroinvertebrate and EPT taxonomic richness than assemblages at reference sites (Clements, 2004). A similar approach was used by Courtney and Clements (2002) to reveal that macroinvertebrate abundance and diversity were reduced at metal-contaminated sites compared to substrates placed at reference locations. Furthermore, colonization of metal-contaminated substrates by invertebrates in the reference stream tended to be reduced when compared to colonization of clean substrates within the reference stream. These results supported a causal linkage between metal contamination of the substrate and water column with reduced benthic macroinvertebrate abundance (Courtney and Clements, 2002) and showed that emigration rather than death was the mode of action in these metal-contaminated streams (Clements, 2004).

Other colonization substrates, such as nutrient diffusing substrates, glass discs, or ceramic tiles, have been employed in gradient designs or reciprocal transplant studies between reference and affected rivers. For example, periphyton biomass collected from nutrient diffusing substrates have been used to identify nutrient limitation below sewage and pulp mill outfalls (Scrimgeour and Chambers, 2000; see Chapter 24). Researchers found that periphyton production was phosphorus-limited upstream of these outfalls, and nutrient limitation was reduced below the outfalls. Ivorra et al. (1999) performed a reciprocal transplant study of algal assemblages between a metal-contaminated and reference stream. They reported that metal concentrations in the transplanted biofilms corresponded to those in the local biofilm within the duration of the study. In addition, composition of algal assemblages changed toward the local assemblages supporting the argument that metal contaminants are one determinant of the algal assemblage.

14.3 Whole Ecosystem Studies

Whole ecosystem studies that expose large areas to a candidate cause would seem to be the ultimate approach for providing insights into when and how causes produce effects. Whole ecosystem studies can provide information on exposure levels associated with effects and modes and mechanisms of action.

Schindler (1998) argued that whole ecosystem studies provide greater environmental realism related to effects on community and ecosystem processes, thereby providing managers with superior knowledge to make more effective management decisions. However, an important drawback of this approach is the problem of establishing adequate replication. Indeed, Carpenter et al. (1998a,b) suggested that instead of replicating whole ecosystems, researchers should use the replicate ecosystems to test alternative, contrasting hypotheses. In addition, ecosystem studies can be logically complex, expensive, and can lead to difficult ethical questions related to damage caused by experimental manipulation which may complicate environmental permitting.

For example, Kidd et al. (2007) used a BACI style approach to investigate cause–effect linkages between exposure to synthetic estrogens in the water column and feminization of fathead minnows (*Pimephales promelas*) in the Experimental Lakes Area, Ontario, Canada. They monitored fish before and after addition of synthetic estrogen to one "impacted" lake and two reference lakes. Levels of vitellogenin (VTG) mRNA and protein were similar among lakes in male and female minnows before estrogen addition, but post-treatment levels of these biomarkers increased substantially in both sexes in the treated lake relative to reference lakes. Approximately 50% of the males from the treated lake developed ova–testes and primary-stage oocytes, and the fathead population in the estrogen-treated lake exhibited year-class failures after two seasons. Year-class failures were not observed in a monitored reference lake. Results from this experiment supported previous laboratory findings and upstream–downstream comparisons that had suggested a link between male feminization and estrogen exposure. Furthermore, the whole ecosystem study helped to illustrate the biological impacts that estrogen exposure would have on reproductive success and the long-term sustainability of this fish population (Kidd et al., 2007). Recent surveys of aquatic systems around the globe have frequently found ambient concentrations of estrogens higher than reported in the Kidd et al. (2007) study (~5 ng/L), suggesting the possibility of ecological impacts on fish populations in these systems.

Whole ecosystem manipulations are unlikely to be performed to support a site-specific causal assessment. They are more likely to be used to address the sufficiency of an exposure to induce an effect or to indicate the types of effects that might occur. For example, the study by Kidd et al. (2007) could be used as supporting evidence if a fish population was diminished and vitellogenin levels were elevated to link that symptom with the population-level effect.

14.4 Summary

Field study designs and in situ methods can enhance the potential for assessment-specific study results to contribute strong, high-quality, and relevant evidence to a causal assessment. These approaches provide the opportunity to document biological responses to stressors under environmentally relevant conditions, while reducing the influence of confounding factors and reducing biological variability (see Table 14.1). Disadvantages include the lack of method standardization, difficulties associated with complex

TABLE 14.1

Summary of the Advantages and Limitations of Some Field Methods Used to Investigate the Fate and Effect of Stressors in Aquatic Environments

Field Study Method	Advantages	Limitations	Examples of Use
Caging techniques	Environmental realism Single to multispecies assemblages Localize stressor exposure	Capture, handling, transporting, and maybe feeding organisms Technique artifacts	Heavy metal impacts (Allert et al., 2009) Bioaccumulation of metals and pollutants (Bervoets et al., 2009)
Colonization substrates	Environmental realism Use indigenous species Natural succession rates	Lack of initial assemblage structure Technique artifacts	Heavy metal effects (Clark and Clements, 2006) Trout farm effluents (Roberts et al., 2009)
Whole ecosystem studies	Environmental realism Single to multispecies assemblages	Lack of replication Suitable reference sites Expense Environmental-permitting Ethical questions	Endocrine disruption (Kidd et al., 2007)

Source: (Adapted from Liber, K. et al. 2007. *Integr Environ Assess Manage* 3 (2):246–258; Crane, M. et al. 2007. *Integr Environ Assess Manage* 3 (2):234–245; Baird, D. J. et al. 2007a. *Integr Environ Assess Manage* 3:259–267; Baird, D. J. et al. 2007b. *Integr Environ Assess Manage* 3:275–278; Wharfe, J. et al. 2007. *Integr Environ Assess Manage* 3 (2):268–274.)

Note: Selected examples of use are provided for each method.

methodological logistics, the potential for high costs, and the lack of suitable replicate reference or affected sites.

Field studies can be specifically designed to elucidate mechanisms, demonstrate concentrations, and other conditions that are necessary for effects to occur. Results of studies conducted at sites other than the one under investigation can be applied when it can be claimed that the environmental setting, stressors, and biota are similar enough that the same results would occur. These assumptions are not necessary if the study is performed at the site under investigation. Then, observed relationships between the cause and effect are undeniably relevant. The cost and logistics sometimes may make field tests impractical until a decision is made to remediate. Then, they can be used to confirm or refute the cause by studying the results of the management action.

15

Laboratory Experiments and Microcosms

Alexa C. Alexander, Joseph M. Culp, and Robert B. Brua

This chapter discusses the role that laboratory experiments and micro-cosms play in causal assessments. By controlling conditions, laboratory experiments provide an important opportunity to isolate the effects of causes. However, they may not realistically reflect environmental conditions.

CONTENTS

Laboratory tests complement field tests and observational studies by providing the opportunity to tightly control exposures and the influence of potentially confounding variables. Laboratory experiments can show that a stressor is capable of causing an effect and provide insights into mechanisms and modes of action. They are used to establish exposures that are sufficient to inhibit responses in test organisms (e.g., growth, reproduction). Evidence for site-specific causal assessments is derived by comparing the levels of stressors and responses in test results to those observed at the affected site. This comparison is most frequently used to evaluate whether stressor concentrations reach sufficient levels to produce the observed effect (Culp et al., 2000a). Other responses such as biomarkers, physiological signs and symptoms, and behavioral responses can also be compared with measurements and effects observed at the site.

Laboratory toxicity tests are particularly useful for evaluating candidate causes that can be acutely toxic or are present at high concentrations (La Point et al., 1996). They are often part of regulatory monitoring programs such as the U.S. EPA NPDES, or Canada's Environmental Effects Monitoring (EEM) program and are used to assess compliance with environmental protection regulations, measure consistency in effluent quality, determine effectiveness of wastewater treatment, and estimate ecological risk (see also Box 15.1). In this chapter we provide a brief review of several types and applications of

---

BOX 15.1   TOXICITY TESTING IN THE NORTHERN RIVERS BASINS STUDY

The Northern River Basins Study was a broad multidisciplinary program to identify and quantify the impact of multiple complex stressors (e.g., nutrient additions, contaminants, and flow changes) on large, northern rivers within Canada. Before the program, relatively little was known about the patterns or processes in large northern rivers, and consequently, this work continues to serve as a baseline for ongoing research efforts into northern and arctic rivers. One of the project's objectives was to evaluate the effects of pulp mill effluent (PME) originating in the Athabasca, Wapiti, and Smoky Rivers. PME is a complex mixture of potential toxicants, nutrients, and particulate organic matter, and a toxicity testing approach was used to quantify its effects on biota. By incorporating laboratory testing, researchers were able to add context to the field program and help determine effect levels for further study in mesocosms. Combined, these methods contributed as part of a weight-of-evidence approach enabling the determinations of PME quality between different mills, the sufficiency of PME treatment, and to better understand the potential impacts of multiple discharges on instream fauna and flora. The study used the three standard test organisms, which provided an overview of where direct effects were likely to occur. For instance, does the PME inhibit plant growth, invertebrate reproduction, or fish development? The selected study organisms were: the freshwater alga (*Selenastrum capricornutum*; Environment Canada, 1992b), common water fleas (*C. dubia*; Environment Canada, 1992a), and fathead minnows (*P. promelas*; Environment Canada, 1992c) which were evaluated with respect to a dilution series that mirrored the dilution in different parts of the rivers of interest. The toxicity testing confirmed that the primary effect of the PME was not a toxic response but rather an enhancement effect due to the presence of nutrients at low effluent dilutions. These findings helped explain field observations that indicated that PME discharge coincided with the increased growth and abundance of benthic invertebrates and stream fishes. Therefore, the toxicity tests conducted for the Northern Rivers Basins Study helped link effects observed in the field to the causative agent, nutrients in the pulp mill effluent (see case study in Chapter 24).

laboratory experiments useful for causal assessment, including standard toxicity tests (see Section 15.1), species sensitivity distributions (see Section 15.2), environmental media quality assessments (see Section 15.3), toxic identification evaluations (see Section 15.4), and laboratory microcosm approaches (see Section 15.5). Advantages and limitations of these types of tests are summarized in Table 15.1.

15.1 Standard Toxicity Testing

Conventional toxicity tests have been used to derive exposure–response curves for individual chemicals. In turn, the curves have been used to delineate the range of acceptable concentrations for exposure to the chemical in the environment (see Figure 15.1). These curves are used to predict the concentration at which 50% of animals die, the median lethal concentration (LC_{50}), the median effective concentration (EC_{50}) for a nonlethal quantal effect (e.g., immobilization), or the 50% inhibition concentration (IC_{50}) for reductions relative to the maximum of a measure of performance (e.g., growth). The lowest-observable-effect concentration (LOEC) and the toxicant concentration at which effects are no longer detected (i.e., no-observable-effect concentration, NOEC) are commonly derived using hypothesis testing statistics. A preferred approach to deriving a benchmark value is to estimate a low effect concentration (e.g., 10%) from the exposure–response curve.

In causal assessments, toxicity test results are used to evaluate whether concentrations observed at the affected site are present at levels expected to cause lethal or chronic effects. This can be accomplished by using the entire curve to evaluate the level of effects that correspond to the observed concentrations, or by comparing concentrations to effect concentrations (e.g., EC_{50}s or LOECs). Detailed protocols for conducting tests are developed by governmental entities and groups like the ASTM (e.g., http://www.astm.org/Standards/E1850.htm). Adherence to proper procedures is a major part of judging the quality of test results. The choice of test organism(s) is critical as these must be sensitive to the stressor of interest, relatively easy to culture and maintain, available, ecologically relevant to the study site, and capable of producing a consistent response to the compound(s) being evaluated. Test results are most relevant for causal assessments when the test organisms are one of, or closely related to, the affected taxa examined at an appropriate life stage.

Often, laboratory tests predict field effects accurately only when a realistic matching of exposure between the laboratory and the field can be established. Acute tests are short and evaluate toxicity to concentrations high enough to rapidly cause death or other severe effects. They are useful for the screening of potential contaminant effects from spills, accidents, and treatment failures.

TABLE 15.1

Summary of the Advantages and Limitations of Various Laboratory Methods Used to Investigate the Fate and Effect of Stressors in Aquatic Systems

Method	Advantages	Limitations	Examples of Use
Standard toxicity testing	Screens and ranks chemicals Establishes concentrations that cause effects Identifies potential effects, symptoms, and effective body burdens Is simple, cost-effective	Usually evaluates one species and one contaminant Evaluates direct effects only Durations are short	Alexander et al., 2007
Species sensitivity distributions (SSD)	Represent the heterogeneity of species Can estimate the proportion of communities affected and the taxa most likely to be affected	The set of tested species may not represent the exposed community The input data are single-species toxicity tests (see above)	De Zwart and Posthuma, 2005; De Zwart et al., 2006; Coffey et al., 2014
Effluent and ambient media testing	Determines the toxicity of complex mixtures in the actual contaminated media	Can be laborious to collect, process, and test effluents, contaminated waters and sediments Difficult to reproduce treatments due to variance in space and time Relationship between sediment chemistry and biological effects is often poor	Long and Chapman, 1985
Toxicity identification evaluations (TIE)	Determines the toxicants within a medium that contribute to the measured response	As above Fractionation and treatments may alter the chemistry of the media and their toxicity	U.S. EPA, 2007a
Microcosm approaches	Can include more species and more relevant species and interactions while controlling potential exposures Approaches conventional laboratory tests for reproducibility and control	Evaluates only a few species or a very simple food web Less realistic than field ecosystems	Scrimgeour et al., 1991, 1994

Note: Selected examples of use are provided for each method.

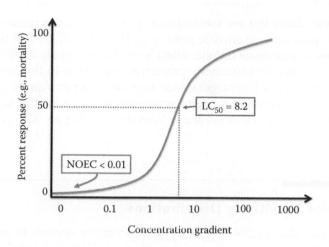

FIGURE 15.1
The log–logistic curve (the s-curve) fits the responses of organisms/populations (y-axis) to stressors along a concentration gradient (x-axis). The characteristic s-shaped curve indicates that more organisms respond (in this example, a mortal response) due to exposure to some toxicant along a concentration gradient (0 to 1000). The median lethal concentration (LC_{50}) is the concentration at which 50% of the animals die. The no-observable-effect concentration (NOEC) is the concentration at which no further response can be detected. The accuracy of both LC_{50}s and NOECs is subject to the quality and variability of the data.

However, acute tests are unlikely to adequately assess the impact of low-level environmental effects, which are becoming more common due to improved wastewater treatment and environmental regulation. For example, several studies indicate that laboratory toxicity test predictions can be poorly correlated with responses of benthic invertebrates (Giddings et al., 1994), algae (Heimbach et al., 1994), and fish (Birge et al., 1989). In contrast, Diamond and Daley (2000) found a relationship between standard effluent toxicity tests using *Ceriodaphnia dubia* and *Pimephales promelas* and results of benthic invertebrate assemblage surveys for 250 effluent discharges across the United States.

Chronic* tests are longer and evaluate responses to lower exposure levels than acute tests. These tests are likely a better proxy for most environmental effects because long-term, sublethal exposure may be responsible for species losses and other effects observed in the field. Chronic tests are particularly useful in the study of effects that take time to develop such as reproduction and multigenerational effects. Chronic toxicity tests also lend themselves to studies of metabolism of toxic compounds through uptake and elimination studies. Due to the time and expense required to conduct longer, lower exposure experiments, acute studies remain the norm. Because chronic tests often demonstrate

---

* Note that the term chronic can refer to the duration of the exposure (e.g., >2 weeks), the tested life stage (embryos and larvae), or the endpoint of the test (i.e., nonlethal effects). In some situations, acute responses may result in delayed lethal effects (e.g., cancer) which are also referred to as chronic effects.

effects at lower doses but are less commonly available, some researchers and agencies calculate acute to chronic ratios (e.g., Hoff et al., 2010) that use results from acute tests to estimate chronic-effect levels (Ahlers et al., 2006). However, these ratios may not be sufficiently protective, especially in the case of compounds like pesticides. Nontarget effects may be better explained by pairing existing safety factors with a mode-of-action concept, thus identifying the species or functional groups that are at risk (Van den Brink et al., 2006a).

15.2 Species Sensitivity Distributions

SSDs aggregate the results of toxicity tests on many species to predict the potentially affected fraction of species in the field (Vaal et al., 1997; Admiraal et al., 2000; Newman et al., 2000; Maltby et al., 2005; Posthuma and De Zwart, 2006). By modeling the responses of many species, SSDs incorporate the underlying heterogeneity of species' responses (Hoy et al., 1998). Methods and issues in developing SSDs are discussed in Posthuma et al. (2002) and summarized in Suter (2007).

SSDs can be used in site-specific causal assessments by comparing the concentrations of chemicals observed at the affected site to the SSD curve. Chemical concentrations corresponding to effects in the SSD well above those observed at the site weaken the argument for that chemical. The comparison can be refined by comparing the reduction in richness observed in the field to the potentially affected fraction of species estimated by the SSD. For example, in the Long Creek case study (see Chapter 22) (see Figure 15.2), storm-flow measurements of zinc concentrations and the reduction in EPT taxa were compared with an SSD constructed for invertebrate species. The reduction observed in EPT taxa observed at site LCN .415 (40%) was well above that predicted from the SSD (approaching 0), weakening the argument that zinc in storm runoff reduced EPT taxa richness.

SSDs have been combined with mode-of-action modeling to predict the response to mixtures (Ashford, 1981; Escher and Hermens, 2002; De Zwart and Posthuma, 2005). This approach has been used to link biological monitoring survey results to likely causes of observed adverse effects. De Zwart et al. (2006) summed the potentially affected fraction from individual chemicals to estimate the response to the mixture. They used these results to estimate the relative contribution of several groups of stressors including habitat, effluent inputs, and toxic chemicals to fish assemblages in Ohio, USA. In a related application, SSDs were implemented in a Geographic Information Systems (GIS) framework for preliminary screening of stressors using multiple independent methods of identification (Kapo et al., 2008).

SSDs based on laboratory results have been criticized since the models are only as good as the toxicity data used to generate them and sensitive species

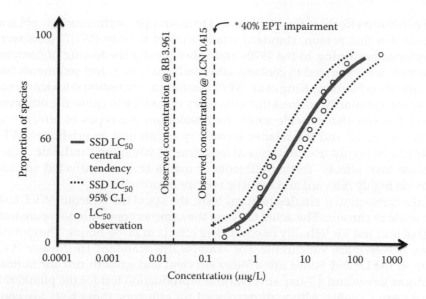

FIGURE 15.2
An SSD based on macroinvertebrate zinc $LC_{50}s$ values. Vertical lines show maximum storm-flow zinc concentrations observed at the Long Creek case study sites RB 3.961 and LCN .415. EPT taxa richness was reduced 40% at LCN .415, a greater reduction than would be expected based on the SSD. U.S. EPA's criterion maximum concentration (CMC) also shown. (Adapted from Ziegler, C. R. et al., 2007a. *Causal analysis of biological impairment in Long Creek: A sandy-bottomed stream in coastal Soutern Maine.* U.S. Environmental Protection Agency, National Center for Environmental Assessment. EPA/600/R-06/065F.)

may not be included. Therefore, SSDs are subject to the same drawbacks, specifically realism and relevance, as individual toxicity tests (Forbes and Calow, 2002). The determination of which species to include in the SSD is a matter of debate. European scientists tend to include all test results (including microbes), whereas others aggregate based on taxonomic category or based on knowledge of the stressor's mode of action (Suter, 2007). In the Long Creek case study, separate SSDs were developed for fish and macroinvertebrate assemblages. SSDs can also be constructed using stressor–response relationships developed from field observations studies (see Section 12.2.4).

15.3 Toxicity Tests of Effluents, Ambient Waters, and Sediments

Toxicity tests of individual chemicals can poorly represent the effects of the complex mixtures that are released or that occur in the environment due to the combined effects of the mixture and to matrix properties such as pH and anoxia. For that reason, standard whole effluent toxicity (WET) tests were developed beginning in the 1970s and subsequently the testing of complex materials was extended to contaminated ambient waters and sediments (see the history in Norberg-King et al., 2005). Similar to conventional toxicity tests, these tests primarily address the sufficiency of toxicity to cause the observed effect, but can also provide some information on the types of effects that could be caused and the relative sensitivity of fish and invertebrates. WET tests can be highly useful in causal assessments when the candidate causes include toxic effects. Testing effluents or media from the affected site can provide highly relevant and realistic toxicity information.

Like conventional single-chemical tests, the standard aqueous WET tests are acute or chronic. The acute tests are the same as conventional acute tests in that they test for lethality or equivalent effects in 24–96 hours. The chronic tests, however, are subchronic. For example, the standard freshwater WET tests in the United States are a 7-day survival and growth test for fathead minnow larvae and a 7-day survival and reproduction test for the planktonic crustacean *C. dubia*. Although developed for effluents, these tests are commonly applied to contaminated ambient waters.

Sediment quality testing is used to determine effects and bioavailability of sediment contaminants. Investigations may examine contaminant interactions, determine spatial and temporal distribution of contaminants, evaluate hazards of dredge materials, rank areas for cleanup, as well as monitor remediation and management efforts (Burton, 1991). Investigations have tested pore water (interstitial water), elutriate (the water extracted after mixing sediment and water), or whole sediments (Ankley et al., 1996). Of these approaches, whole sediment tests are thought to be the most realistic and provide a direct measure of toxicity.

Although tests of contaminated site media can provide highly useful evidence, assessors must be aware of potential problems. These include modification of the medium or contaminants during collection, processing, and storage. Also, the tested samples may be unrepresentative if contamination is variable in time or space. Finally, the tests are of short duration and use few species, so relevant toxic effects may not be detected. Assessors must be aware of these potential limitations when interpreting results as evidence, but they do not negate the inherent advantages of testing the site media.

Tests of effluents or contaminated media from another site provide useful evidence when they address similar contaminants (e.g., bleached kraft mill effluent at both sites) or when the matrix is judged to be more relevant than available conventional toxicity test (e.g., a sediment from a similar ecosystem which is contaminated by one of the candidate causes at the affected site). Media tests have also been used to set national or regional benchmark values. However, use of contaminated sediment tests to derive sediment quality guidelines has been widely debated in the literature (Swartz, 1999) due at least in part to the complexity of the mixtures found in sediments, complex

---

BOX 15.2   THE SEDIMENT QUALITY TRIAD

The sediment quality triad (Long and Chapman, 1985) was developed to improve sediment testing by explicitly defining what is needed to determine whether contaminated sediments are causing effects. The triad approach incorporates chemistry, toxicology, and ecology in a weight-of-evidence approach (Cassee et al., 1998; Culp et al., 2000a; Lowell et al., 2000; Preston and Shackelford, 2002). The original sediment quality triad used chemical analysis of sediment, benthic invertebrate surveys, and lab tests of sediments. Variants on the triad approach are common, such as systems where laboratory-reared organisms are deployed in field conditions (*in situ*) and where organisms are exposed to a range of natural gradients (e.g., sediments and turbidity) as well as any contaminants that are present.

---

partitioning at sediment–water–biota interfaces, and the site-specific nature of sediment contamination (Driscoll and Burgess, 2007) (see Box 15.2).

15.4 Toxicity Identification Evaluations

Although testing of effluents and contaminated ambient waters and sediments can determine whether these complex materials cause a specific toxic effect, this approach is often insufficient. It may be necessary to determine which constituents are responsible. In particular, rather than shutting down an effluent or dredging sediment, effluent treatment or sediment remediation may be designed to target specific chemicals or classes of chemicals.

The toxicity identification evaluation (TIE) method (Norberg-King et al., 2005; U.S. EPA, 2007a) determines the causes of toxicity in effluents and ambient waters. In TIE, the toxic constituents of a mixture are identified by removing components of the mixture and testing the residue, fractionating the mixture and testing the fractions, adding components of the mixture to background media and testing them, and other techniques. TIE methods for sediments have been developed based on the previous work on TIE for effluents and receiving waters (e.g., U.S. EPA, 1991, 1992, 1993). Because WET testing usually precedes TIE, the standard WET tests are usually used in TIE. For example, Maltby et al. (2000) combined TIE evaluation with standard WET tests to determine that chlorine was the primary contributor to sediment toxicity below a bleaching mill discharge.

Among the drawbacks of TIE for sediments is the difficulty in extracting sufficient pore water for toxicity testing, preventing the use of larger species.

Also, the extensive testing required to track effects over time can be costly and laborious because successive investigations are needed to determine which compound(s) are responsible for the toxic effect. The advantage of the TIE method is that the effluents or ambient media from areas of interest are used directly in the testing, and therefore, the potentially toxic components are characterized in a directly relevant and meaningful way.

15.5 Laboratory Microcosms

Laboratory microcosms range from collections of microbial species in test tubes to aquaria with multiple trophic levels. These "mini-ecosystems" can be used to simulate aquatic systems, thereby making it possible to conduct experiments to determine fate and effects of contaminants under highly controlled conditions without confounding factors. Pond microcosms often include complex food webs (Traas et al., 2004; Van Wijngaarden et al., 2004), while laboratory stream systems tend to use more simplified ones (Lamberti, 1993; McIntire, 1993). Laboratory microcosms have been used to evaluate toxic effects that are mediated by species interactions, such as competition and predation (Scrimgeour et al., 1991, 1994), the effects of multiple pesticides on pond organisms (Van Wijngaarden et al., 2004), and the interactive effects of nutrient enrichment and contaminants on planktonic and benthic components of freshwater food webs (Brock et al., 1995; Van Donk et al., 1995). Laboratory microcosm results have been combined with models to examine causal scenarios and predict indirect stressor effects (Traas et al., 2004). Design of laboratory microcosm experiments requires a tradeoff between control of environmental conditions and departure from environmental realism. Thus, investigators must consider that microcosm results may be constrained by factors such as oversimplification of species assemblages, and the potential for indoor microcosm systems to become impoverished over time due to their isolation from natural biological colonization (Lamberti, 1993; Traas et al., 2004).

When using microcosms in causal assessments, assessors must consider whether the conditions created in the microcosm system are important for understanding relationships in the affected ecosystem (e.g., multiple species or trophic levels). Otherwise, any evidence derived from the microcosm is no higher quality than evidence from conventional toxicity tests.

15.6 Summary

The strength of laboratory tests is the ability for investigators to control exposures and potentially confounding variables. In cases of suspected direct toxicity, laboratory toxicity tests are especially important, as these tests remove the "noise" so often found in observational field studies. Their weakness is the relevance of test conditions, exposures, organisms, and responses to the effects under investigation. The ability of laboratory tests to explain ecological effects in the field is dependent in part on the validity of test assumptions. These include the presumption that biological complexity is adequately simulated, that highly controlled laboratory conditions are representative of more variable environmental conditions, that test species will express similar responses to the contaminant as will native organisms, and that exposures in the laboratory (potentially short, acute) are representative of longer term, multiple exposures in the environment. Forbes and Forbes (1994) cautioned that single-species data have a limited capacity to predict community responses because they are measures of effects in the absence of an ecological context. Standard laboratory methods also do not account for bioaccumulation of contaminants, temporal changes in exposure or response, or multiple stressor effects (Waller et al., 1996). Not surprisingly, a growing body of literature suggests that sublethal and indirect effects are not only more common, but are also more difficult to predict from single species bioassay data (Brock et al., 2000; Fleeger et al., 2003; Rohr and Crumrine, 2005; Rohr et al., 2006). Clearly, the use of laboratory testing in helping to establish causal relationships could be strengthened by conducting tests that include species that are representative of the affected organisms and which incorporate any affected functions as endpoints (Kersting, 1994). For example, the use of multispecies laboratory microcosms offers an opportunity to combine components of the ecosystem that are believed to interact in the induction of the effect. Although laboratory testing is a powerful tool that provides significant benefit when attempting to establish causal relationships, the effective use of these techniques requires the investigator to have a thorough understanding of the advantages and limitations of the techniques employed. For these reasons, we recommend that evidence from laboratory tests be used in combination with other sources of evidence in causal assessment (Barbour et al., 1996; Hall and Giddings, 2000; Cormier and Suter, 2011).

16

Mesocosm Studies

Joseph M. Culp, Alexa C. Alexander, and Robert B. Brua

This chapter discusses the use of mesocosm studies for causal assessment. Mesocosm studies combine some of the natural conditions and interactions captured in field studies with some of the controls afforded by laboratory experimental systems. These methods are more likely to be used when the costs are justified by the need for more confidence in the assessment conclusions and when results can be applied to many site-specific assessments.

CONTENTS

Mesocosms are outdoor or indoor facilities with controlled physicochemical conditions and sometimes standardized biological assemblages. They are used to simulate complex exposure dynamics under simulated but realistic field conditions (Culp and Baird, 2006). Mesocosms are variously defined, but here we follow the general classification of Boyle and Fairchild (1997). They describe mesocosms as facilities that range in size from ponds or large experimental streams with defined physical dimensions and water quality, to smaller semicontrolled limnocorrals, tanks, and streams, to small (<1 m^3) tanks or recirculating streams (i.e., sometimes labeled microcosms) with strictly controlled biological assemblages and physicochemical conditions.

Mesocosms allow some control while approaching near-natural environmental conditions. They provide the opportunity to investigate effects that cannot be studied in smaller laboratory settings, such as structural (e.g., diversity) and functional (e.g., production) changes in communities, direct and indirect effects of stressors on food web components, and lethal and

sublethal effects of suspected causative agents under more realistic conditions (see Table 16.1) (Lamberti and Steinman, 1993; Culp et al., 2000b).

Mesocosm experiments are logistically more complex and usually more
expensive than observational field studies or laboratory tests. These shortcomings are countered by the production of high-quality information that can
reduce uncertainty in causal assessments. Advantages of mesocosms compared to field observational studies include replication and control of exposure. Compared to laboratory tests, they provide more realistic environmental
conditions. Mesocosms have been criticized for their smaller spatial scale and
temporal duration compared to whole-ecosystem studies (Carpenter, 1996,
1999; Schindler, 1998). However, mesocosms provide the means to simulate
natural processes without harming natural systems as a result of experimental
manipulation (Guckert, 1993; McIntire, 1993; Lawton, 1996; Boyle and Fairchild,
1997; Drenner and Mazumder, 1999; Clements et al., 2002).

Mesocosms are most useful when they realistically simulate the causal
processes and effects under investigation (Lamberti and Steinman, 1993;
Culp and Baird, 2006). Mesocosm tests may be conducted in later stages of
an assessment to confirm a cause or to better understand a complex process. Results from mesocosm experiments conducted at other locations for
other purposes may be available early in the assessment process but must be
judged relevant to the investigation. Criteria to determine relevance include
similarity in physicochemical environment simulated (e.g., light and temperature regime), comparable biotic communities, and the effects and stressors
examined.

This chapter describes how results can be used in causal assessments
(see Section 16.1) and reviews types of freshwater aquatic mesocosms (see
Section 16.2).

16.1 Mesocosm Study Applications for Causal Assessments

Mesocosm studies can be used to tackle an extensive range of topics potentially relevant to causal assessments. They distinguish the contribution of different stressors under simulated site conditions and can be used to develop
stressor–response relationships that can be compared to observations from
the site. They can be used to better understand the link between laboratory
results and field observations and capture more complex causal processes
(e.g., food web dynamics) than can be easily tested in the laboratory.

Mesocosm studies have been used to discriminate and identify the contribution of individual stressors that covary at the affected site. For example,
riverside mesocosm experiments were used in the Athabasca River, Alberta,
Canada, to separate the effects on benthic food webs of nutrients and contaminants contained in pulp mill effluent (see Chapter 24; Culp et al., 2000b).

TABLE 16.1

Summary of the Advantages and Limitations of Various Mesocosm Methods Used to Investigate the Fate and Effect of Stressors in Aquatic Environments

Mesocosm Methods	Advantages	Limitations	Examples of Use
Constructed ponds or ditches	Potential to follow long-term recovery after disturbance Includes complete food web	High set-up cost Potential for high replicate variation	Insecticide fate (Webber et al., 1992) Insecticide effects (Brock et al., 2009)
Limnocorrals	Similar to ponds but lower cost	Biofilm on walls may affect water quality Restricted horizontal and vertical mixing Costly to maintain for long periods	Antisapstain biocide effects (Liber et al., 1994) Nutrient enrichment effects (Forrest and Arnott, 2006)
Littoral enclosures	Similar to limnocorrals Includes shoreline biota	Similar to limnocorral High predator variability Risk to treatment integrity as water depth increases	Insecticide fate (Heinis and Knuth, 1992) Insecticide effects (Lozano et al., 1992)
Fabricated tanks	Lower cost than larger systems Highly controlled environment Simple to replicate No environmental release of pollutant	Natural colonization highly restricted High potential for wall biofilm to affect water quality Difficult to include large predators	Insecticide fate and effects (Rand et al., 2000) Insecticide and pH effects (Relyea, 2006)
Large-stream mesocosms	Potential to follow long-term recovery after disturbance Ability to include complete food web	High set-up cost Potential for high replicate variation	Review of large-stream systems (Swift et al., 1993) Next generation facility (Mohr et al., 2005)
Small-stream mesocosms	Lower cost than larger systems Highly controlled environment with natural communities Simple to replicate No environmental release of pollutant	Recolonization by downstream drift absent High potential for wall biofilm to affect community structure Difficult to include large predators	Insecticide effects (Alexander et al., 2008) Metal effects (Clements et al., 2002) Pulp mill and sewage effluent effects (Culp et al., 2000a)

Note: Selected examples of use are provided for each method.

The environmentally relevant effluent concentrations did not produce measurable toxicity in insects or algae, but effluent-associated phosphorus increased algal biomass and insect abundance. In a series of experiments, Clements (2004) exposed benthic macroinvertebrate assemblages to zinc separately and in combination with other metals (cadmium and copper). He was able to establish relationships between metal concentration and several responses of biological structure and function (e.g., invertebrate richness and drift, community respiration) and demonstrate that macroinvertebrate responses to the three-metal mixture was greater than either that of zinc alone or zinc plus cadmium.

Mesocosm studies have been used to quantify stressor–response relationships for specific contaminants or effluents (Bothwell, 1993; Clements et al., 2002; Dubé et al., 2002; Culp et al., 2003; Clements, 2004). Stream mesocosms have been used to identify toxicity thresholds of invertebrates or algae for petrochemical effluents (Crossland et al., 1992), consumer product chemicals (Belanger et al., 1994), and insecticides (Alexander et al., 2008). They have been used to develop quantitative nutrient thresholds for autotrophic production in streams (Bothwell, 1993). Chambers et al. (2000) employed this approach to demonstrate phosphorus limitation and recommend nutrient-loading limits in the Athabasca River (see Chapter 24).

Mesocosm studies can be used to establish when results from laboratory tests are relevant to field conditions (Crane et al., 1999; Hanson et al., 2003). Mesocosms have been used to study pond and lake ecosystems for a variety of chemical and physical stressors (e.g., pesticides, pharmaceuticals, nutrients, pH). These studies tackle an extensive range of topics including the indirect effects of stressors on food webs (deNoyelles et al., 1994; Relyea, 2006), recovery responses of macroinvertebrate assemblages after short-term exposure to stressors (Brock et al. 2009), and pesticide effects in the environment (Van den Brink et al., 2006b). Crane et al. (1999) employed mesocosms to demonstrate the effectiveness of laboratory bioassays to evaluate the bioavailability and toxicity of insecticides to invertebrates, and Hanson et al. (2003) demonstrated that laboratory and mesocosm studies could produce similar ranges in variation of macrophytes endpoints.

Mesocosms have contributed to understanding complex interactions and processes that cannot be simulated in laboratory tests. For example, mesocosms have been used to demonstrate the importance of spatial refugia in the recovery rate of macroinvertebrates to insecticide spray-drift exposure (Brock et al., 2009). In a review of 36 mesocosm studies using pesticides as the stressor, deNoyelles et al. (1994) concluded that mesocosm studies can track indirect effects of insecticides or herbicides as different components of the food web are altered (deNoyelles et al., 1994; Relyea, 2006). Similarly, Relyea (2006) strongly advocated tracking food-web changes in mesocosms because potential indirect effects of stressors, such as pesticides, cannot be predicted from laboratory experiments. Finally, mesocosm results can be combined with predictive modeling to better link laboratory and field results (Van den Brink et al., 2006b).

In sum, these studies show that mesocosm studies are a way to judge the relevance and accuracy of laboratory studies for deriving evidence in causal assessments. Evidence relevant to a specific case can be derived by comparing mesocosm study results with observations from the case, including stressor and effects levels, specific community alterations, and measurements reflecting exposures, mechanisms, or modes of action.

16.2 Mesocosm Methods

The following sections provide a brief overview of mesocosm designs that have been used to simulate aquatic systems. Understanding the challenges and considerations involved in developing different mesocosm types aids the design of effective studies for the site under investigation and the interpretation of results from other available studies.

16.2.1 Mesocosm Methods for Lakes and Ponds

Investigators have used various lake and pond (i.e., lentic) mesocosm systems to evaluate the fate and effects of stressors and to quantify cause–effect relationships in a multispecies context. The physical size of lentic mesocosms ranges from artificially constructed systems (fabricated tanks, constructed ponds, ditches) to isolated subsections of the natural habitat (limnocorrals, littoral enclosures) (Boyle and Fairchild, 1997). The choice of mesocosm method can depend upon study question formulation (e.g., inclusion of fish predators), but the large investment of time and money required for mesocosm research often directs researchers toward the use of smaller mesocosms. Fortunately, smaller systems can demonstrate causal associations similar to that generated in larger systems at a substantial cost savings. For example, Howick et al. (1994) found that fiberglass tank experiments produced identical effects of insecticides on invertebrates as earthen pond experiments and were 80% less expensive. A summary of the advantages and limitations of various mesocosm methods used to investigate the fate and effects of stressors in lentic environments is detailed in Table 16.1.

Pond-like mesocosms have a long history of use in assessing the ecological risk of pesticides. Constructed, earthen ponds range in size from 0.01 to 0.1 ha and have been used in higher tier-testing procedures required by the U.S. EPA for pesticide registration (Graney et al., 1994). Pond mesocosms are normally less than 3-m deep (Christman et al., 1994; Howick et al., 1994; Johnson et al., 1994). Plastic or clay liners may be necessary to improve water retention capacity (Howick et al., 1994). When establishing pond mesocosms, one must decide either to allow the occurrence of biological colonization through natural dispersal processes or to introduce sediments from well-established ponds. To reduce replicate variability and facilitate rapid establishment of biological communities, Christman et al. (1994) suggest inoculation of biota with lower dispersal rates (e.g., some macroinvertebrates and macrophytes). In contrast, Ferrington et al. (1994) and Howick et al. (1994) demonstrated that the addition of mature pond sediments to mesocosms rapidly produced biological communities with similar biodiversity, abundance, and seasonal phenology to mature ponds.

Lentic mesocosm methods also include limnocorrals and littoral enclosures. Limnocorrals isolate replicate subsections of the aquatic environment using dividing sidewalls anchored to bottom sediments and filled with filtered (e.g., Forrest and Arnott, 2006) or unfiltered pelagic water (e.g., Thompson et al., 1994). Limnocorrals have similar advantages to other mesocosm approaches (e.g., replicated design, standardized physicochemical environment, multispecies interactions) (see Table 16.1). Limitations that must be considered are wall effects that result from restricted vertical and horizontal mixing. Wall effects can affect nutrient and chemical dynamics, physicochemical properties, and interspecies interactions (e.g., predation), resulting in conditions that diverge with time relative to the surrounding water body (Graney et al., 1995). Littoral enclosures differ from limnocorrals in that they have a natural shoreline and three plastic walls embedded in the sediments (Graney et al., 1995). While they have comparable advantages to limnocorrals, their limitations include high replicate variation in top predator density and the possibility that fluctuations in water depth will compromise enclosure integrity (see Table 16.1).

The most controlled lentic mesocosms are fabricated tanks where the sediment and water source are the same among replicates and the biological communities are carefully manipulated (Rand et al., 2000). Tanks from 2000 to 20,000 L are large enough to be representative of lentic food webs (often excluding fish) and have ambient environmental conditions of temperature, light, wind, etc. (Graney et al., 1994). The primary advantages of tank mesocosms are lower cost and the ability to prevent contamination of the natural environment with study compounds. The small size of mesocosm tanks often limits the inclusion of larger predators and can result in undesirable wall effects. Of particular importance is the choice of sampling technique because destructive sampling (e.g., sediment grab samples) can affect experimental results in these small systems (Graney et al., 1995).

16.2.2 Mesocosm Methods for Streams

Stream (i.e., lotic) mesocosms have been used in ecological and ecotoxicological research for over 50 years (McIntire, 1993). They range in size from large, constructed channels (>100 m³) with once-through flow, to smaller (<1 m³) systems consisting of fabricated tanks with partial recirculation or once-through flow (Swift et al., 1993; Culp and Baird, 2006).

Large stream mesocosms (>50 m in length; >100 m³) are very rare (Swift et al., 1993). The high costs of construction and maintenance greatly limit the number of replicate streams and potential for variety in experimental design (see Table 16.1). Their operation has largely been restricted to government agencies or large consortia due to cost and logistical complexity. Recognizing the need for larger, replicated systems that are ecologically realistic, Mohr et al. (2005) have developed a mesocosm facility that offers a highly flexible design which can be arranged as eight replicate streams (up to 106-m long each) or joined into a single stream approximately 850-m long.

Small stream mesocosms have been used widely by ecologists and ecotoxicologists (e.g., Lamberti and Steinman, 1993; Culp et al., 2000b; Dubé et al., 2002; Schulz et al., 2002; Crane et al., 2007). These systems range from flow-through flumes (Bothwell, 1993), to transportable streamside mesocosms (Culp and Baird, 2006), to greenhouse systems with naturally colonized substrates (Clark and Clements, 2006). Water velocity in the systems is controlled by stirring mechanisms, water pumps, or the slope of the experimental unit. The establishment of benthic communities varies and includes natural colonization, seeding of benthic communities from reference riffles (Alexander et al., 2008), and the use of trays of substrates that have been colonized in reference streams (Clark and Clements, 2006). Because of their small scale, researchers generally conduct shorter term experiments (<30 days). Shorter studies decrease the possibility of wall effects and undesirable divergence from the reference stream composition as a result of the absence of biological processes such as invertebrate drift from upstream habitats. Despite these limitations, small-scale stream mesocosm experiments have proved to be particularly informative when combined with field and laboratory studies (Culp et al., 2000a; Clements et al., 2002; Clements, 2004).

16.3 Summary

Mesocosm experiments offer several key benefits, including control over stressor exposure and duration under relevant ecological complexity and environmental conditions, appropriate treatment replication, elimination of potential confounding factors, and the ability to evaluate stressor–response relationships in a multispecies setting (see Table 16.1). They provide greater realism and environmental complexity than laboratory testing (Forbes and Forbes, 1994). They also facilitate the study of chronic exposure of food webs at a more logistically and financially feasible scale than whole ecosystem experiments (Kimball and Levin, 1985; Shaw and Kennedy, 1996).

17

Symptoms, Body Burdens, and Biomarkers

Glenn W. Suter II

Symptoms, body burdens, and biomarkers are observations that can be used to confirm that the cause has produced characteristic effects or that biologically relevant exposure has occurred.

CONTENTS

This chapter discusses evidence that supports the practice of diagnosis, including symptoms, body burdens, and biomarkers.

Symptoms are effects that are characteristic of a particular causal agent or a group of similarly acting agents.* Symptoms demonstrate that the cause has altered biota in characteristic ways. Body burdens and biomarkers can confirm that a biologically relevant interaction has occurred. Because they provide evidence of uptake or of a process by which organisms respond to exposure to a particular agent or group of agents, they can serve the function of symptoms.

---

* Many veterinarians and physicians reserve the term "symptom" for subjective evidence of disease as observed by the human patient (e.g., a headache). They would likely use the term "signs" for the features described in this chapter. We are following the common language definition of symptom.

17.1 Body Burdens and Biomarkers

Body burdens and biomarkers reflect the underlying mechanistic processes by which organisms are internally exposed to chemicals. For example, for a chemical to produce an effect in an organism, it must be taken into the organism, leading to accumulation in organs or induction of metabolic enzymes. Therefore, body burdens, chemical concentrations in target organs, or enzyme levels provide evidence that biologically relevant exposure has occurred. That is, the accumulation of a chemical or a proximate biochemical response indicates that the candidate cause not only co-occurred with the organisms, but was in a bioavailable form and was taken up.

Body burdens are commonly used as evidence of whether contaminants are causing organism-level ecological effects. In particular, they are commonly used as evidence that organisms have actually been exposed to a chemical. However, body burdens are useful for only those chemicals that are accumulated and not internally regulated by organisms. For example, most organic chemicals are readily metabolized or excreted. Internal concentrations of nutrient elements are regulated, but regulation of micronutrient metals may be overwhelmed by high exposures. Also, aquatic arthropods are capable of sequestering metals in nontoxic forms, so total concentrations are poor predictors of effects (Rainbow, 2002). Metabolites of organic compounds are sometimes easier to measure than the parent compound (Cormier et al., 2000, 2002).

Body burdens may also be used in exposure–response models to associate tissue residues with relevant effects. Such relationships have been summarized for aquatic organisms by Jarvinen and Ankley (1999, 2009) and for wildlife by Beyer and Meador (2011).

Biochemical and physiological biomarkers have been the subjects of research for decades and have been applied to numerous environmental cases (Bartell, 2006; Forbes et al., 2006; Amiard-Triquet et al., 2012). For example, metabolic enzyme levels have been used as biomarkers of exposure to organic chemicals, blood vitellogenin levels in male fish have been used as biomarkers of estrogenic compounds, and DNA adducts have been used as biomarkers of exposure to specific mutagens. Because these responses are more specific than the usual field monitoring metrics (presence/absence and abundance of taxa), biomarkers provide useful symptoms in specific cases. However, biomarker research has emphasized developing sensitive measures of effect rather than on determining the causes of observed higher–level effects. Few biomarkers or sets of biomarkers have been shown to be symptomatic of a particular causal agent. A recent review of biomarkers for fish lists nine potential uses for biomarkers in field studies and identifying causes is not included (Schlenk et al., 2008). This situation is changing, particularly in Europe, where toxicity profiling is being developed as a method for causal assessment (Hamers et al., 2013). Toxicity profiling involves applying a battery of tests of a chemical or mixture to laboratory organisms or to organisms exposed in the field to identify responses that constitute a toxicological fingerprint.

17.2 Symptoms

Symptoms observed in the affected biota provide evidence that a cause contributed to the effect. For example, observing bright red gills in fish at a site under investigation provides evidence that cyanide may have contributed to a fish kill (Meyer and Barclay, 1990). The reliability of a symptom is judged based on how consistent or exclusive the symptom is. Consistent symptoms always appear when a particular agent is acting. Exclusive symptoms appear with exposure to only one particular agent and not with others. For example, bright red gills are associated with respiratory blockers and respiratory membrane irritants, in general, but not with narcotics or metals.

Symptoms used in site investigations are typically based on previously conducted work and published in the literature or manuals. Examples include investigations of fish kills (Meyer and Barclay, 1990; Roberts, 2012), incidents of aquatic toxicity (Norberg-King et al., 2005), and wildlife investigations (Friend and Franson, 1999; Braun, 2005; Huffman and Wallace, 2011; Cooper and Cooper, 2013). Guides to symptoms for distinguishing air pollution effects and diseases in plants have been particularly well developed for decades (Skelly et al., 1990; Flagler, 1998). Others such as guides for diagnosing causes of effects on corals are under development (Raymundo et al., 2008).

Symptoms can be based on observations from controlled laboratory settings, semicontrolled field enclosures (e.g., fish hatcheries), or field observational studies. Symptoms therefore carry the advantages and limitations of the source of observations used to develop them. Symptoms identified in laboratory studies have the advantage of being observed under controlled conditions, but the studies may lack realism or may be based on irrelevant organisms. Symptoms identified in field observational studies have the advantage of realistic exposures, but may be influenced by other stressors. Ideally, symptoms would be based on a combination of studies, using Koch's postulates or other criteria to verify that exposure to a cause consistently produces a symptom. For example, a combination of field observations and laboratory isolation and infection studies have been used to determine the symptoms of five diseases in corals (Sutherland et al., 2004). The symptoms of white plague Type II disease are diagnostic of *Aurantimonas coralicida*.

Symptoms are often defined at a lower level of biological organization than the effect of interest. In the cyanide example, the observation of oxygen-saturated hemoglobin (i.e., red gills) is used to provide evidence concerning the cause of an organism-level effect (i.e., mortality). The following discussion organizes symptoms by the level of biological organization associated with the effect of interest at the organism, population, and community level.

17.2.1 Symptoms of Organism-Level Effects

Symptoms used to evaluate organism-level effects include features that can be observed in organisms in the field (e.g., gross pathologies and behaviors).

Gross pathologies are the symptoms that anyone can see if they examine a dead, dying, or injured organism. They include deformities, lesions, tumors, and other physical traits that are visible in the field or during necropsy and that are indicative of disease, toxicity, or injury. They are often used in investigations of wildlife or fish kills (see Table 17.1), other wildlife kills (e.g., marine mammals, birds), forest declines, or other system-wide effects like loss of eel grass or coral decline. The kit fox case study used conventional veterinary diagnostic symptoms to exclude diseases and used characteristic injuries to identify coyote kills (see Chapter 25). Some gross symptoms are characteristic of a set of causal agents. For example, pigmented salmon syndrome (yellow pigmentation of the ventral surface and around the gill arches) results from combined exposure to diesel oil and resin acids, but not from either separately (Croce and Stagg, 1997).

Behavior provides useful symptoms in a few cases. When dissolved oxygen is low, fish gasp at the surface. In response to neurotoxicants such as cholinesterase-inhibiting pesticides exposed animals exhibit convulsions, and at high mercury exposures, mammals and birds exhibit ataxia.

TABLE 17.1

Examples of Symptoms for Chemically Induced Fish Kills

Symptom from Fish Kills	Possible Causal Agent
White film on gills, skin, and mouth	Acids, heavy metals, trinitrophenols
Sloughing of gill epithelium	Copper, zinc, lead, ammonia, detergents, quinoline
Clogged gills	Turbidity, ferric hydroxide
Bright red gills	Cyanide
Dark gills	Phenol, naphthalene, nitrite, hydrogen sulfide, low oxygen
Hemorrhagic gills	Detergents
Distended opercules	Phenol, cresols, ammonia, cyanide
Blue stomach	Molybdenum
Pectoral fins in extreme forward position	Organophosphates, carbamates
Gas bubbles (fins, eyes, skin, etc.)	Gas supersaturation

Source: Adapted from Norberg-King, T. J. et al. 2001. *Toxicity Reduction and Toxicity Identification Evaluations for Effluents, Ambient Waters, and Other Aqueous Media.* Pensacola, FL: Society of Environmental Toxicology and Chemistry.

17.2.2 Symptoms of Population-Level Effects

There are few examples of population-level symptoms. Munkittrick and Dixon (1989a, b) proposed that the causes of declines in fish population could be diagnosed as caused by one of a set of standard causal mechanisms based on a set of metrics commonly obtained in fishery surveys. This method was subsequently refined and expanded (Gibbons and Munkittrick, 1994), applied to assessments of Canadian rivers (Munkittrick et al., 2000), and incorporated into the causal analysis component of the Canadian Environmental Effects Monitoring Program (Hewitt et al., 2005). Numerous metrics contribute to the set of symptoms, but they are condensed to three response categories: age distribution, energy expenditure, and energy storage. The types of causes that can be diagnosed are: exploitation, recruitment failure, multiple stressors, food limitation, niche shift, metabolic redistribution, chronic recruitment failure, and null response.

17.2.3 Symptoms of Community-Level Effects

Many investigators have attempted to identify changes in taxa composition (i.e., presence, absence, or abundance) that are symptomatic of particular causal agents. Most efforts to date are based on associations of species with particular agents or sources in field observational studies. For example, the Hilsenhoff Index combines the presence of aquatic invertebrate taxa with their tolerance ranking to compute an index that is strongly associated with organic loading characteristic of poorly treated sewage (Hilsenhoff, 1987). Most attempts to develop community-level symptoms have been based on multimetric indices and primarily on taxonomic traits (i.e., on the presence, abundance or relative abundance of species or higher taxa).

The frequencies of traits other than taxonomic composition may also be used to determine the causes of changes in communities (Culp et al., 2010). In theory, traits make better symptoms, because they are aspects of the effect, and can reflect the underlying mechanisms by which effects are produced. For example, the Ohio EPA uses declines in fish that require clean sediments for spawning (simple and lithophilic spawners) as evidence that excess silt is contributing to the decline in their fish biotic index (Yoder and DeShon, 2002). Also, trait-based symptoms may be more broadly applicable because many traits occur in all ecosystems of a particular type, while particular taxa with particular traits may be restricted to a region. An example is provided by a study comparing the percent of macroinvertebrates that cling with percent EPT as indicators of sedimentation of streams (Pollard and Yuan, 2010). The results indicated that the trait was more generally useful than those taxa.

Community-level symptoms have also been based on the tolerance of species for the causal agents. This concept originated with indicator taxa such as oligochaetes, or more specifically, enchytraeid worms as indicators of organic pollution and low dissolved oxygen. Most tolerance categories (e.g., tolerant/intolerant) or tolerance values (i.e., the level of exposure tolerated by a species) are still based on tolerance of organic wastes (SWCSMH, 2010). However, tolerance of any agent could be used, based on data from either the laboratory or field. For example, laboratory test data on tolerance of species to pesticides, organics, and salinity have been incorporated into the SPEcies At Risk (SPEAR) method (Liess and von der Ohe, 2005; Beketov et al., 2009; Liess, 2014). Species sensitivity distributions (SSDs) also provide tolerance values, relative rankings of tolerance, or at least, categories of tolerance that can serve as community level symptoms (see Section 13.2.4). That is, if an impaired site contains species from the upper end of the SSD for a candidate cause (tolerant species) but not those at the lower end (intolerant species) that finding is symptomatic of the candidate cause (Coffey et al. 2014).

Although taxa or trait occurrences may be strongly associated with particular causes, many taxa respond in similar ways to different agents. For this reason, symptoms of community-level effects may not be exclusive and may not be able to differentiate many stressors. The most useful community-level symptoms demonstrate that the different taxa are present and absent with different causes. For example, indices that differentiate organic pollution, acidification, and low flow were developed in Britain (Clews and Omerod, 2009). The case studies of Clear Fork and Pigeon Roost Creek used community-level symptoms as one type of evidence to distinguish some of the causes of degraded stream macroinvertebrate assemblages (Coffey et al., 2014; Gerritsen et al., 2010; see Chapter 23). Other attempts to use taxa patterns to differentiate agents have been less successful (Chessman and McEvoy, 1998; Norton et al. 2000, 2002b; Riva-Murray et al., 2002; Yoder and DeShon, 2002).

17.3 Using Symptoms, Body Burdens, and Biomarkers in Causal Assessments

In most site-specific causal assessments, symptoms, body burdens, and biomarkers will be combined with other types of evidence in order to reach a causal conclusion. The observation of a symptom provides evidence that a particular agent produced a characteristic alteration and was involved in producing an effect. The observation of body burdens provides evidence that biologically relevant exposure occurred. The observation of a biomarker can indicate that a particular mechanism or mode of action occurred.

Symptoms can be combined with other supporting information into diagnostic protocols for determining the cause of commonly encountered conditions (see Chapter 3). Diagnostic protocols for nonhuman animals and plants are not nearly as well developed as for humans, but some are available in the veterinary; ecotoxicology; and fish, wildlife, and plant pathology literatures. For example, diagnostic criteria for lead poisoning in waterfowl include a hepatic lead concentration of at least 38 ppm and at least one characteristic symptom (see Box 17.1).

In some cases, sets of symptoms (i.e., symptomologies) have been identified as a result of ecoepidemiological studies. Perhaps the best known case is the Great Lakes embryo mortality, edema, and deformity syndrome (GLEMEDS) (Gilbertson et al., 1991). This symptomology has been identified in multiple species of fish-eating birds and has been associated with dioxin-like compounds, but it is characterized by more symptoms than the diagnostic effect of dioxin in the laboratory, chick edema syndrome.

The use of symptoms and body burdens in the Coeur d'Alene River is shown in Box 17.1. This case illustrates the development of diagnostic criteria. The most direct method is to perform a toxicity test and record the body

BOX 17.1   SYMPTOMS AND BODY BURDENS IN WATERFOWL OF THE COEUR D'ALENE RIVER

Kills of waterfowl were frequent in the Coeur d'Alene River watershed beginning in the early 1900s following contamination by lead mining and smelting. Attention was particularly focused on tundra swans. Because poisoning by lead shot is common in waterfowl, symptoms were well established. Blus et al. (1991) found that 46 swans found dead or moribund in the Coeur d'Alene River all showed multiple symptoms of lead poisoning, "notably enlarged gall bladders containing viscous dark green bile" and blood lead levels above the conventional benchmark of 0.5 µg/g in whole blood. Swans from uncontaminated areas showed neither symptoms nor elevated lead concentrations. The authors attributed the mortalities to exposure to sediments containing up to 8700 µg/g lead. Lead shot was eliminated as a source by x-ray imaging.

Due to the high potential cleanup costs for the Coeur d'Alene River, additional studies were conducted to supplement and confirm the initial ecoepidemiological study. They included feeding studies with contaminated sediment that related lead ingestion to blood lead and then blood lead to physiological and histological injuries (Beyer et al., 1998a, 2000; Day et al., 2003). Field studies of sediment ingestion rates confirmed that the birds were receiving toxic doses by that route. Detailed field studies relating sediment lead levels to blood lead levels and hematological symptoms in ducks and geese (Spears et al., 2007). Simultaneously, laboratory and field studies of poisoning by lead shot were ongoing. These studies led to the development of diagnostic criteria for waterfowl lead-poisoning surveys consisting of a liver lead level above 38 ppm dry weight and at least one characteristic lesion (Beyer et al., 1998b).

burdens, lesions, behaviors, and other pathologies associated with death or other distinct effects. However, it is desirable to confirm the body burdens and other symptoms in the field. They may vary among species and exposure rates, durations, and conditions. For example, perching birds do not display the same symptoms of lead poisoning as waterfowl, and mallards tend to have fewer lesions than geese or swans. Alternatively, field studies may be used to identify symptoms. Ideally, field-derived symptoms should be confirmed by laboratory studies. Care must be taken to distinguish those symptoms that consistently appear when organisms are exposed from those that appear only in association with death or other effects of interest. For example, engorged gall bladder was seen in 80% of lead-poisoned waterfowl but also in nearly half of waterfowl that are exposed but not diagnosed as lead poisoned (Beyer et al., 1998a).

The Coeur d'Alene case suggests that reliable diagnosis is based on controlled studies that have been confirmed in the field and include symptoms of both exposure and effects. However, symptoms that do not meet those standards are still useful. For example, Beyer et al. (1998a) concluded, based on a single large field study, that hepatic lead levels alone provide a defensible criterion for lead-poisoned waterfowl, but they would not consider it diagnostic.

17.3.1 Weighting Evidence from Symptoms

The weighting of symptoms is discussed here, but, even if they are not considered symptomatic, body burdens and biomarkers would be weighted using equivalent considerations.

Relevance: What properties of a symptom developed from one species, population, or community would suggest evidence of causation in another? As with toxicity data, evidence of symptoms is most relevant when the species or communities are similar to those under investigation. A symptom is likely to be broadly relevant if it has been reported in diverse species and communities. Relevance may also be judged on the basis of knowledge of the generality of the mechanisms involved in producing the symptoms.

Strength: The strength of symptoms can be evaluated in several ways. The first is by relative frequency of the symptom compared with background levels. We tend to think of symptoms as occurring only rarely and only in cases of disease, toxicity, or injury. However, some symptoms may occur at low frequency under normal circumstances and some, particularly the population or community attributes, occur to some degree in all systems. As a result, the evidence is weighted with respect to strength primarily by comparing symptom frequencies in the case under investigation with frequencies under background exposures. Second the degree of manifestation of a symptom relative to background is indicative of the strength of the symptoms as evidence (e.g., the size of lesions or the severity of ataxia). Third, if more than one symptom is associated with a cause, strength increases as more symptoms are observed. For example, dark gills and gasping at the surface are both symptoms of exposure to low dissolved oxygen in fish. Observing both increases the strength of the evidence.

Some symptoms occur at levels below those that produce the effect of concern. For example, some biomarkers are better indicators that some exposure has occurred, rather than indicators that a relevant effect has occurred.

Reliability: As with other evidence, reliability of a symptom results from various properties that make it more convincing (see Chapter 19). For example, evidence of symptoms generated by a laboratory study that clearly demonstrates the causal relationship is more reliable than evidence from a field study. Symptoms that are consistently observed with exposure to an agent are more reliable, as are symptoms that are exclusive to an agent.

17.4 Summary

Symptoms, body burdens, and biomarkers can provide evidence of the characteristic signs expected of a cause. These signs may be a more specific manifestation of the effect (usually at a lower level of biological organization) or reflect the underlying mechanisms by which a cause produced the effect. Because the expectations are developed based on observations from field studies or laboratory experiments, they carry the strengths and limitations of the source of information used to develop them.

Although a set of symptoms may be sufficient to diagnose a cause, we recommend performing a full causal assessment in any case. One would want to document that the diagnosed cause and effect co-occurred in space and time and determine that other available evidence was consistent with that cause. In addition, one would want to identify any other causes that may also be acting in the affected system.

18

Simulation Models

Glenn W. Suter II

Simulations models have the advantage of providing estimates for processes or conditions that are difficult to observe. However, they typically must be calibrated or refined before being applied for a particular investigation.

CONTENTS

Mathematical simulation models allow us to use our understanding of the components and processes that make up a system to describe how the system responds to perturbations. The equations used in simulation models are based primarily on physical, chemical, and biological understanding rather than empirical relationships. Although the models are based on theory, they incorporate results from laboratory tests and field observations to provide case-specific parameter values. Simulation models are routinely used in environmental management to determine harvest levels for forests, fisheries, and wildlife. In environmental science, they are routinely used to estimate fate and transport of chemicals given release rates and environmental conditions. In population viability analysis, they are used to estimate the degree of protection required to restore endangered species.

Simulation models have many different potential uses for causal assessment. Fate and transport models can be used to estimate the levels of stressors that are difficult to measure or are no longer present or to estimate the contribution of different sources (see Section 18.1). Stressor–response models can be used to estimate biological response from exposure to different agents and are especially useful for evaluating combinations of agents (see Section 18.2). The U.S. EPA's Council for Regulatory Environmental Monitoring (CREM) is a good source of generally useful models that have been extensively tested and reviewed (U.S. EPA, 2013c).

The primary alternative to simulation models is empirical models that use observed associations to describe relationships between variables (e.g., Chapter 12). Empirical models have the advantage of realism, but they are often inadequate because the necessary data for the region, effects, or candidate causes are not available. In addition, relationships in the field may be so confounded that causal models cannot be derived from field data. Simulation models have the flexibility to address effects on different spatial and temporal scales and to reflect processes and results that are difficult or not possible to observe in the field.

18.1 Fate and Transport Models

In most causal assessments, exposure levels are estimated using measurements of chemical concentrations, habitat properties, or other measures of the intensity and spatial and temporal distributions of potentially causal agents. However, in some cases such measurements are unavailable or unreliable. When sources are known, it is possible to estimate exposure using transport and fate models as is commonly done in risk assessments for proposed sources (Schnoor, 1996; Mackay and Mackay, 2007). Further, even when measurements of exposure are available, models may be more reliable. In particular, episodic exposures such as aqueous exposures to pesticide applications are difficult to measure but can be modeled (Acevedo et al., 1997; Morton et al., 2000). For aquatic ecosystems, watershed modeling systems such as BASINS (U.S. EPA, 2013d) and HSPF (U.S. GS, 2014b) simulate a variety of candidate causes including sediment, nutrients, and toxicants.

The results of fate and transport models are used to generate causal evidence by combining them with measures of the biological response. The evidence can be used to judge whether candidate causes occur at sufficient levels at the same time and place as effects. Exposure estimates can also be combined with observed responses to evaluate whether biological responses increase with exposure or to produce a stressor–response model. Fate and transport models can also be used to provide evidence that a source is capable of producing observed levels of an agent. In these cases, the model would be used to estimate levels of the candidate cause at the location of the effect and that result would be compared to measured exposures to determine whether the hypothesized sources and pathways are credible. Receptor models run in the opposite direction of conventional transport and fate models.

That is, they use the exposure levels and environmental characteristics to determine the relative contributions of sources to the exposure, rather than using source characteristics to estimate exposure (Gordon, 1988; Scheff and Wadden, 1993). Receptor models play an important role in evaluating different management options.

18.2 Stressor–Response Models

Mathematical models can be used to simulate the responses of organisms, populations, or communities to exposures to contaminants and other agents (Barnthouse, 2007; Bartell, 2007; Giddings et al., 1981; Dixon, 2012).

Like empirical stressor–response models, simulation models can be used to determine whether the exposure was sufficient to cause the observed effects. Modeled results that are similar to responses observed at the site would support the argument for the candidate cause. Stressor–response models can also be used to simulate the pattern of effects that would be produced by different candidate causes, such as the relative abundance of species or of life stages in response to different candidate causes.

18.2.1 Ecotoxicological Models

Environmental toxicology is becoming increasingly mechanistic, which provides an opportunity to better estimate the combined effects of multiple stressors.

The simplest mechanistic ecotoxicological models are the exposure additivity and the effect additivity models for chemical mixtures. If chemicals have the same mode of action, it may be assumed that they act together to induce their effect, differing only in their relative potencies. Hence, the corresponding mixture models are exposure additive (i.e., you can add their toxic doses or concentrations to estimate effects). When one chemical of a set of concentration-additive chemicals occurs in a mixture at half (0.50) of its avian lethal dose and another occurs at two-thirds (0.66) of its lethal dose, then the mixture contains 1.16 lethal doses. That mixture could account for a bird kill, but not either chemical alone. In contrast, when chemicals have independent modes of action, their effects are additive. That is, when two effect-additive chemicals appear together at concentrations sufficient to kill a fraction of exposed organisms, some are killed by Chemical A, some by Chemical B, and a few could be killed by both.

A more sophisticated example of an ecotoxicological model is the biotic ligand model (BLM) (Paquin et al., 2002; Niyogi and Wood, 2004). BLM combines an aqueous metal speciation model with a model of competitive binding of nutrient and toxic cations to ligands on the respiratory surfaces of fish and aquatic invertebrates. Binding by toxic metals results in a cascade of physiological effects ultimately resulting in organismal responses. This model can provide better estimates of effective levels of some aqueous metals than laboratory toxicity data alone. Ultimately, simulations of the effects of other chemicals (known as toxicokinetic and toxicodynamic models) will provide better models of toxicological causation and a better basis for estimating mixture effects and identifying symptomatic effects.

The BLM illustrates the potential for mechanistic models to move beyond simple additivity models for mixtures to represent interactions. The combined toxicities of aqueous divalent metals have been estimated by applying exposure additivity models to concentrations normalized by the biotic ligand model. That is, if two metals occur at half their median lethal concentrations as estimated for the receiving water using the BLM for each, then lethality should ensue. However, the metals not only combine to disrupt calcium or sodium uptake, but also interact by competing for ligands, so more accurate models of metals' mixture toxicity include that competition (Jho et al., 2011). Consideration of hydrogen ions (pH) illustrates even greater interactive complexity. They are important determinants of metal bioavailability, acting directly on the metals and indirectly through their effects on the metal-binding characteristics of dissolved organic matter and the speciation of inorganic carbon, as well as weakly binding to sodium channel ligands. Hence, the overall effect of pH on metal toxicity is interactive rather than concentration-additive. Software is available for performing the relatively complex BLM calculations (HydroQual, Inc., 2007).

18.2.2 Population, Community, and Ecosystem Models

Ecological models of populations, communities, and ecosystems can be used to estimate the combined effects of multiple causes. For example, the rapid collapse of the lake trout population in Lakes Huron and Michigan appeared to have been due to more-than-additive combined effects of harvesting, lamprey parasitism, and chlorinated organic chemicals, which could be modeled when assumptions were made about compensatory and depensatory capabilities of the population (Gentile et al., 1999).

Population models have been developed and used primarily for the management of fisheries and wildlife resources (Quinn and Deriso, 1999; Haddon, 2001; Starfield, 1997). Because these models address the consequences of harvesting, they are most directly applicable to agents that act by killing organisms, such as power plant cooling systems. More recently, population models have been used in the management of rare or declining species. These population viability models simulate the response of species to various management actions such as increasing habitat extent and quality (Beissinger and McCollough, 2002; Morris and Doak, 2002). Population models developed for either of these purposes could be adapted for causal analyses.

The kit fox case study (see Chapter 25) provides an example of the current utility of simple population-level models. Demographic data from the site was used to calculate demographic parameters and to create a model of the kit fox population on the Elk Hills. Simple demographic models consisting of an age- or stage-specific survival, fecundity, and abundances are used to project the size of a population given data or assumptions concerning the demographic parameters. The kit fox model showed that the observed high level of mortality, particularly in young-of-the-year foxes, was sufficient to account for the 30% per year population decline and that reduced fecundity was not a contributing factor. That result, plus data on the cause of death, showed that predation by coyotes was the primary proximate cause. This model was made possible by the years of work on the site monitoring the population. Population modeling for ecological risk assessments is described by Forbes et al. (2011) and Barnthouse (2007).

Ecosystem models simulate transfers of biomass, energy, and materials between different compartments of an ecosystem (Starfield and Beloch, 1986; Bartell, 2007). They are particularly useful when causal processes cascade through several different trophic groups. For example, an ecosystem model can indicate whether an herbicide that reduces algal production could account for a decrease in fish abundance. The U.S. EPA has developed a general aquatic ecosystem model for this purpose called AQUATOX (Park et al., 2008; U.S. EPA, 2013e). Examples of the application of AQUATOX to assessments of contaminated sites can be found at the EPA website. Other uses of system models in ecological assessments can be found in Bartell (2007).

Hybrid population and ecosystem models have been developed to simulate population management while accounting for the ecosystem context (Walters et al., 2008; Fulton et al., 2011). Such models may be particularly useful for determining the cause of population declines when habitat and species interactions are implicated.

18.3 Evaluating Simulation Models

To determine how much weight to assign to evidence from models, several considerations must be evaluated. Organism, population, and ecosystem models are worth the effort they require if they deliver realism and relevance (Bartell et al., 2003). Realism refers to the extent to which the model represents the components and processes of the system being considered, particularly those that characterize the candidate cause. For example, if we hypothesize that the toxicity of copper to nitrogen-fixing cyanobacteria is responsible for the adverse effects on invertebrates in a stream, the model must distinguish cyanobacteria from other phytoplankton and must represent changes in cyanobacterial functions and their consequences for the ecosystem. Relevance refers to the ability of a model to provide the needed output. For example, a bioenergetic model will estimate changes in biomass, but we may be trying to determine the cause of demographic changes, such as mortality of larval fish. Further, quality questions must be asked concerning the data used to implement the model, the sensitivity of the model to assumptions, the degree to which model parameters are defined by calibration rather than from data, and similar issues with respect to implementation.

18.4 Summary

Simulation models can be used to explore scenarios and ask "what if" questions such as "could the short-term release of the effluent of a factory account for the observed effects?" Transport, fate, and exposure models for contaminants are well developed and are likely to be useful for cases in which contaminants are implicated and the time and resources are available to model them. Simulation models of effects have yet to be widely used in causal assessments. For example, a review of the use by state water quality regulators of the U.S. EPA's CADDIS system found that simulation modeling was the only type of causal evidence that has not been used (Harwood and Stroud, 2012). It is likely that multiple good reasons exist why simulation models are not used. The most important is probably the lack of sufficient data and other information for the candidate causes and affected systems. Also, we have found that applied ecologists often lack training in mathematical modeling, are cautious in going beyond the data, and are skeptical of modeling results. However, simulation modeling is well established in environmental fields such as climate change assessment and the management of forests and populations of fish and wildlife. In those fields, the models that have been developed for predictive management purposes can readily be adapted to explanatory causal assessments.

Part 2C

Forming Conclusions and Using the Findings

The last phase of a causal assessment brings together the evidence to form a conclusion in a way that is useful for a decision.

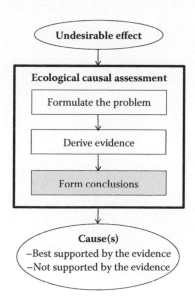

19

Forming Causal Conclusions

Glenn W. Suter II, Susan M. Cormier, and Susan B. Norton

> This chapter describes how evidence is brought together to identify the best supported cause or causes.

CONTENTS

Having formulated the problem and collected and analyzed the evidence, you are ready to use the evidence to determine causation. By now it should be evident that assessing causation is not just a matter of statistically analyzing the degree of association of candidate causes and effects. The body of evidence should be analyzed to infer which candidate cause is most strongly supported by the evidence (i.e., the most likely cause). We recommend that this can be done by weighing the evidence.

Colloquially, weighing evidence refers to determining the relative degree of support for two sides of an argument. The expression is from jurisprudence. Evidence is placed in each of two pans of the scales of justice which tips the scale to one side or the other. The evidence itself is described as having weight, a metaphor for how much it influences the outcome. Like any metaphor, it simplifies the complex reality, but its common usage suggests that it has been a useful way to portray the deliberative process.

We recommend weighing evidence to ensure that all relevant evidence is considered to the degree that it deserves. Inference to the most likely cause is performed by assembling all available and relevant evidence for and against each candidate cause, weighting each piece of the evidence on the basis of its relevance and quality, strength, and reliability, and weighing the body of evidence for each candidate cause against the others (see Figure 19.1). If those basic steps are insufficient to derive a conclusion, various techniques may be used to revise the list of candidate causes, add evidence, and reassess the body of evidence until the assessment is complete and the findings can be reported.

A conclusion may be reached in at least three different ways. First, the candidate cause with the greatest degree of support can be accepted as a cause, with due caution. For example, the illicit discharge identified in the Willimantic River case study (see Chapter 1) had the greatest degree of support, so it was remediated by the state. The due caution disclaimer is needed because a possibility always exists that the evidence is incomplete in a significant way or has unrecognized quality problems. Ideally, diverse evidence all points to the same causal conclusion, and weakening or refuting evidence removes all alternative explanations. However, even when one cause is supported by relevant, strong, diverse, and consistent evidence, other causes may also be supported to some degree. A second option, therefore, is to identify all the well-supported causes. This is a common outcome. For example, in the Long Creek case study (see Chapter 22), support was found for dissolved oxygen, altered flow, woody debris, and ionic strength, and all were found to potentially contribute to the altered invertebrate assemblage. A third option is to take a step back and consider the candidate causes that have some merit and hypothesize new alternatives that account for the evidence. This can be illustrated by constructing a new conceptual model that shows how the

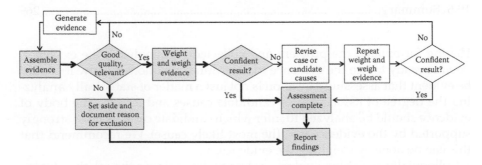

FIGURE 19.1
Schematic of the process in which adequate-quality, relevant evidence is evaluated and re-evaluated until the assessment is complete and the findings can be reported. The basic steps and decisions of the weight-of-evidence process are shaded and with bold arrows. Reiterating the assessment is shown in white and with thin arrows.

proximate cause is affected from multiple interacting causes and factors. For example, in the kit fox case (see Chapter 25), candidate direct causes became indirect causes.

Although other approaches for inferring causation are available and potentially useful, the flexible approach presented here is applicable to typically encountered environmental evidence. A flexible approach is needed because site-specific causal investigations often include a mixture of site-specific observations, regional observations, laboratory tests, mechanistic studies, and symptoms. Our approach brings disparate pieces of evidence together in a way that can be consistently applied even when different types of evidence are available for different candidate causes.

Inference by weighing evidence has been criticized for being nonquantitative, ad hoc, and subjective because it is most often based on the assessor's informal and unstructured judgment (Weed, 2005). Much of this criticism can be addressed by using a well-described method that is identified in advance. In this way, the assessor can avoid being accused of having reached a predetermined conclusion and then defining a method that gives that sought-after conclusion.

A formal method has two important advantages. First, it increases the likelihood of arriving at the right answer. We all have biases and lapses in judgment, and a formal method provides discipline to ensure that evidence is not ignored or inappropriately discounted. Second, it provides transparency. Anyone reading the assessment can see what was done and why it was done. They can evaluate whether the method was rigorously applied, whether they disagree with particular judgments, and whether different judgments would lead to different conclusions. This is particularly important for peer review.

The approach presented in this chapter is not a cookbook method. The array of causal assessment problems is too diverse to allow that. Simpler problems and smaller sets of evidence allow simpler systems for evaluating the evidence. Also, an assessor needs to feel comfortable with whatever system is used. However, each causal assessment should describe and use a clearly defined method, not just a general approach.

19.1 Assemble and Organize the Evidence

The success of an assessment depends on the skill of the assessors in finding evidence or generating it from data, observations, and knowledge. Chapters 9 through 18 describe the various sources of evidence for a causal assessment, for example, field and laboratory studies performed for the case or reported in the literature.

It is likely that evidence will require sorting into groups or categories because it is diverse. Using some sort of organization makes it easier to develop and apply scoring systems for comparing weights across the different candidate causes. For example, it is relatively easy to compare and summarize the degree to which results from a set of laboratory tests support a candidate cause and then compare the relative strength of laboratory results to those from a field observational study. It would be more difficult to weigh all field and laboratory studies in one step.

Categories of evidence may be selected based on the amount and diversity of evidence and the preferences of the assessors. We provide two sets of categories. First, evidence may be categorized by the characteristic of causation that it illustrates, that is, evidence of antecedence, co-occurrence, sufficiency, interaction, and alteration (see the discussion in Chapter 4 and examples of their use in Chapter 23; Coffey et al., 2014; Haake et al., 2010a; Wiseman et al., 2010a). Second, it may be organized by the type of evidence that it represents, such as laboratory tests of site media or mechanistic plausibility (e.g., see Chapters 4, 21, 24, and 25; Hicks et al., 2010). However, the categorization of evidence should be defined in a way that facilitates the assessment. For example, when there is little evidence, it may be sufficient to simply categorize it as laboratory- or field-based.

The degree to which evidence is aggregated also depends on the circumstances of the assessment. A highly controversial assessment that involves disagreements among agencies and stakeholders about the interpretation of evidence may benefit from a highly disaggregated assessment. That is, one might weight each individual piece of evidence and aggregate them into small categories (see U.S. EPA, 2011a, Appendix A for an example). In other circumstances, it may be more important to succinctly lay out the broad categories of evidence and the bases for weighting them. When all is said and done, the presentation of evidence should have three components: a description of the evidence, the weight, and a justification of the evaluation.

19.2 Weight the Evidence

Although the idea that some evidence is more influential than others is not controversial, the practice of explicitly weighting evidence is. Some explicit form of weighting is needed; otherwise the inference is drawn from a free-form mental narrative, which is prone to errors and biases (see Chapter 5). Different assessors will subjectively assign different weights to the same piece of evidence, and even the same assessor will find a piece of evidence more compelling on one day than another. Explicitly weighting does not solve the problem of subjectivity, but it does make judgments apparent. To minimize subjectivity, it is important to use criteria for weighting (i.e., a scoring system) and to apply them consistently.

**BOX 19.1   RELEVANCE AND QUALITY ARE USED
TO DETERMINE INCLUSION AS EVIDENCE**

Relevance includes similarity of the agent, organisms, conditions,
and measured responses to the system and effects of concern.
Quality includes considerations such as study design and
performance.

The considerations that are used to judge reliability (see Table 19.3)
are equivalent to those for relevance and quality. Relevance and quality
are used for a dichotomous judgment of adequacy rather than as attributes that can make the evidence worthy of greater weight.

Evidence should be included only when it meets minimum standards
of relevance and quality (see Box 19.1). When it meets those standards, it
is given a minimum score that also indicates its logical implication for the
candidate cause (supporting, weakening, or ambiguous). The evidence may
then be given additional scores based on strength and reliability (see Boxes
19.2 and 19.3).

When evidence is organized into categories, an assessor might still weight
each piece of evidence individually. However, when the evidence within a
category is reasonably consistent, one might skip weighting each piece, and
simply assign weights to each category as a whole.

We recommend using a system of symbols. We prefer a system of +, −,
and 0 symbols developed for epidemiology by Susser (1986), and adapted for
environmental assessment (Fox, 1991; Suter, 1998; U.S. EPA, 2000a). However,
other symbols or even colors can be effective as long as they are well defined.
For example, *Consumer Reports* uses circular symbols and colors to assign
scores to different attributes of consumer goods. Restaurant reviews and

---

**BOX 19.2   THE STRENGTH OF A PIECE OR TYPE OF
EVIDENCE EVALUATES THE DEGREE OF DIFFERENCE
FROM BACKGROUND OR REFERENCE**

Magnitude—degree of difference between the amount of exposure or symptomatic response at an affected site and at unaffected sites
Association—degree to which variation in a variable representing
a cause explains variation in a variable representing the effect
Number—the number of elements of a set (e.g., of symptoms or
steps in a causal pathway) that are reported

BOX 19.3   RELIABILITY RESULTS FROM FACTORS THAT MAKE EVIDENCE CONVINCING BEYOND ADEQUATE RELEVANCE AND STRENGTH

Quality—evidence that has higher quality than is required for inclusion (see Box 19.1) is more reliable

Abundance—evidence from numerous data is more reliable

Minimized confounding—evidence is more reliable when extraneous correlates are controlled by the sampling design or data analysis

Specificity—evidence (e.g., a symptom or set of symptoms) that is specific to one cause or a few related causes is more reliable

Potential for bias—practices that reduce bias, and thereby increase reliability including blind study designs, random sample study designs, and acknowledgement of the sources of funding and purpose of the study

Standardization—a standard method decreases the likelihood that the evidence is biased and that analyses are inaccurate

Corroboration—using models, indicators, or symptoms that have been verified by many studies and are accepted technical practice can greatly increase reliability

Transparency—completeness of the description methods and inferential logic and availability of data for reanalysis increases probability that a report is reliable

Peer review—an independent peer review increases reliability of a source of information

Consistency—the degree to which an association does not vary in repeated instances within a study (e.g., across years, locations, study teams, or methods) is an indicator of reliability

Consilience—evidence that is consistent with prior knowledge is more reliable

travel guides use stars or diamonds for different attributes. The use of symbols avoids the implication that there is a measureable or countable attribute of evidence that determines how much weight it should be given in the assessment.

+++ or − − −	Convincingly supports or weakens
++ or − −	Strongly supports or weakens
+ or −	Somewhat supports or weakens
0	No effect (neutral or ambiguous)
NE	No evidence

We do not concur with some assessors who recommend numerical scores (Linkov et al., 2012). Numerical scores have the apparent advantage of being arithmetically combined as in the Massachusetts system for contaminated sites (Menzie et al., 1996). However, this can also be a disadvantage. Weights for different types of evidence cannot literally be added to weigh the body of evidence. For example, a strong (++) toxicity test is not equal to two weak (+) field studies.

19.2.1 Relevance and Quality

Evidence is judged to be admissible when it meets minimum relevance and quality criteria (see Box 19.1). Poor quality or irrelevant pieces of evidence are documented as such. Either they are not included in the body of evidence or they are given a score of "0" as having no influence on the body of evidence.

When the evidence is relevant and based on good data and analyses, it is recorded and assigned a +, 0, or − depending on its logical implication. That is, does it support the candidate cause, weaken it, or not clearly go one way or the other? A score of + indicates support for the candidate cause and − indicates weakening the candidate cause, and 0 indicates that its implications are neutral or ambiguous. This scoring of logical implication is generally straightforward. For example, as relevant evidence of co-occurrence, dissolved copper is greater than regional background or it is not. As relevant evidence of sufficiency, the amount of dissolved copper measured at the affected site has caused the effect in laboratory tests or has not.

As far as possible, criteria regarding relevance and quality for including evidence should be defined in advance and should be clearly articulated. For example, field data might be limited to studies within a particular region (e.g., Central Appalachia), of certain types of ecosystems (e.g., first- through third-order streams), and with certain quality attributes (e.g., chemical analyses performed by EPA methods). Peer review is a common quality criterion, but data from a study by a state agency that uses clearly documented protocols, staff training, and quality assurance plans could be higher quality than data from many academic studies in peer-reviewed journals, especially the burgeoning open-access journals (Bohannon, 2013). Funding sources may be scrutinized to determine whether the research is likely to be unbiased, but once again, judgment with justification must be documented.

Having criteria to judge relevance and quality helps to maintain consistency. The required degree of relevance and acceptable quality will vary from case to case and among iterations of the same case. For example, when the effect involves trout, and the toxicity data for salmonids is abundant for a candidate cause, then data for fathead minnows would not be sufficiently relevant to include. However, when fathead minnows were the only tested fish, those data would be sufficiently relevant to include. Similar judgments would be involved in judging adequate quality. Early on, the data may be sparse and of questionable quality, but analysis of such data can lead to collection of better information until the body of evidence is composed of evidence of a quality that is adequate for drawing a conclusion.

19.2.2 Strength

The strength of evidence refers to the degree to which it demonstrates a large difference or a high degree of association between a cause and effect relative to background levels or degrees of association (see Box 19.2). Strength is a property of evidence in the case rather than inherent in the type of evidence.

The criterion for weighting strength has various manifestations depending on the type of evidence. When comparing affected sites or treatments with reference sites or treatments, it is the magnitude of difference (e.g., difference in ambient concentrations or body burdens). When relating potentially causal variables to response variables, strength may relate to measures of the cause–effect association such as the correlation coefficient or the slope. When weighting evidence of symptoms or steps in a causal pathway, strength lies in the number of elements of the set included in the evidence (e.g., the number of symptoms in a set or of steps in a causal pathway that have been documented). In all of these criteria, strong evidence implies a clear distinction from background, reference, control, or below-threshold conditions. For example, strong evidence of co-occurrence worthy of a second plus could be that the concentrations of the candidate cause at the affected site are 10 times greater than at reference sites in the region.

The strength of the evidence is one weighting criterion that lends itself to standardization. For example, the following scores were used in the causal assessment of salinity measured as specific conductivity and the extirpation of genera in Appalachia (see Table 19.1). The category was correlations of potentially causal agents and biological responses.

These scores are based on experience with field data and judgment. For example, a correlation coefficient (r) between 0.25 and 0.75 is supportive, but does not get an extra plus for strength. However, $r > 0.75$ is considered to be relatively strong for a correlation between a water quality measure and a biological response from a regional data set, so it gets the extra plus.

TABLE 19.1

Weighting the Strength of Correlations (r) and Noting the Logical Implication

Assessment	Logical Implication and Strength	Score
The sign of the correlation coefficient depends on the	$r > \|0.75\|$	++
relationship. For toxic relationships such as the correlation	$\|0.75\| \geq r \geq \|0.25\|$	+
between conductivity and number of *Ephemeroptera*, the	$\|0.1\| < r < \|0.25\|$	0
sign should be negative. Weak or positive correlations	$r < \|0.1\|$	−
weaken the case for that candidate cause	r has the wrong sign	− −

Source: Adapted from Cormier, S. and G. Suter II. 2013b. *Environ Toxicol Chem* 32 (2):272–276.

Consistency and a general reasonableness of the scores are more important than the precise values chosen for the cutoffs, because they are used to compare candidate causes. Other examples can be found in CADDIS (U.S. EPA, 2012a), Wiseman et al. (2010 a,b), Haake et al. (2010a,b), and the case studies in Section 3 of this book.

19.2.3 Reliability

Some evidence is more trustworthy than others. Some attributes of evidence can increase even a skeptic's confidence that the evidence truly represents a causal relationship and not a chance association, a confounding factor, research bias, or measurement error. For example, a large difference (i.e., strong evidence) may not be convincing because it is influenced by correlated factors. In contrast, a small difference may truly indicate a causal relationship when it reflects a highly relevant change, such as an increase above a threshold response. Reliability is the property that makes evidence convincing.

Several factors increase confidence that evidence represents a true causal relationship (see Box 19.3). Evidence is more convincing when other factors that might influence causation are minimized or controlled. For example, confidence that the evidence reflects a causal relationship is increased when experiments manipulate only one variable and field studies minimize potential confounders by careful design or analysis. Symptoms are more trustworthy when they are specific. Specific individual symptoms or sets of symptoms (i.e., symptomologies) are reliable when they consistently predict only one or a few causes. Symptoms may be reliable if they have been verified in many studies or if they are unverified but come from a very high-quality study. In the Clear Fork case (see Chapter 23), evidence using local community symptoms was used to attribute causes to sites with low biological index scores. For some candidate causes, the error rate of the model is low. This evidence could warrant a plus for reliability in that situation. In others, the model cannot discriminate causes well. In such cases, there may be reason to assign a score for logical implication, but not for reliability.

Consistency is a special criterion that applies at all levels of weighting evidence. It is important here because it addresses the consistency of data used in a piece of evidence across years, locations, methods, or sampling teams. Weighting is also done with respect to the consistency of evidence within a category of evidence (see Section 19.2.4) and the consistency of the entire body of evidence for a candidate cause (see Section 19.3).

The attributes of strength and reliability are not entirely independent. If a strong association is derived from a very confounded model, strength and reliability may cancel each other out, leaving a score of a single plus or minus or an ambiguous zero. Likewise, a cause with many specific symptoms has a greater chance of attaining three pluses or minuses because there are more symptoms to observe and more possibility that the effects are specific because there are more of them. For these reasons, it is always advisable to record how

> ## BOX 19.4   WEIGHTING CATEGORIES OF EVIDENCE
>
> *Number and consistency*—more pieces of evidence within a category increases the reliability, as long as each piece is generated independently and they are consistent. Inconsistencies, particularly multiple inconsistencies, diminish the degree to which evidence is convincing.
>
> *Diversity*—pieces of evidence based on results from different methods, investigators, and end points are more likely to be similar when they were produced by the same cause.

the scoring is done, so it will be consistent across all candidate causes and to ensure that the assessor has critically examined the scoring that is used.

Some types of evidence have been considered inherently more reliable. In reviews of medical interventions, clinical trials are given greater weight than observational studies, and anecdotal evidence is given the least weight (Pope et al., 2007). Similarly, because of their inherent relevance, some investigators give biological surveys more influence in ecological assessments than toxicity tests, even when the surveys have "major limitations" (McPherson et al., 2008).

19.2.4 Weighting Categories of Evidence

The weight of a category composed of several pieces of evidence (e.g., toxicity tests with different relevant species or from different laboratories) is assigned on the basis of the quality, strength, and reliability of the constituent pieces of evidence and properties of the category as a whole: number, consistency, and diversity (see Box 19.4). Weight is increased when numerous results are consistent. Consistency is convincing because chance associations are less likely to repeatedly occur; therefore, the evidence is likely to be the result of a truly causal relationship. Inconsistencies, particularly multiple inconsistencies, diminish the influence of evidence. However, situations may occur where inconsistencies are explainable and therefore should not influence the weight. For example, one of the toxicity tests may have been conducted with a species that is known to be tolerant. Diverse sources and methods also increases confidence that results are not due to chance or a common error.

19.3 Weighing the Evidence

After weights are applied to evidence and perhaps to categories of evidence, the evidence for each candidate cause is weighed to evaluate the degree to which the body of evidence supports it. As with categories of evidence, the weighing of a body of evidence takes into consideration the properties of the evidence and the number, diversity, and consistency of the combined evidence.

In addition, when the body of evidence is not consistent, it may be weighed with respect to its coherence. That is, can all the evidence be logically fitted together so as to account for apparent for inconsistencies? For example, inconsistent evidence concerning metal toxicity might be made coherent by considering the influence of dissolved organic matter on bioavailability. Hence, coherence is consistency achieved by post hoc inferences from the evidence. Note that a more extreme form of post hoc inference is the reconsideration of the list of candidate causes and the formation of new causal scenarios (see Section 19.4.3).

Weighing can be done in one step or multiple steps, depending on the amount of evidence, how it has been categorized (see Section 19.1), and on the preferences of the assessors. When there is little evidence, or when it is all of one type, one may simply evaluate the body of evidence after weighting each piece of evidence (see Figure 19.2a). When there is little difference among pieces of evidence within categories, one may skip weighting the pieces and begin by weighting the categories of evidence (see Figure 19.2b). For example, when there are numerous acute lethality tests for aquatic invertebrates, it is appropriate to weight them as a set. When evidence is abundant and diverse, one may weight the pieces of evidence, aggregate them into categories and determine weights for them, and finally weigh the body of evidence (see

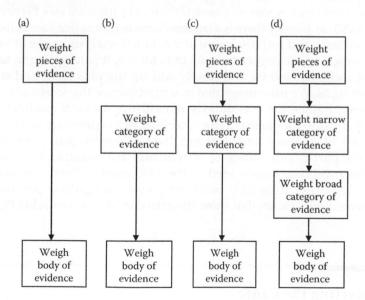

FIGURE 19.2

Four alternative approaches (a–d) to weighting and weighing evidence which are described in the text. The choice depends on the amount and diversity of evidence, the circumstances of the assessment, and the preferences of the assessors.

TABLE 19.2

Evidence Available for the Willimantic River Case Study

Type of Evidence	Candidate Cause[a]			
	Embedded Sediments	Low DO	Heat	Toxic Mixture
Spatial co-occurrence	− −	−	−	+
Manipulation of exposure				+++
Covariation of the stressor and the effect from the affected site and nearby sites		+	+	+
Stressor–response from other field studies	− −		+	
Stressor–response from laboratory studies		−	+	
Causal pathway	−	−	+	+

[a] Evidence for only four of the candidate causes are shown for this simplified example. Blank cells indicate no data available.

Figure 19.2c). At the extreme, one might be very methodical with a very large and diverse set of evidence in a contentious case. For example, one might weight each piece of evidence, then weigh finely defined types (e.g., laboratory acute invertebrate toxicity tests, laboratory chronic invertebrate toxicity tests, laboratory acute fish toxicity tests, etc.), then weigh broader types (e.g., laboratory toxicity tests), then combine the weight for that type with other types of evidence of sufficiency (e.g., models or field tests), and finally weigh the body of evidence across all characteristics of causation (see Figure 19.2d) (U.S. EPA, 2011a). Clearly there is a tradeoff between simplicity and efficiency, on the one hand, and full disclosure of a detailed analysis on the other.

When weighing a category or body of evidence, it is tempting to take the weighing metaphor too far and simply add up the plus signs and subtract the minus signs. We discourage that practice because the scores are usually not additive. Rather, the overall body of evidence for each candidate cause is evaluated using the same criteria described in the previous section. Is the overall body of evidence consistent and diverse? Are the pieces or categories of evidence particularly strong or convincing? For example, the scores for four of the candidate causes used in the Willimantic River case study are shown in Table 19.2. This table facilitates pattern recognition but should be supplemented with tables that show the actual evidence (see Table 19.3).

19.4 Inferring the Cause

The final causal conclusions are based on the collective body of evidence for each candidate cause and are formed by comparing the evidence among the candidate causes. We have identified and used three methods for inferring

TABLE 19.3

Weighing of Evidence for the Willimantic River Case Study

Characteristic	Score	Type of Evidence
Toxic Mixture—Most Likely		
Co-occurrence	+	*Spatial co-occurrence*—gray mixture episodically released where the effect began
	+++	*Manipulation of exposure*—taxa diversity increased after repair of the pipe and cessation of release of the mixture
	+	*Covariation of the stressor and the effect*—Zn was strongly correlated and Fe was moderately correlated with EPT diversity. Cr concentration was 10 times greater than at upstream site
Antecedence	+	*Causal pathway*—the mixture traced to a broken waste pipe permitted to discharge organic matter, ammonia, and metals to the waste treatment plant
Heat—Ambiguous		
Co-occurrence	–	*Spatial co-occurrence*—temperatures were warmer at the upstream comparison site
	+	*Covariation of the stressor and the effect*—Temperature was moderately correlated with non-EPT taxa diversity
Sufficiency	+	*Stressor–response from laboratory studies*—Temperatures > 20°C are reported to adversely affect some invertebrates in laboratory studies (Cox and Rutherford, 2000; Panov and McQueen, 1998)
	+	*Stressor–response from other field studies*—Below impoundments 5°C increase reduced number of taxa (Lessard, 2000)
Antecedence	+	*Causal pathway*—More impervious cover, which can heat storm water, occurred at the affected site
Low Dissolved Oxygen—Unlikely		
Co-occurrence	–	*Spatial co-occurrence*—Dissolved oxygen was moderately high, 8.9 mg/L
	+	*Covariation of the stressor and the effect*—Dissolved oxygen levels correlated weakly with taxa richness
Sufficiency	–	*Stressor–response from laboratory studies*—Dissolved oxygen levels greater than reported levels that affect sensitive taxa (Nebeker, 1972)
Antecedence	–	*Causal pathway*—Biological oxygen demand and total phosphorus were unchanged from upstream and well below benchmarks for effects. A dam and riffles provide aeration, increasing oxygen
Embedded Substrate—Unlikely		
Co-occurrence	– –	*Spatial co-occurrence*—The proportion of substrate composed of sand was half (12%) of the upstream comparison site (25%)
Sufficiency	– –	*Stressor–response from other field studies*—More than 12% sand and fines have been associated with decreased taxa (Gerritsen et al., 2010; Cormier et al., 2008) but these levels were not observed at the affected site
Antecedence	–	*Causal Pathway*—The bank stability score was unchanged from upstream. Armored riprap and granite wall was prevalent at both the affected and upstream site

the cause of an effect from the weighed evidence. They are (1) choosing the candidate cause that best explains the evidence (see Section 19.4.1), (2) identifying those causes that are supported by the evidence (see Section 19.4.2), or (3) identifying a new alternative scenario that explains the body of evidence for all supported causes (see Section 19.4.3). In the first method, the evidence for candidate causes is compared and the one cause that best explains the evidence is identified. In the second method, the evidence for each candidate cause is appraised to determine whether it is adequate to infer that the candidate cause may be acting either independently or in combination with other well-supported causes. The last method, the revision of the case or candidate causes, involves a shift from the original scope of the investigation. As a result, that method may involve a second iteration of the weighing evidence, additional data generation, or an iteration of the entire process (see white boxes in Figure 19.1).

19.4.1 Identify the Best Supported Causal Explanation

As discussed above, we can usually identify the candidate cause that is best supported by the evidence as the explanatory cause, although we cannot prove causation (Lipton, 2004). For example, in the Kentucky River bourbon spill, the evidence was overwhelming that low dissolved oxygen from bacterial respiration was the cause of the on-going fish kill. In the kit fox case, predation by coyotes clearly had stronger and more consistent evidence than the alternatives (see Chapter 25). In the Willimantic River case, the body of evidence for an episodically released mixture was strong and consistent, whereas the other causes were weak or ambiguous (see Table 19.3).

Table 19.3 summarizes the body of evidence in a way that makes it easy to identify the most compelling evidence, for example, the recovery of the macroinvertebrate assemblage after the repair of the pipe in the Willimantic River case study. These pieces or types of evidence would be highlighted when communicating the findings (see Chapter 20).

Another approach for reaching conclusion from the body of evidence is to use the characteristics of causation (see Chapter 4) and ask, "Is there evidence of each of the characteristics of a causal relationship?" In the best cases, a body of evidence for a cause demonstrates all of the characteristics of causation. For example, in the Kentucky River bourbon spill example (see Table 6.2), the evidence relevant to asphyxiation due to deoxygenated water is from diverse sources and represents all of the causal characteristics. The body of evidence has great weight because it is derived from high-quality data that are from the case or clearly relevant to the case, and the effects of the deoxygenated water could be distinguished from the other candidate causes. The explanation is consistent because it is all supportive, and all of the causal characteristics are logically connected to low dissolved oxygen and the death of the fish. By comparison, evidence of the causal characteristics for the other two candidate causes is absent, very weak, or contrary to the cause.

In many cases, identifying the best supported cause is enough to inform a management decision. When the result is not contentious, the manager may decide to remediate, shut down, or otherwise address that apparent cause. An assumption here is that there is enough evidence to be confident in the conclusion (otherwise, see Section 19.5).

19.4.2 Identify Well-Supported Causes

In some cases, it may appear that multiple candidate causes have enough supporting evidence to conclude that each may be playing a role in producing the effect, either alone or by interacting with others. Other candidate causes may have enough weakening evidence to eliminate them from further consideration or may have too little evidence to decide whether they are a cause. In the kit fox case (see Chapter 25), although predation was a clear explanatory cause, road accidents were also supported, but toxicity and disease could be eliminated. In the Long Creek case (see Chapter 22), evidence supported dissolved oxygen, altered flow, woody debris, and ionic strength as causes of the altered macroinvertebrate assemblage. Altered food sources were not supported, and evidence for fine sediments was ambiguous.

A body of evidence can be adequate even when evidence is uneven across candidate causes or unavailable for some categories. For example, evidence for time order is rarely available in ecoepidemiological cases, because effects and causal events are seldom anticipated and frequent monitoring is rare. When the evidence for time order is not available but the candidate cause is present at levels that are sufficient to cause the effect and the effect is specific to that cause, then even when there are other causes with supporting evidence, that candidate cause may be accepted as a cause.

It is desirable to not eliminate a candidate cause too quickly or easily. In the Kentucky bourbon spill example, ethanol toxicity was rejected because there was no temporal co-occurrence of the kill with ethanol exposure. That one piece of evidence alone was enough to refute ethanol-induced narcosis as the cause, but it is even more compelling when the symptoms are consistent and when a strongly supported cause is identified. The entire case becomes coherent when both the causes and noncauses are reasonably explained.

19.4.3 Revise the Case or Candidate Causes

In some cases, a clear conclusion is not reached after evidence is evaluated. However, it may be possible to review the evidence to generate a new alternative that does explain the evidence.

One approach is to review the inconsistencies in the assessment results and ask what could explain them. Potentially useful approaches include refining the definition of the effect and adding, combining, or otherwise restructuring the list of candidate causes or the causal network (see Section 8.3). The approach is based on knowledge gained during the assessment as well as the apparent inconsistencies in the evidence that must be resolved.

Inconsistencies may be resolved by refining the effect of concern. In particular, it is often valuable to make the effect more specific. For example, the effort to determine the cause of decline in peregrine falcons was unsuccessful until the effect was redefined as reproductive failure due to egg shell thinning. That redefinition eliminated several candidate causes, such as egg collecting and shooting. Redefinition may, as in that case, clarify the sort of causal mechanism that is appropriate, thereby reducing the list of candidates. Specificity can be enhanced by reducing the spatial scale, by identifying specific species or other lower level taxa that are particularly affected, or as in the peregrine falcon example, by identifying a specific symptomology.

The list of candidates may also be refined by clarifying or combining one or more of the candidate causes. For example, it may become clear that two of the candidate causes are acting jointly and therefore should be combined. That is, neither of the pair may be consistently sufficient to cause the effect where it is observed, but where either is insufficient alone, the other is present and they are sufficient in combination. It is important to ensure that the combined agents are actually joint proximate causes and not a cause and a confounder or a cause and part of the preceding causal network.

Refining a candidate cause may also involve describing it more specifically. For example, inconsistent evidence concerning copper as a cause of aquatic effects may be resolved by defining it more specifically as dissolved ionic copper. That will resolve a discrepancy due to the fact that the copper in toxicity tests is nearly all dissolved in the free ionic state, but copper in the field is often complexed by dissolved organic matter.

The third approach to resolving inconsistencies is to combine the candidate causes that have not been eliminated into a new causal network which may include new assumptions or overlooked information. The presumption of this approach is that inconsistencies that include some clearly positive evidence should be explained. A generally useful approach is to create a new conceptual model that combines credible elements of the prior conceptual models into a new synthetic structure. That is, remove the box–arrow combinations that are not supported. With the remaining box–arrow combinations that are supported by evidence, look for ways that they can be combined to create a new conceptual model that is more consistent with the evidence. Refined candidate causes and effects can contribute to this approach.

An example of revising candidate causes is provided in the kit fox case study (see Figure 19.3 and Chapter 25). Strong and consistent evidence was found for predation by coyotes. Accidents also had consistently positive, though much weaker, evidence. However, those two candidate causes could be combined as independent additive causes of acute lethality. Inconsistent evidence for habitat disturbance and reduced prey abundance could be accommodated by making them indirect causes.

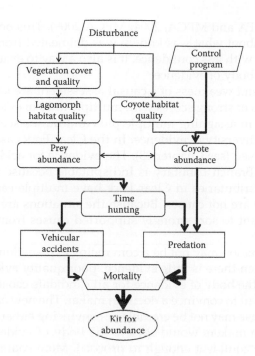

FIGURE 19.3
The final conceptual model showing the final conclusions of the kit fox case study (see Chapter 25).

Revised candidate causes are assessed by weighing the evidence for and against them, as discussed above. However, the process should be simpler and quicker because of all the previous work. The candidate causes have already been assessed, and the evidence for the new relationships has, for the most part, been assembled and weighted. This reinterpretation in light of the revised relationships may be the best causal explanation of the effect.

Alternatively, the new relationships will suggest some attribute of the environment that has not been noted or observed, and finding that evidence should decide the case. This is the classic hypothetico-deductive method of science devised by Francis Bacon.

19.5 Evaluating Confidence

An assessment is a scientific study intended to inform a decision (Suter and Cormier, 2008). It hinges on the confidence of the final conclusion. The case studies that we have noted throughout the book include cases in which even screening assessments provided information for decision-making (Bellucci et al., 2010; U.S. EPA and MPCA, 2004; MPCA 2009). This occurred because the information about what was known was delineated from what was not known or known with less confidence. It is also a way to evaluate the overall confidence in the body of evidence.

The strengths and weakness of a causal assessment are identified by recognizing patterns of strong scores with multiple lines of evidence, strong or large differences in assigning multiple plus and minus scores, corroborating studies, and diversity of evidence. In the Clear Fork case study, several streams are assessed (see Chapter 23). The evidence for acid mine drainage in the Stonecoal Branch tributary is indisputable because the evidence is so strong. Other tributaries in Clear Fork have multiple causes and many agents occur that are not causes. Because the situations are more complex, it is very important to sort strongly supported causes from those that are weak or refuted.

The best explanation may not be a convincing explanation and it may not even be true. When there is little evidence, poor quality evidence or inconsistent evidence, the body of evidence for all candidate causes may be inadequate or it may fail to convince a decision maker. The new candidate causes for the post hoc case may not be credible or convincing either. Ideally, knowledgeable decision-makers would review the body of evidence and decide whether they are confident enough to proceed. More commonly, assessors must make a judgment concerning adequacy based on experience and any guidance provided by policies and precedents. Assessments that do not confidently identify the cause can still serve as screens. Screening assessments eliminate candidate causes with refuting evidence so that further efforts can be focused on the remainder.

19.5.1 Generate More Evidence

When weighing the evidence or revising the relationships does not give a reasonably confident conclusion, the best course is to generate new evidence. By this point in the assessment process, it should be possible to infer what critical evidence would distinguish among the remaining causal relationships. That is, infer consequences of the remaining candidate causes that are observable and characteristic of only that candidate cause. Then, perform the necessary studies. For example, when a pathogen is the cause, it should be possible to identify it in the affected organisms, and when low dissolved oxygen is the cause, as in the bourbon spill, air-breathing organisms should be unaffected. When two causes are acting together, the combined level of effect predicted from a test of mixtures of those agents should be similar to the level of the observed effect.

When the evidence really is critical and the proper studies are well performed, those results will resolve the case. Otherwise, the new evidence should be weighted and the bodies of evidence for the remaining candidate causes should be reweighed (see Figure 19.1).

19.5.2 Adapt the Management Action

When the results of a causal assessment are ambiguous, and particularly when multiple causes seem to be contributing, the assessors may turn things over to the engineers. That is, rather than continue to analyze the case until causation is resolved, a remedial action may be designed that deals with that ambiguous situation, thus solving the question of causation and the biological effect all at once. Alternate management strategies are described in Chapter 21.

19.6 Summary

In the real world, causation is seldom characterized by any one piece of evidence or any one analysis. However, many types of evidence are potentially available that can increase confidence that a relationship is causal and others are not. Hence, the assembling and weighing of evidence is a practical necessity. The process varies depending on the circumstances, but it should include the following activities:

- Assemble the evidence and define the weighting scheme.
- Weight all relevant evidence for each candidate cause before weighing the body of evidence.
- Use a formal method for weighting and weighing to strengthen the credibility and impartiality of the assessment.
- Fit the amount of formal detail to the needs of the decision, thereby enabling a timely decision to take appropriate action.
- When there is more than one likely cause, show how they relate to one another.
- Revisit the case definition and the candidate cause list when the conclusions are inconclusive.
- If it is possible to derive additional information, use the results of the assessment to determine what information is likely to provide critical evidence for the next round of assessment.

New insight is always gained from an assessment; when you have not been able to identify the cause, make new insights the main message of your conclusion.

Comparative conclusions are more readily obtained and defended. That is, it is easier to say why the evidence is more convincing for A than for B that to say that the evidence for A is convincing. This is the inherent advantage of the method of multiple working hypotheses (Chamberlin, 1995;

Lipton, 2004). Going through a deliberate evaluation of the evidence provides a rigorous foundation for presenting the explanation to others. The next chapter describes how findings can be effectively communicated in narrative and diagrammatic forms so that your audience will be engaged and motivated to act.

20

Communicating the Findings

Susan M. Cormier

Effective communication ensures that an assessment's findings are used. This chapter discusses the content and style of assessment products and provides perspectives for communicating with different audiences.

CONTENTS

Why do some assessments make a difference and others gather dust on the shelf or land in the recycling bin? How can you be sure that yours leaves a legacy? It is not just about the quality of the work. It is also about getting the results to the people that can take the information and change the way things are done. The outcome depends on how the assessment is shared.

20.1 Roles of an Assessor: Scientist, Synthesizer, and Advisor

Producing and communicating a causal assessment requires someone who has the skills of a scientist but also the skills of an astute synthesizer and a policy advisor all in one (Suter and Cormier, 2012). A scientist investigates natural phenomena to reveal new truths and gain a better understanding of nature. A synthesizer organizes, analyzes and evaluates the results of relevant scientific studies with the intention of informing a decision. A scientific advisor considers options and, when asked, recommends management or policy actions based on the assessment, scientific understanding, and professional experience. During the course of a causal assessment, one person may need to perform all three functions: doing high-quality data gathering, recognizing relevant science, assembling it into an informative whole, and articulating the consequences of choosing among management or policy decisions. This three-part responsibility requires a special type of person, someone who is good at the different responsibilities of a scientific explorer, scholarly and analytical synthesizer, and a diplomatic advisor, and who is mindful of these distinct roles and responsibilities.

Some personal qualities are required to perform and communicate assessments. Foremost is personal and scientific integrity. Without them there is no credibility and the ripple effect of unethical behaviors by a few diminishes the good names of all assessors. Credibility accrues when assessors consistently exhibit both open-mindedness and skepticism. Such a person is able to see solutions and connections while simultaneously protecting their audience from misguided actions. This healthy combination of openness and wariness of potential bias it not enough. An assessor also needs to possess and maintain a broad range of physical, chemical, and biological knowledge. Environmental assessments by their very nature involve evaluations of complex interactions between the physical world and living things. Critical thinking is a must, and skepticism must be tempered with a willingness to accept some uncertainty while recognizing the best scientific explanation (Suter and Cormier, 2013b).

When presenting the technical results of an assessment, it is not sufficient to merely make the results clear and complete. Assessors should also present results in a way that prepares the ground for good decision-making and be cognizant of the possible reactions of stakeholders to different situations. That is, it is important to clearly and objectively describe the scientific results as evidence for or against each candidate cause while staying true to the science and how it best informs the decision. The inclination to please organizational superiors can be strong. Therefore, it is important for assessors to maintain their integrity and scientific position when drawn into an advisory role. It takes some attention to routinely recognize the transition from technical consultant to advisor and to find a clear but unobtrusive way to ensure that decision-makers are aware of the switch from facts to values.

Some assessors are downright charismatic. They know how to tell a story so well that people are easily convinced by their explanation. It is a dangerous gift because people want the world to make sense. One way this happens is when lots of details are provided (Heath and Heath, 2008). A story can be strong that it over shadows the facts, producing "a narrative fallacy" (Taleb,

2010) Great story tellers need to consider the influence they can have and then balance what is known with what is inferred with what might be advised.

An assessment is a scientific study that provides the scientific basis for informing a decision. Its success hinges on a clear, concise, and compelling executive summary backed by a scientifically strong report and appendices. The ultimate audience is the decision-maker, but the immediate audience is often a scientific one; namely, the scientists that advise the decision-maker and the stakeholder groups that influence the decision. Therefore, the bulk of the assessment is for scientific readers who are trained to be detail-oriented, critical, and logical thinkers, but the more accessible the assessment is to everyone the better.

The conclusion summarizes the relative weight of evidence for the candidate causes, the results, and the strengths and weaknesses of the assessment. The causal characteristics can be used as an organizing structure for this summary, either for all the candidate causes (see Table 6.1, Kentucky River case) or just the cause that is best supported by the evidence (see Table 20.1). Details can be included in tables that display the evidence and associated scores (e.g., see Tables 22.4 and 22.15 in the Long Creek case study and Table 23.3 in the Clear Fork case study). In the text, the causes are assigned concluding categories: for example, likely causes, unlikely causes, uncertain causes, and causes lacking information that were deferred or not analyzed.

Conceptual models can be organized by candidate cause or as an explanation of multiple interacting causes, indicating the most important pathways

TABLE 20.1

Summary of Evidence for Lead as a Cause of Mass Mortality of Tundra Swans in the Coeur d'Alene River Watershed

Causal Characteristic	Evidence
Co-occurrence	Swan kills occurred in lead-contaminated lakes and wetlands and not elsewhere in the region
Sufficiency	Mortality occurred in laboratory tests at lead doses and body burdens seen in dead or moribund swans in the field
	Consistent mortality in the field at blood lead levels >0.5 µg/g
Time order	No evidence—no pre-mining information on swan mortality
Interaction	Dead and moribund swans had high blood and liver lead levels
	Lead-contaminated sediments were found in swan guts and excreta
Specific alteration	Swans had pathologies characteristic of lead, particularly, enlarged gall bladders containing viscous dark green bile
Antecedence	Spills of lead mine tailings and atmospheric deposition from smelters account for the high sediment lead levels

Source: Based on information from URS Greiner, Inc., and CH2M Hill. 2001. *Remedial investigation report for the Coeur d' Alene basin Remedial Investigation/Feasibility Study.* Seattle, WA: U.S. EPA, Region 10. URS DCN: 4162500.06200.05.a2.

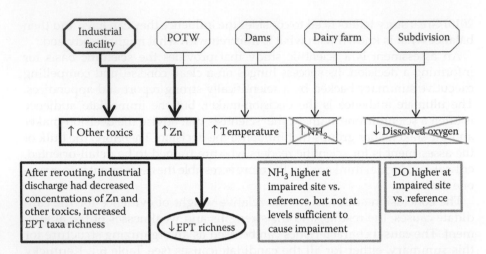

FIGURE 20.1
Conceptual model with associated evidence indicating that toxicants including metals such as zinc from a ruptured effluent pipe at an industrial facility were the cause of reduced numbers of EPT taxa in the Willimantic River case study. Episodic toxicity was the dominant cause for the reduced number of taxa (bold lines). The strongest piece of evidence supporting the episodic toxic mixture as the cause was a manipulation of exposure by removing an illicit point source (see evidence in box with bold outline). After the discharge was removed, the benthic macroinvertebrate assemblage downstream recovered to a condition that was judged to be acceptable by state standards. Two causes are crossed out indicating that they were rejected based on the evidence in the gray outlined boxes. (Adapted from Bellucci, C., G. Hoffman, and S. Cormier. 2010. *An Iterative Approach for Identifying the Causes of Reduced Benthic Macroinvertebrate Diversity in the Willimantic River, Connecticut.* U.S. Cincinnati, OH: U.S. Environmental Protection Agency, Office of Research and Development, National Center for Environmental Assessment. EPA/600/R-08/144.)

with bolder lines (for the kit fox case study, see Figure 25.3; Suter and O'Farrell, 2008). Pathways may be annotated with evidence that supports or weakens the argument for a candidate cause (see Figure 20.1).

20.2 Providing Options or Advice

Before communicating results and, even more importantly, before giving advice, review the evidence to assess confidence in the conclusions. An assessment is intended to inform a decision. Is there enough evidence to inform the decision that needs to be made?

In general, scientists tend to point out the weaknesses and uncertainties of an argument. The last paragraphs in many scientific papers conclude with— "more research is needed." Such a statement is given for continued scientific advancement but a near failure for an assessment, because decision-making is impeded by uncertainty and failure to clarify the options (Lehrer, 2009; Kahneman, 2011). In fact, creating uncertainty and controversy is a ploy for delaying actions. Tobacco executives have been quoted as saying: "Doubt is our product since it is the best means of competing with the 'body of fact' that exists in the minds of the general public. It is also the means of establishing a controversy" (Michaels, 2008). Therefore, it is important to be strong when the evidence is strong while being honest about the uncertainties. One option is to emphasize what is known rather than what is uncertain.

An influential causal assessment lays out the potential next steps. When confidence in results is low, an option is to plan studies to obtain critical evidence. When confidence in the results is high and a probable cause is identified, options may include finding the source of the causal agent and estimating the relative contributions from multiple sources. This is sometimes called a source assessment. Another option is to recommend performing a risk assessment to estimate the risks and benefits of options for reducing the level of the causal agent.

To enable decision-making, it helps to reduce the clutter of information and choices. An influential assessment provides enough information to enable a decision to be made but not so much as to create confusion. The compelling parts of an assessment need to be brought forward and highlighted. The assessor needs to anticipate specific technical questions. The assessor also needs to be ready for open-ended questions such as "how will your assessment be challenged?" and "who will be affected by these findings?" This should not be a problem, because a good assessor should have already asked and answered the relevant technical questions during the assessment.

To motivate action, an advisor has to be willing to state that the science is good enough to guide informed decision-making. It is good to remember that the advice can be no action, more data collection, assessment of remedial or protective options, or strategic action.

As you can guess, this is not a job for the faint of heart. But it is worth it. A career as an assessor is an important opportunity to make a difference in a very direct way. It is a way to provide important contributions as a technical consultant and advisor to ensure that environmental decisions have the benefit of the best science, insights, and wisdom of the scientific community.

20.3 Communicating to Different Audiences

Although the ultimate audiences for a causal assessment are the decision-makers and the stakeholder community, peer reviewers and professional communicators are additional audiences that assessors are likely to encounter. Understanding the needs of each audience improves the chances that an assessment will be both useful and used.

20.3.1 Peer Reviewers

Review by independently selected, competent scientific reviewers increases the credibility of an assessment and often its quality. Peer reviewers are asked to be critical and skeptical. They are reading with the intent of finding weaknesses to improve a scientific product or to block poorly done work. The author can ease the assessment through the peer-review process by proactively writing for this audience. When the assessment is complete, transparent, scientifically sound, and clearly written, it is likely to pass the peer-review gauntlet.

Most peer reviewers are scientists. Using familiar styles and formats focuses reviewers on the information rather than the presentation. Although some reviewers are assessors, many are academics who are more familiar with the typical scientific journal format of abstract, introduction, methods, results, and conclusions. In assessment terminology, this translates to an executive summary, a problem formulation, evidence derivation, and conclusions. The similarity between the two formats can be made apparent with additional sections and headers to call attention to the differences between an information-rich assessment and journal articles which tend to focus on one cause and a narrow range of evidence.

To show that the assessment is scientifically sound, it must be well documented. As a report, the number of tables and data are unlimited. Appendices and on-line materials are handy for ancillary material, long tables, and raw data. Bottom line: show everything so that the reviewers will know it was considered. It is best not to leave them guessing.

The lack of experiments or statistical hypothesis tests will unsettle some. So, it is important to explain how the statistics are being used and how the assessment relies on the overall weight of the evidence not on one statistical product. It may be useful to explain how the causal assessment informs the next steps in problem solving and how it is different from basic science that usually focuses on only a part rather than the whole system (Walters and Holling, 1990).

Being complete means the work is transparent. This fosters credibility and trust. Being complete does not mean being verbose. Make it easy for the reviewers to get the information without wading through unnecessarily complex or wordy sentences. Liberally use topic headers. A well-organized assessment implies clarity of thought and by association a thoughtful conclusion. A concise assessment increases the chance that the readers will make the connection between the evidence and inferences and that their attention will be maintained to the conclusion.

Peer reviewers frequently want to expand the scope of the assessment. Clearly explain the scope in terms of space, time, taxa, levels of organization, and types of causes that are of interest to the decision-makers. Then, explain how the scope is determined by the problem to be solved and the context, including legal and policy constraints.

Finally read the assessment as a reviewer would. The author knows the weaknesses better than anyone. Give those weaknesses a hard look. Do not bury them. Avoiding the issue does not help advance science or contribute to a good decision. Instead, do the needed analyses and then indicate wherever appropriate how much uncertainty it creates, what was done to address it, what could be done to remove that uncertainty, or why it is not influential to the overall conclusion. Finally, the conclusion of the assessment should be the final paragraph so the reviewer does not miss it.

20.3.2 Decision-Makers

Decisions-makers want the bottom line. The executive summary provided at the front of the document is for them. Executives are unlikely to read a scientific report that is sufficiently technical and detailed to pass peer review. They are unlikely to read an entire assessment. The detailed assessment and appendices are there as support, and executives rely on staff or peer reviewers to make sure the bulk of the document is complete, transparent, and scientifically sound. Therefore, the executive summary must present the case, key evidence, conclusions, and important certainties and uncertainties. Technical terms are appropriate but may need brief definitions. Jargon is out. Figures and maps can help orient and summarize more quickly than text and tend to stick in memory. Keep in mind that about one in every 10 men is color blind, so check to make sure maps and figures are accessible to everyone (Reimchen, 1987; Niroula and Saha, 2010). Executive summaries can run from 1 to 30 pages depending on the complexity of the assessment and on how important the assessment is to the decision-maker. When the assessment is a high-visibility product, the executive summary may be printed separately, often with more photographs and explanatory figures so that it will be accessible to a wider audience.

How the message is structured will vary depending on whether it is written for a business, government, or advocacy group. For example, many causal assessments are performed because businesses need to know when they are responsible for an adverse environmental effect. This has potential impacts on company's reputation and therefore stock value. Obviously, all businesses hope that the evidence will show that they are not causing harm, and when the assessment shows this, executive officers and managers will want to know what evidence frees them from culpability. If their business is the source of some or all of the causes of ecological damage, they want to understand the evidence and the options to address the source. This might be as simple as repairing a broken effluent pipe. Other solutions might be more difficult and expensive, so the business would want stronger evidence to justify costly decisions. The report or briefing that is prepared for them is more helpful when it provides next steps even when it is only to describe the next assessment or the need to improve any predictive models.

Decision-makers do not want to be handed a problem with no ideas for dealing with it. Usually the decision-maker wants to know the identity of the cause, its magnitude, its source, and a segue to an assessment of options that will address the problem.

Assessments performed by governmental entities are ultimately devoted to advancing public welfare and immediately devoted to implementing their legal mandates. Assessments by governmental entities may be self-initiated, but are often begun because scientists, politicians, or advocacy groups have raised a concern. As such, the assessment must speak to those concerns, either allaying fears or moving toward resolution of the problem. Government executives need to juggle priorities, work with limited resources, and respond to the legal, economic, social, and political ramifications of decisions. There may be distinct assessments to evaluate these issues in addition to the technical causal assessment.

Assessments performed for environmental advocacy groups will usually reflect the interests of the segment of the public that comprises their membership. Depending on the issue, advocacy groups may not have an a priori position on a causal assessment (they may just want a resource restored) or they may have a particular responsible party in mind (they may be an antimining group). In the former case, their causal assessment may resemble that of a resource agency. In the latter case, it may resemble a prosecutor's case against a defendant. Either way, the assessment is important as a foundation for marketing the findings to like-minded individuals, lobbying for greater visibility and support, and pressuring industry and government.

Business owners, government representatives, and advocacy groups all need a sound causal assessment to plan a course of action. Therefore, the most important need for decision-makers is a complete, well-balanced, scientifically sound causal assessment. Whenever possible, leave the spin to the public relations staff.

20.3.3 Communicators

Public relation experts and the news media get the word out on many causal assessments. Assessors, because of their deep involvement with the assessment, can get wrapped up in technical issues, thus detracting from the main message (Anderson, 2013). There is a good reason why the White House has a press secretary providing written press releases and answering questions. It is not just to save time. It is about getting the message out to the public. "Obviously, the president's message doesn't get far unless it passes successfully through the filter of the press" (Mike McCurry, Press Secretary to President Bill Clinton*). Lasting legacies can be achieved by

---

* http://www.whitehousehistory.org/whha_press/index.php/backgrounders/press-secretary-recollection.

making science and assessments interesting and accessible, as attested by the long life of programs like the Public Broadcasting System's NOVA and the impact of the film, *An Inconvenient Truth* narrated by former vice president Al Gore.

Although there is growing pressure on journalists to provide entertainment or to validate the opinions of their audience, the Society of Professional Journalists recommends that "Analysis and commentary should be labeled and not misrepresent fact or context" (SPJ, 1996). In practice, journalists are looking for a story. When scientists provide a sellable story, there is no need for a journalist to embellish it. By managing the content of the narrative and not getting off track, the scientific content of the story can be aligned with the content of the assessment. So adopt the practice of many public relations experts and craft the narrative to be unambiguous providing a consistent and scientifically defensible message.

In the attempt to give voice to opposing points of view, journalists sometimes mask the consensus of the scientific community by giving equal time to a few entertaining outlier scientists. One way to reduce potential bias is to routinely provide reporters with concrete evidence of peer reviews or provide graphically compelling polls showing the consensus of scientists. Another is to preemptively address uncertainty without making it the dominant message. Zehr (2000) suggests noting which scientific uncertainties have been accounted for and explained, including uncertainty bounds when appropriate, and describing how confidence will be increased with any future activities, and then returning to and reemphasizing the findings and the message.

In addition to telling the story about assessments performed by others, the documentary film and news media develop and present their own assessments (i.e., investigative journalism). A solid piece of reporting is condensed, resembling an executive summary, and the detail is provided by links to sources or on-line material.

20.3.4 Stakeholders

The information needs of stakeholders are as different as the stakeholders themselves. However, they all have one thing in common. They want their concerns to be considered when it is time for decision-making. Generally, stakeholders' knowledge comes from information filtered by the public relations experts, the media, and a stakeholder's preferred advocacy groups. However, websites and public briefings also provide key information to people who are interested in the assessment and provide access to the detailed assessment if they want it. Some stakeholders have their own staff scientists or consultants to perform their own assessments or critique yours. Ultimately, an assessor wants stakeholder support, so sound decisions will be made that will be accepted by as many people and groups as is possible. The public at large is the broadest stakeholder group. Communicating to a broad audience requires having the knowledge and training to share information while also being receptive to the information that stakeholders want to impart. A professional communicator can be an indispensable coach, even crafting and delivering the message.

20.4 Typical Assessment Outline

Assessments include some special elements. The case studies provided in Part 3 of this book begin with a summary and then are organized by the three major steps of causal assessment: problem formulation, evidence, and conclusions. This is one suggestion, but not the only way the assessment of record can be organized.

The executive summary is an extended abstract or condensed assessment. It summarizes the case, the conclusion, and the critical evidence. It is the only section many people will read.

The problem formulation section is a combination introduction and methods section with several distinct parts that are often called out with separate headers. It includes sections on the impetus for the assessment, the geographic setting, the specific effect, the causes that were considered, and a methods section that documents the sources and quality of information and provides a short description of the approach used to form conclusions.

The evidence section describes the data and information sources and the analytical methods and results, typically in tables or figures as well as text.

The conclusion section describes the rationales for the conclusions made by weighing the evidence for individual causes and considering interactions among causes. The conclusions section shows how the individual pieces of evidence build a coherent case (see Chapter 19; Suter and Cormier, 2011; see also Box 20.1).

BOX 20.1 COMPARISON WITH OTHER ASSESSMENTS

If not already included as evidence, a comparison with other assessments of the same effect is worth describing. In the Touchet Case Study, previous studies had concluded that some of the same agents were causing the biological effects, and those causes were corroborated by the new causal assessment (Wiseman et al., 2010a, b). In the Groundhouse River assessment, the screening assessment (U.S. EPA and MPCA, 2004) was corroborated by the subsequent assessment (MPCA, 2009) with improved information on less dominant causes. In the Kit Fox case study, the new assessment revealed a very different cause than a previous assessment and that difference required explanation (see Chapter 25).

20.5 Suggested Tips and Good Practices

1. Respect the audience; find out what they know and what technical information will need to be explained.

2. Pay close attention to decision-makers' needs and interests. Give them the information they need when possible. Otherwise, acknowledge their needs and suggest how they can obtain that information.

3. Take a tip from journalists, "Don't bury the lead." Make sure the final conclusion or recommendation is in the abstract, executive summary, and introduction. That way it is easier for the audience to follow the arguments and to be critical in a constructive way.

4. Avoid jargon and minimize acronyms.

5. Use a concise style and avoid repetition.

6. Make the logic clear and complete.

7. Liberally use section headings and subheadings to help direct the reader's attention.

8. Present the key evidence.

9. Do not present poor quality, irrelevant, or ambiguous evidence, because it detracts from the message. Document that this evidence was considered by putting it in an appendix along with the reasons for not using it in the assessment.

10. Describe the certainties and uncertainties, without making them the centerpiece or last statement.

11. Each figure should tell a story by showing relationships, not just restating a table or text. Write detailed figure legends. Some people just review the figures and skip the text.

12. Make figures attractive with relationships that are obvious. It may be helpful to use annotated conceptual models.

13. Provide enough background and technical information in the text for the reader to understand what was done and why.

14. Provide detailed background information and technical details that only a specialist would want to read in appendices. This makes the narrative flow more easily and keeps the assessment short enough that it might be read.

15. Include everything that is relevant by taking advantage of appendices or on-line material. Transparency and full disclosure increase trust.

16. Keep good metadata and records. A tough audience may demand those materials. Good records make it easier to defend or revise the assessment or to update the assessment after a period of inactivity.

17. Read the report from the perspective of a peer reviewer. Then add analyses, text, figures, and tables so that there will be fewer questions later on. Even for a journal article, it may be better to be a little long and show the depth of thought about the issues. The extra detail can be removed or shortened if requested by the editor.

18. Revisit the original impetus for the causal assessment and ultimately what options may exist to address those issues.

19. Get a friendly reviewer to raise technical issues before proceeding to the formal review.

20. Enlist an editor to improve the style of the text.

21. Document responses to peer review and public comments.

22. In consultations or open meetings, listening to others is as important as delivering the message.

23. Get help from a professional communicator.

20.6 How Do You Know You Did it Right?

A causal assessment is successful if it prompts management action that corrects the adverse effect. When a causal assessment targeted the location of a toxic source in the Willimantic River, the Connecticut Department of Environmental Protection went to the location and discovered the source (Bellucci et al., 2010). The broken effluent pipe was repaired. They also went on to study and address the statewide problems with stressors associated with impervious cover and urban land use. When DDT was shown to be the cause of population collapses in high-trophic-level birds and when granular carbofuran was shown to cause frequent bird kills, both pesticides were banned in the United States and many other countries followed suit. Following the DDT ban, populations of the affected birds recovered.

You know the assessment was effective when it changes what people do and, more importantly, when it leads to an improved environment.

21

After the Causal Assessment: Using the Findings

Susan B. Norton, Scot E. Hagerthey, and Glenn W. Suter II

> This chapter discusses what comes after an ecological causal assessment is completed. Causal assessments describe the causal relationships that must be altered for beneficial changes to occur. Optimally, the assessment's findings will be used to guide subsequent actions taken to improve biological condition and contribute to the scientific understanding of causal processes in ecosystems.

CONTENTS

As discussed in Chapter 20, a successful causal assessment influences management decisions. When those decisions result in action that improves biological condition, that is evidence that our understanding of the causal processes, while never complete, is adequate to guide effective management actions. Management actions provide opportunities to evaluate the accuracy of the causal assessment and the underlying understanding of ecological and physiological processes. In this way, assessments of their outcomes can inform and improve future actions, future causal assessments, and the knowledge base of ecological causal processes.

There are many documented examples of biological recovery following management actions. Removal of the Quaker Neck Dam in North Carolina, USA, expanded the range of striped bass and shad (Burdick and Hightower, 2006). Reducing nutrient inputs decreased the occurrence of bluish-green algal blooms in Lake Washington (Lehman, 1986). Implementing a fishing moratorium increased Chesapeake Bay striped bass populations (Richards, 1999). Improvements in POTWs stemming from the Clean Water Act reduced fin anomalies in fish near POTW outfalls (e.g., Setty et al., 2012). Experiences like these provide evidence that our ecological knowledge can be applied to improve the environment.

Unfortunately, there are also many examples of management actions that did not improve biological condition. Dam removal can increase fish and invertebrate mortality by exposing sediments, facilitating the invasion of nonnative plant species, and transporting fine sediments downstream (Stanley and Doyle, 2003). Violin et al. (2011) reported that in-stream biota did not improve after a physical habitat restoration project: the only measureable difference between restored and reference stream reaches was that the riparian forest was removed at the restored reach. Experiences like these provide a reminder that biological condition does not always improve in expected ways and that management actions can have unintended consequences that harm the very resources that are valued.

The outcome of management actions can be used to provide feedback to causal assessment only if there are strong linkages between the causal assessment, the selection of a management action, and the assessment of the action's success. For this reason, this chapter begins by describing an ideal sequence of activities and how they are linked together (see Section 21.2). Then we discuss how causal assessments can contribute to the evaluation of management options (see Section 21.3) and the evaluation of the outcomes of actions (see Section 21.4). Finally, we discuss how the knowledge gained from the sequence of assessments can be synthesized and communicated to improve the scientific basis for making environmental decisions (see Section 21.5).

21.1 Assessment Sequences

This chapter revisits the sequence of assessments introduced in Chapter 1. One possible sequence of assessments addresses the questions: Is there an undesirable condition? What caused the undesirable condition? What is the best course of action? Did the action work? (see Figure 21.1; Norton et al., 2004; Cormier and Suter, 2008). Each question is addressed using a specific type of assessment (e.g., of condition, cause, options, and outcomes, respectively). The sequence draws on, and contributes to, subject area knowledge of ecological causal processes (depicted by the large gray background in Figure 21.1).

In an ideal world, the products of each assessment direct the design and conduct of assessments that follow. The causal assessment identifies the cause of the observed biological effect; an assessment of management options (e.g., risk assessment) evaluates the possible outcomes of alternative actions; the management action addresses the cause of the biological condition; and an outcome assessment evaluates the actions taken to improve the biological condition that initiated the whole sequence.

Sometimes the linkages are clear, enabling smooth transitions between assessments. For example, the State of Connecticut Department of Environment Protection was involved throughout the sequence of activities to address degraded biological condition in the Willimantic River. They conducted the monitoring that identified the degraded biological assemblage and led the causal assessment that identified the cause as a broken pipe that discharged intermittently into the stream. The management action, rerouting the pipe into the municipal waste treatment system, was clear, feasible, and taken immediately. Sampling conducted by the department over the subsequent 2 years verified that the biotic assemblage improved (Bellucci et al., 2010).

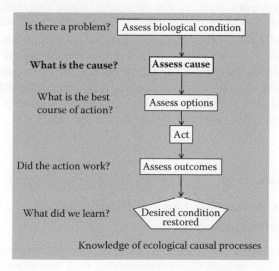

FIGURE 21.1
An idealized sequence linking assessments of biological condition to cause, management action, and the assessment of outcomes. The ultimate goal is to restore biological condition to a desired state (Adapted from Norton, S. B. et al. 2004. In *Ecological Assessment of Aquatic Resources: Linking Science to Decision-Making*, edited by M. T. Barbour, S. B. Norton, H. R. Preston, and K. W. Thornton. Pensacola, FL: SETAC Press; Cormier, S. M., and G. W. Suter II. 2008. Environ Manage 42 (4):543–556.)

Weak linkages between the steps in the sequence increase the risk of a missed connection that ultimately reduces the chance of restoring biological condition. Weak linkages may arise due to different assessments being conducted by different groups, under different programs, and with different objectives or priorities. Weak linkages can result from significant lags between each assessment because time is needed to plan, collect data, and conduct analyses. The frequency of such disconnections between steps of river restoration projects was noted as a major factor reducing the likelihood of achieving intended results (Bernhardt et al., 2007). As a result of missed linkages, management actions may address causes that are known to be capable of causing effects, but may not be the most important factor influencing biological quality at a particular location. Monitoring for effectiveness may not target the right indicators or be designed to properly evaluate success. There may be a desire to show that management action improves biological condition, even when biological degradation was not the reason for taking management action. Or, the linkage to biological degradation is lost by the time action is taken, and success is instead defined in terms of public perception. Missed linkages become missed opportunities to improve biological condition.

Strong linkages between causal assessment, management option assessment, and management outcome assessment are not a panacea for all of the challenges of restoring biological condition. Much still needs to be learned about how ecosystems recover. For example, some degraded ecosystems have an inherent resilience that makes them resistant to restoration efforts (Gunderson and Pritchard, 2002). Still, causal assessments describe the relationships that must be altered if management actions are to produce a desired result. Understanding these relationships is a prerequisite for realizing when and why management actions succeed. The next section describes ways causal assessment results contribute to the evaluation of management options (see Section 21.2) and outcomes (see Section 21.3).

21.2 Causal Assessments and the Evaluation of Management Options

In the sequence shown in Figure 21.1, the selection of a management option is informed by the results of an assessment that identifies the causes of degraded biological condition. For example, the removal of the Kent and Monroe Falls Dams on the Cuyahoga River was preceded by a causal assessment that determined that the dams caused slow flows and low oxygen levels, embedded sediments, and impeded fish passage, that, collectively, contributed to degraded fish assemblages (Tuckerman and Zawiski, 2007).

In contrast, reviews of stream restoration rarely identify a separate causal step. Instead, concepts associated with condition and causes are frequently lumped together under broad categories of "goal setting" or "motivation" (Downs and Kondolf, 2002; Bernhardt et al., 2005, 2007; Palmer et al., 2005).

A separate causal-assessment step increases the likelihood that a management action is directed at the causes of biological degradation. The products of causal assessment help to define the goals and the targets of management action (see Sections 21.2.1 and 21.2.2), the amount of change needed (see Section 21.2.3), the means to effectively intervene (see Section 21.2.4), and the prioritization of actions (see Section 21.2.5).

21.2.1 Describing the Desired Management Outcome

Ecological degradation is frequently cited as the reason for undertaking a river restoration project (Bernhardt et al., 2007; Rumps et al., 2007). Such broad statements, however, do not identify the specific desired outcome required to develop an effective management action. Translating the concern over ecological degradation into a more precise description that guides management decisions is informed by the products of causal assessment.

In their standards for river restoration, Palmer et al. (2005) discuss the importance of beginning restoration projects with a specific guiding image of the most ecologically desirable system possible for a given location. The image provides the restoration goal and is constructed using many different approaches, including historical information or the characteristics of undisturbed or less disturbed sites. During the "define the case" (see Chapter 7) step of a causal assessment, the identification of comparison sites provide a concrete and pragmatic description of the biological conditions that are achievable when the causes have been addressed. Realistic goals may be described in increments toward the much more difficult or impractical goal of achieving historical or regional reference conditions.

21.2.2 Identifying the Target of Management Action

We agree with Palmer et al. (2010) that "managers should critically diagnose the stressors affecting an impaired stream and invest resources first in repairing those problems most likely to limit restoration." Most river restoration and management projects target physical and chemical attributes that have the capability of causing effects. Indeed, river restoration goals in the United States are frequently stated in terms of reduction of stressors; for example, enhancing water quality, managing riparian zones, improving in-stream habitat, stabilizing banks, modifying flow, and providing for fish passage (Bernhardt et al., 2005, 2007). Tomer and Locke (2011) summarized the results of agricultural conservation practices in terms of improved water quality, most frequently sediment and nutrient loads. U.S.

EPA's TMDL* program targets pollutants degrading water and sediment quality (U.S. EPA, 2013f).

However, management actions may proceed before confirming that they will address the specific causes of biological degradation at the locations being restored. For example, projects to improve habitat frequently proceed before evaluating whether habitat is indeed the dominant causal factor (Roni et al., 2008; Palmer, 2009; Feld et al., 2011; Haase et al., 2013). This view is so prevalent that it has been dubbed the "Field of Dreams" hypothesis (e.g., Palmer et al., 1997) alluding to the 1989 film's motto of "If you build it, they will come." Recent reviews have concluded that habitat enhancement projects in streams have a low overall success rate, presumably because causes other than habitat degradation dominate. The greatest degree of success was reported from projects in small streams in forest settings where causes other than habitat may be less influential (Stewart et al., 2009; Miller et al., 2010).

When the causal assessment identifies one dominant cause, the target for management action is clear: manage for the dominant cause. However, when the causal assessment is unable to clearly distinguish the relative contributions of multiple causes, it may be more efficient to design a management action that either targets many of the potential stressors or that uses monitoring to provide feedback on whether conditions are improving. Several options are:

Remediate a dominant and potentially sufficient cause. In many cases, a causal agent, acting independently, is clearly sufficient to cause adverse effects but may mask the effects of less severe agents that may be contributing to the effect, or causing effects that do not become apparent until the dominant stressor is remediated. Hence, the dominant stressor may or may not be sufficient to cause all of the negative effects. In the Willimantic River case, the benthic invertebrate community recovered to a legally unimpaired condition, based on Connecticut biocriteria status, following removal of the dominant stressor, a point-source discharge. However, the quality of the formerly impaired reach still did not equal that of high-quality reference conditions, suggesting that additional causes were influencing the stream assemblage (Bellucci et al., 2010). In the case of the Yakima River, Washington, USA, reduction of suspended solids laden with legacy pesticides and PCBs successfully reduced fish tissue concentrations of bioaccumulative toxicants (Joy and Patterson, 1997). However, the reduction of suspended solids also clarified the water, allowing excess nutrient loads to spur thick star grass growth, which has since become problematic (Wise et al., 2009).

Remediate a necessary cause. In some cases, one of a set of multiple causes may be necessary but insufficient alone to induce the effect. Remediation of any necessary cause should eliminate the effect. This situation may occur because of interactions among the causal agents. For example, the Long Creek

---

* TMDLs identify the amount of a pollutant that a water body can receive and still achieve water quality standards.

case study concluded that low flow and low dissolved oxygen concentrations were likely interacting to cause effects. Levels of low flow and dissolved oxygen that individually would not be sufficient to cause the observed effect may be sufficient when acting together. Since both are necessary, remediation of either would be expected to improve condition.

Remediate all plausible causes. Multiple causes are not a problem if a feasible management action will remediate them all. For example, a stream channel restoration could improve habitat structure, temperature, and dissolved oxygen concentrations. An agency might decide to proceed with the management action, leaving the issue of relative contributions of the causes unresolved, but monitor to verify that the management action indeed was effective.

Be pragmatic. In some cases, the relationship among multiple causes may not be clear, but remediation of one cause may be much more feasible. For example, when one of a set of causes is illegal or is a bad practice that is readily corrected, it should be remediated and the results monitored. In other cases, the pragmatic choice may be based on a political judgment. In the case of the decline in the striped bass fishery in the Chesapeake Bay, both harvesting and toxicity appeared to be important causes (Barnthouse et al., 2000). While it would be difficult, expensive, and time-consuming to track down and remediate the sources of toxicity on the many tributaries, harvesting could be readily and quickly managed at relatively little cost. Hence, a moratorium on striped bass harvesting was implemented, and the population subsequently recovered. Elimination of toxicity in the tributaries where striped bass spawn also might have been sufficient and may eventually occur. Such action might allow more intensive harvesting. When a remedial strategy is chosen without first identifying the cause or the relative contributions of multiple causes, the environmental manager is gambling. The remediation may or may not improve biological condition.

21.2.3 How Much Change is Needed to Improve Biological Condition?

By the end of the causal assessment, the direction of change needed to improve conditions should be clear: for example, increase the concentrations of dissolved oxygen, decrease the deposition of fines, or decrease the concentrations of toxic chemicals. When the management options are dichotomous, such as to remove a dam or not, the direction of the required change may be enough to guide the decision. In many cases, remedial options are continuous. For example, a treatment facility may be scaled up to remove an increasing proportion of dissolved organic matter in a waste stream. In these cases, estimates of the amount of stressor reduction will improve biological condition that are needed to guide the choice and design of a management action.

Predicting the amount of stressor reduction required to improve biological condition usually involves a model. Ideally, the model would quantify the changes that occur under different intervention scenarios. Changes are often modeled in two stages: (1) models of the relationship between the degree of intervention or source reduction and the level of stressors and (2) models of the relationship between stressors and biological response.

Some progress has been made on models of the first stage, modeling the amount of stressor reduction expected under different management scenarios. For example, Tomer and Locke (2011) report good progress in projecting the results of agricultural conservation practices on hydrological changes; however, predicting effects on water quality and sediment yields have proven to be more difficult.

The stressor–response models developed during the causal assessment are a starting point for the second stage, estimating the amount of stressor reduction needed to improve biological condition. Yet these models will be imperfect predictors of change. Stressor–response models developed using experiments have the advantage of being the result of a controlled manipulation, which isolates the response to the cause. However, the conditions of the experiment may not reflect the conditions of the management action or the exact actions that are being implemented on the ground. In particular, results from small pilot treatments may be very different when applied to longer and more extensive temporal and spatial scales.

Stressor–response models that are developed using observational data have the advantage of realism, but are a product of the system of initial conditions and causal processes that operated when they were produced. Most field-derived stressor–response models are developed using an association between biological condition and levels of stressors at different locations to mimic the response expected with increased degradation over time. Observations along the gradient are usually a result of different degrees of degradation; very few models are based on observations that were produced by restoration. The reaction of the degraded community to a decrease in stressor levels could be quite different than the reaction of the original community to the imposition of stress. In some cases, the degraded biological condition may reflect an alternative stable state that may be resilient to efforts to alleviate the problem (Gunderson and Pritchard, 2002; Suding et al., 2004; Folke et al., 2004; Scheffer and Carpenter, 2003). For example, prairies that have become dominated by woody vegetation because of the lack of fire may not revert back to prairie after reintroducing a natural fire regime, because the new vegetative cover responds differently to burning (Suding et al., 2004). In streams, it has been suggested that the dominance of stream invertebrates tolerant of acidic conditions may prevent sensitive invertebrates from recolonizing after the acidic conditions subside (Monteith et al., 2005).

Field-derived stressor–response models may also inaccurately reflect responses to management actions that deliberately break the causal processes that led to the original association. The words of Box (1966) capture the reason well, "to find out what happens to a system when you interfere with it you have to interfere with it (not just passively observe it)." This is particularly problematic for models that relate land-use to biological response. For example, many investigators have found strong statistical associations between percent impervious surfaces (e.g., parking lots, roofs) and indices of biological condition (e.g., Kennen, 1999; Morley and Karr, 2002; DeGasperi et al., 2009). Slowing runoff and enhancing percolation using drainage swales, pervious pavement, and rain gardens are common best management practices in urban areas (Booth and Jackson, 1997). These remedies are expected to be especially effective because they disconnect some of the processes (e.g., reducing biological exposure to peak flows of surface runoff) likely to be mechanisms behind the original statistical association. The original models relating percent impervious surface to biological condition that used observations from historical data and conventional developments are likely to be inaccurate predictors of the benefits of these newer management actions.

The solution to the issues with field-derived stressor–response models is to improve the understanding of biological recovery under different management scenarios. Eventually, the results of enough management actions will be available to develop empirical models of recovery, including the identification of conditions under which biological recovery would or would not be expected given mitigation of physical or chemical stressors. Until then, models based on reactions to degradation should be considered to be a best-case scenario for response to decreased levels of physical and chemical stressors. They should be applied with caution, considering the proximate causes expected to change with the management action, the state of the biological systems being restored, and the causal processes expected to produce the response.

21.2.4 Identifying the Intervention(s)

Causal assessment has proved useful for identifying the appropriate intervention required to improve poor biological conditions. This was the case for the Kent Dam removal project on the Cuyahoga River, where dam removal was a clear (but not the only) option for increasing dissolved oxygen levels.* However, the best intervention is not always obvious. For example, Wilson et al. (2008) found that sediment loads in five US streams were not derived from erosion of surface soil, as expected. Instead, 54–80% of the sediment loads were derived from channel sources. Consequently, management actions directed at reducing erosion from agricultural fields would have little effect.

Causal assessments help identify an effective intervention in three ways. First, possible interventions can be identified using the conceptual models developed during the course of the causal assessment because they identify what is known about sources and human activities that produce proximate causes. Jansson et al. (2005) recommends the use of conceptual models to

---

* An alternative considered was to require a wastewater treatment plant to meet more stringent pollutant limits.

evaluate river restoration projects. Mika et al. (2010) embraced this idea, using a conceptual model to guide restoration of in-stream and riparian structure and function in the Upper Hunter River, Australia.

Second, if a causal assessment reduces the number of likely causes, then the number of causal pathways that need to be traced is also reduced. Only the pathways associated with the likely causes need consideration. Management actions may be directed at many different points along the path from human activity to biological effect. These may include managing human activities, such as limiting a pesticide use or putting a moratorium on fishing or managing sources of stressors, such as reducing phosphorus loads by reducing runoff from agricultural fields. Management actions may also be directed at interacting stressors or environmental factors. For example, efforts to increase fish populations in trout streams have included application of lime to increase pH, thereby reducing exposure to monomeric aluminum, the proximate cause (e.g., Allan and Castillo, 2007).

Third, the process of causal assessment encourages investigators to clarify the means by which exposure and effects occur. In turn, this knowledge can suggest creative ways to intervene to reduce effects. For example, the granular formulation of Diazinon was cancelled after it became clear that birds were ingesting the granules when foraging for food and grit (U.S. EPA, 2004a, b). Jenkins and Boulton (2007) identified that the length of time a stream is desiccated was the predominant factor influencing invertebrate mortality in floodplain lakes in Australia. This finding was then used to recommend water allocations to support ecological functions. The prevalent cause of bat mortality from wind turbines was found to be from the rotating turbine blades inducing a sudden change in barometric pressure resulting in hemorrhaging in the bats (i.e., barotraumas; Baerwald et al., 2008). That finding combined with the knowledge that bats fly most during periods of low wind speeds suggested that increasing the operational wind speed could decrease bat mortality. Indeed, these adjustments have been shown to reduce bat mortality up to 93% with less than 1% loss in annual wind energy produced (Arnett et al., 2011).

The success of management actions is contingent upon identifying the stressor source. In their review of agricultural best management practices, Tomer and Locke (2011) noted that "making assumptions about contaminant sources without data-based evidence can lead to ineffective recommendations and loss of stakeholder trust in the process of water quality management." Yet, investigations of the sources of stressors can be every bit as involved and complicated as investigations of the proximate causes of biological degradation. Because source assessments are often used to allocate responsibility and legal liability, methods are often referred to as environmental forensics (Murphy and Morrison, 2002). Environmental forensic methods typically emphasize tracer studies and the use of chemical fingerprinting, for example, the use of isotope analyses to determine sources of nitrogen (de Bruyn et al., 2003) and DNA analyses to identify sources of pathogens (Simpson et al., 2002).

Source identification can benefit further from the causal assessment framework described in this book (i.e., formulating the problem by carefully defining the issue requiring explanation and delineating alternative possible explanations, developing evidence from different sources, and forming conclusions by weighing and comparing the evidence). An example comes from the city of Austin, Texas, USA (City of Austin, 2005). Scientists there were confronted with unexpectedly high concentrations of PAHs in sediment samples collected from several urban streams and storm water control structures. Of particular concern was the finding of elevated PAHs in sediment of streams that feed into Barton Springs Pool, a valued local resource and home to an endangered species, the Barton Springs salamander. Elevated concentrations were found in a dry stream bed and were even higher further upstream near the parking lot of an apartment complex (see Figure 21.2). The findings triggered a series of investigations and analyses to determine the source of PAHs.

Possible sources included parking lots, specifically the coal-tar emulsion sealcoat applied to parking lots; buried waste; eroding stream banks; deposition of airborne particulates from diesel engines; and motor oil and tire wear in parking lot runoff. Eroding stream banks and buried waste were ruled out as a source when elevated PAH concentrations were not found in bank sediments and soils. In contrast, concentrations in the scrapings from the parking

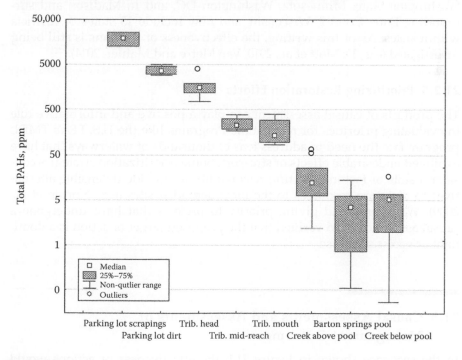

FIGURE 21.2
Total PAHs decreased from a parking lot downstream to Barton Springs Pool. (Adapted from Figure 4.2, City of Austin (2005).)

lot were higher than those in downstream sediments (see Figure 21.2). Further evidence was provided by observational studies that showed 63% of the variation in total PAH concentration could be explained by the percent of sealed parking lot area in the watershed. Modeled PAH loads from sealed parking lots compared closely with measured storm-load events (Mahler et al., 2005). Different types of sealants were analyzed by comparing the chemical pattern of sealant PAH constituents with that found in the sediments. The ratios of specific PAHs in sediment overlapped more closely with PAHs in particles from coal-tar emulsion sealcoat than those from asphalt-emulsion sealcoat or unsealed asphalt and concrete pavement (Mahler et al., 2005).

The investigation was expanded to locations outside of the city of Austin, when it was learned that cities in the western United States used a different type of sealcoat with lower PAH concentrations (Van Metre et al., 2009). Relative to cities in the eastern and central United States, PAH concentrations in dust from seal-coated pavement in the western cities were much lower. PAH concentrations in dust from seal-coated pavement were also much greater than unseal-coated pavement. Tirewear and oil spills would have been expected to occur on both pavement types; hence, this result weakened the case that they were the source of the PAHs. In 2006, the City of Austin banned the use of coal-tar sealants (City of Austin, 2006). Bans followed in Washington State, Minnesota, Washington DC, and in Madison and surrounding Dane County, Wisconsin, and now include 15 states or districts within states. As of this writing, the effectiveness of the bans is still being investigated (e.g., DeMott et al., 2010; Van Metre and Mahler, 2014).

21.2.5 Prioritizing Restoration Efforts

The products of causal assessment can play a positive and informative role in evaluating priorities for restoration. Programs like the U.S. EPA's TMDL program face the need to address tens of thousands of waterways that have exhibited undesirable effects or stressors. Some prioritization strategies consider a suite of factors affecting restorability to provide defensible alternatives to directing resources to the most degraded sites (e.g., Norton et al., 2009). We recommend giving priority to projects that have undergone a causal assessment and verified that the proposed target of action is a dominant or necessary cause.

21.3 Causal Assessments and Assessments of Management Outcomes

In the sequence shown in Figure 21.1, the effectiveness of actions would be evaluated based on whether the management action was implemented according to plan, whether it induced the desired change in the cause, and the degree to which it improved the biological effects of concern. However, very few projects collect evidence about all three types of outcomes (Bernhardt et al., 2007; Rumps et al., 2007; Feld et al., 2011).

Causal assessments can contribute to the process by helping set expectations for recovery (see Section 21.3.1) and identifying indicators useful for evaluating incremental success (see Section 21.3.2). In addition, if management actions fail to produce the desired outcome, a causal assessment process can help identify why (see Section 21.3.3).

21.3.1 Setting Expectations of the Degree and Time Course for Recovery

As described in Section 21.2.1, the comparison sites identified during the causal assessment process embody the biological conditions that can be achieved through management action. These same biological conditions define the degree of biological recovery that can be expected if the causes are adequately addressed.

Causal assessments can also help set expectations for the degree of incremental recovery when some, but not all, of the causes of environmental degradation can be addressed. For example, in a partial dredging effort from 2002 to 2006, 68,800 tons of PAH-contaminated sediment were removed from the Little Scioto River. In 2009, the site was listed on the National Priorities List, the Remedial Investigation report was completed in 2013, and the feasibility study completion date is 2014. A causal assessment in 2002 had indicated that some parts of the reach were also impaired by stream channelization (Norton et al., 2002a; Cormier et al., 2002). Those involved with the initial dredging were unaware of that causal assessment. The results of the assessment were brought to the attention of the site managers so that recovery expectations would be realistic (S.M. Cormier, personal communication).

Knowing the causes and sources of effects can help establish the time scale of recovery. Recovery of fish populations and macroinvertebrate assemblages from temporary disturbances including chemical spills or treatments, floods or droughts, construction activity, and invasive organism removal, can range from less than 1 year to 6 years (Niemi et al., 1990; Detenbeck et al., 1992). These time periods are consistent with observed organism reactions following management actions that create abrupt changes. Some biological improvements from dam removal are immediate, for example, increased access of fish to spawning habitat. Riverine taxa tend to replace those adapted to slow-moving water within a year of dam removal (Stanley and Doyle, 2003). The rate of improvement in communities downstream and immediately upstream of dams can take longer, in the order of years, due to the greater volume of accumulated fine sediments inhibiting recovery (Stanley and Doyle, 2003; Feld et al., 2011). Relatively short recovery times for macroinvertebrates have also been observed after excluding livestock from streams (3 years) and treating mine waste in streams (10 years for macroinvertebrates and 3–9 years for fish) (Meals et al., 2010; Spears et al., 2011).

The recovery of fish populations following in-stream habitat restoration has been reported, but questions remain about whether those improvements are sustained over time (Meals et al., 2010; Feld et al., 2011).

Recovery of fish populations and macroinvertebrate communities from long-term disturbances, including channelization, logging, mining and eutrophication, can take 3 years to more than 52 years (Niemi et al., 1990; Detenbeck et al., 1992; Spears et al., 2011). These recovery times are consistent with the time course of physical and chemical improvements after implementation of non-point-source best management practices (BMPs). Non-point-source BMPs intended to reduce stream sediments may take up to 50 years before an effect is observed, practices that adjust fertilizer application rates may take up to 40 years to reduce nitrate concentrations, and efforts to reduce winter road salt applications may take 50 years or more to decrease chloride concentration in groundwater (Meals et al., 2010). It may take longer than 50 years before a restored riparian forest becomes a sustainable source of large woody debris for a stream (Feld et al., 2011). However, much faster improvements in eutrophic lakes (<5 years) were reported by reducing external sources of nutrients in lakes combined with managing internal nutrient sources, such as capping or dredging of sediments and manipulating fish populations (Spears et al., 2011).

21.3.2 Identifying Monitoring Indicators

Monitoring after remediation or restoration is performed inconsistently, unless required by regulation. In a review of 39 river restoration projects, Alexander and Allen (2007) reported what one would expect: most (79%) of the projects collected some form of monitoring data. In contrast, a review of 317 river restoration projects showed that only 59% of the projects used quantitative monitoring data to evaluate project success (Bernhardt et al., 2007). In a survey of over 37,000 broadly–defined river restoration projects, only 10% of project records indicated any form of monitoring (Bernhardt et al., 2005). Reviews of river restoration effectiveness in the European Union have noted an overall paucity of monitoring data, a lack of pre-restoration baselines, and few attempts to improve monitoring through a process of watershed management planning (Feld et al., 2011; Verdonschot et al., 2012, 2013).

In a sobering finding, almost 30% of the river restoration projects that conducted post-project monitoring did not use the monitoring data that had been collected to evaluate the outcome, either because the data were not relevant to project goals or the monitoring design was inadequate to support evaluation (Bernhardt et al., 2007). In addition, little distinction was made between monitoring to evaluate whether the action was implemented as planned, whether causes were reduced, or whether the biological condition improved. As would be expected, programs that focus on reducing stressors in streams (e.g., conservation practices supported by the U.S. Department of Agriculture (USDA) and the U.S. EPA TMDL program) focus monitoring efforts on documenting implementation and the reduction of stressors. For example, the Conservation Effects Assessment Project documents the reductions of nutrients and sediments achieved through agricultural conservation practices (Tomer and Locke, 2011), and the U.S. EPA reports on the number of TMDLs being addressed through permits or watershed management plans.

The products of causal assessment can help identify monitoring parameters that demonstrate incremental progress between management action and biological response. The conceptual models show where different monitoring results from intervention to stressor reduction fit together on the path to biological recovery. For example, management actions to reduce acid deposition in the United Kingdom were evaluated by monitoring the decline in sulfate and base cations, the trends toward more acid-sensitive diatom and macroinvertebrate taxa, and the appearance of juvenile trout (Davies et al., 2005; Monteith et al., 2005).

In addition, the specific biological effect identified while defining the case (see Chapter 7) can be used to identify more sensitive responses to better evaluate the effectiveness of the management action. The macroinvertebrate and fish indices frequently used to categorize biological condition in streams and rivers typically capture the effects of multiple stressors. Therefore, they may be too coarse or generic in nature to detect early reaction to a management action aimed at a single stressor. Feld et al. (2011) highlighted the need to monitor biological responses that are sensitive to specific restoration measures and advocated using better, mechanistic understanding of cause–effect relationships to guide post-restoration monitoring designs in the future. A better understanding of expected specific biological reactions would inform the project manager which components of the biological community would be expected to respond earliest and strongest to the change.

21.3.3 Investigating Why Management Actions Fail

Management actions do not always achieve the intended goal or target. In studies of reach restoration projects in Germany, positive responses from fish were observed only about half of the time (Melcher et al., 2012; Haase et al., 2013). The responses of macroinvertebrates and macrophytes were even weaker (Stewart et al., 2009; Haase et al., 2013). Results for restoration of eutrophic lakes have been more promising: about two-thirds of 43 projects reported nutrient reductions and positive biological response (Verdonschot et al., 2013). However, biological recoveries in lakes following reductions of acid deposition have generally lagged behind aluminum and pH responses (Verdonschot et al., 2013). As the results of environmental actions accumulate, investigations have turned toward identifying the conditions under which expected improvements do and do not occur.

When an intervention does not have the intended effect, it seems reasonable to investigate the reasons why. As discussed in Chapter 5, identifying alternative causes is an important part of formulating investigations in a way that prevents cognitive errors. Many authors have offered explanations for why management actions may not have had the intended effect (see Table 21.1) (see also reviews by Spears et al., 2011; Haase et al., 2013; Verdonschot et al., 2013). A few authors systematically investigate the evidence for and against different hypotheses.

TABLE 21.1

Proffered Explanations for the Underperformance of Management Actions

Candidate Explanation	Example	References
Wrong cause	Habitat restoration projects in urban watersheds, where the impacts of degraded water quality and altered flow regimes likely dominate, showed little positive impact on fish and macroinvertebrate assemblages.	Palmer et al. (2010) Miller et al. (2010) Tullos et al. (2009) Violin et al. (2011)
Wrong intervention	Management actions targeted at reducing soil erosion from agricultural fields did not result in the expected reductions in stream sediments because the primary source of the sediment was from the stream channel and bed.	Tomer and Locke (2011)
Insufficiently bold intervention	Stream assemblage may not react to riparian buffer restoration where the length of the buffer is not extensive enough.	Lorenz and Feld (2013); Sutton et al. (2010)
Intervention not sustained over time	Habitat enhancements were not sustained because watershed sources of sediment and flow regime changes were not also managed.	Moerke and Lamberti, (2003); Entrekin et al. (2008)
Other causes encroach	The effects of new urban development can mask gains from river restoration and conservation practices.	Moerke and Lamberti (2003)
	Mussel harvesting prevented expected eelgrass recovery after nutrients were reduced because the harvesting decreased consumption of phytoplankton stocks by mussels and contributed to sediment resuspension.	Carstensen et al. (2013)
Insufficient time elapsed	Although sulfate from acid rain has decreased significantly in United States since controls were initiated in 1990, calcium levels in soils have yet to increase.	Lawrence et al. (2012)
Recolonization impeded by barriers, distance, or extirpation	Lakes further from source populations of fish recovered more slowly after liming treatments than those close to source populations.	Degerman et al. (1992)
	River restoration sites with intact habitat upstream (providing a source of colonizers) showed greater response to restoration efforts.	Lorenz and Feld (2013)
	Culverts hindered upstream movement of caddisflies into restored reaches.	Blakely et al. (2006)
Species interactions prevent recolonization	Established acid-tolerant macroinvertebrate species may prevent the re-establishment of the full complement of acid-sensitive species despite reductions in acid deposition.	Monteith et al. (2005)

Such investigations sometimes reveal unexpected reasons for failure. In their review of the modest recoveries found after the substantial reductions in acidic precipitation in United Kingdom, Monteith et al. (2005) evaluated the evidence for several potential reasons and concluded that aquatic communities may have changed in a way that makes them resistant to the improvements in water quality. In their investigation of the lack of expected eelgrass response to reductions in nutrient inputs, Carstensen et al. (2013) concluded that the increases in mussel harvesting prevented eelgrass recovery by releasing phytoplankton from herbivory and contributing to sediment resuspension.

Management interventions that do not produce expected effects provide an opportunity to gain valuable insights that can guide the next generation of actions. For example, when agricultural soil erosion control measures were not having the intended effect of reducing overall sediment loading, subsequent analyses determined that the primary sources of sediments were stream banks and beds and led to the recommendation that future sediment control projects also should manage flow regimes (Tomer and Locke, 2011). Although there is much incentive to record success stories (e.g., see U.S. EPA, 2013g), creating incentives to investigate disappointments may be even more valuable for improving future decisions.

21.4 Synthesizing and Communicating Knowledge about Causal Processes

Optimally, the knowledge gained throughout the assessment process will improve our understanding of ecosystems and causal processes. Accurate understanding is required to identify the conditions under which management actions will most likely lead to improved biological conditions and to prevent actions that unintentionally harm ecological systems (Palmer, 2009).

There is also a need to innovate ways to communicate which actions do and do not work within the scientific community, and, more importantly, to the community of practitioners. Scientific publications are effective for communicating among scientists, but are unlikely to serve practitioners well. The disconnect is apparent in the review of 319 river restoration projects by Bernhardt et al. (2007), in which the authors found that only one project used a scientific paper to guide design decisions.

21.4.1 Synthesizing Knowledge from Management Actions

Project reports and case studies provide the fundamental building blocks for reviews and syntheses. Useful information can be gained from project reports even though few projects study or report the entire sequence of events from implementing an action, to monitoring physical and chemical improvements, to documenting biological recovery (Feld et al., 2011). Intensive studies of management actions can be used to pool resources in ways not possible on a routine basis and can be targeted to fill specific knowledge gaps (e.g., Bernhardt et al., 2007).

Reviews of program effectiveness should synthesize knowledge of what works. However, it is difficult to use overall statistics on program effectiveness to evaluate conditions leading to biological improvement (see Box 21.1). Part of the problem is that many environmental actions are taken for reasons other than biological degradation. For example, since human pathogens comprise the greatest reason for listing a stream as impaired (i.e., listed according to Section 303d of the U.S. Clean Water Act), management actions implemented in these streams would not be expected to necessarily improve the condition of aquatic biota (U.S. EPA, 2013f). More surprisingly, perhaps, is that a management action may be initiated for reasons other than either ecological or human health. Fewer than 50% of interviewees reported that river restoration projects were initiated because of a recognized need to address some form of river degradation. Rather, restoration may be motivated by factors like public safety issues, legal mandates, funding availability, aesthetics, and the ease of addressing the problem (Bernhardt et al., 2005, 2007; Rumps et al., 2007). Project success is, therefore, evaluated on the basis of criteria other than biological condition, such as positive public opinion and post-project appearance. In a review of 39 river restoration projects in the midwestern United States, more than half claimed success based on "positive effects on the human community" (Alexander and Allen, 2007).

BOX 21.1   MANAGEMENT ACTIONS AND THE U.S. EPA TOTAL MAXIMUM DAILY LOAD PROGRAM

The results of management actions prompted by the U.S. EPA TDML program are difficult to assess. EPA reviews and approves States' lists of impaired waters and the TMDL or other watershed management plan developed to address the cause of impairment. However, the state-led plans that follow a TMDL can be implemented by a combination of federal, state, and local programs, making it difficult to attribute outcomes to specific program actions. Point sources are managed by EPA through discharge permits, but there is not a parallel authority to control nonpoint sources. States can, but are not required to, regulate nonpoint sources. USDA programs provide greater funding for implementing best management practices, but these are not necessarily linked to water quality improvements related to TMDLs or impaired waters. EPA has limited authority to require new post-TMDL monitoring, data tracking, and reporting (U.S. EPA, 2007b).

Another approach for evaluating the outcomes of management actions is to use the results of large programs devoted specifically to biological monitoring of condition. For example, the U.S. EPA undertakes probabilistic surveys of biological condition under the National Aquatic Resource Survey program, and many states use state-level rotating basin surveys that monitor habitat, water quality, and biota (Yoder and Barbour, 2009). These types of monitoring surveys have many benefits for ecological causal assessment including providing a valued source of data and a broader context within which site-specific results can be evaluated. Over time, these surveys will quantify overall trends in condition that, in theory, could be linked to overall environmental policies and management actions. However, the surveys are designed to quantify trends over coarser geographical scales than the individual site and will reflect environmental changes other than management action (e.g., climate changes, urban sprawl). For this reason, linking these results with management actions taken at specific locations will be difficult.

Reviews of restoration practices like those conducted by Bernhardt et al. (2007), Palmer (2009), Roni et al. (2008), Feld et al. (2011), Tomer and Locke (2011), Melcher et al. (2012), Verdonschot et al. (2013), and Hering et al. (2013) can be used to synthesize knowledge gained from conducting management actions. Approaches being used to develop "evidence-based medicine" provide a useful model for extending synthesis beyond just a review (e.g., Sackett, 1997). A key component of these approaches is to couple the systematic review of the literature with quantitative analysis of results (Higgins and Green, 2008). In addition to quantifying the degree and direction of effects, such approaches can highlight where biases might occur in the literature (e.g., a bias towards publishing positive results). The Eco-evidence methodology (Norris et al., 2012) and reviews conducted by Stewart et al. (2009), Miller et al. (2010), Melcher et al. (2012), and Miller et al. (2013) illustrate how these concepts have been applied to issues of environmental management.

Conceptual models are another way of synthesizing and communicating what is known and to identify where knowledge gaps exist. Sime (2005) used a conceptual model to describe research needed to guide the restoration of the St. Lucia Estuary and Indian River Lagoon in Florida, USA. In their comprehensive review of river restoration projects, Feld et al. (2011) used conceptual models to summarize the evidence for different linkages in restoration projects (see Figure 21.3).

21.4.2 Promoting Shared Understanding

Environmental management actions often impact and involve numerous interested and affected parties (i.e., stakeholders), depending on the project's scale, scope, and societal interests. In many situations, river restoration projects involve as many as eight stakeholder groups representing, federal,

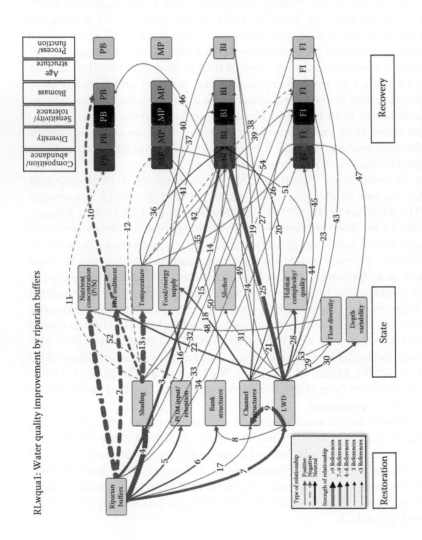

FIGURE 21.3

An example of a conceptual model developed to synthesize river restoration knowledge by depicting literature summarizing water quality improvement by riparian buffers. The numbers within the arrows are labels denoting the group of references that describe the connections being made; two-letter abbreviations refer to organisms: PB, phytobenthos; MP, aquatic macrophytes; BI, benthic invertebrates; FI, fish. (Adapted from Feld, C. K. et al. 2011. *Advances in Ecological Research* 44: 119–209. With permission.)

state, and local government agencies, tribal nations, nongovernmental organizations (NGOS), private landowners, and volunteers (Bernhardt et al., 2007).

A shared understanding of causal processes and their relationship to management outcomes is particularly important for communicating expectations on the time for biological recovery to occur after a decision to take action has been made. Once a decision to take action has been made, the hope is that implementation will begin immediately; however, the reality is that it can take a considerable amount of time to plan and implement actions. For example, a selected action could require significant upgrades to a wastewater treatment plant. Obtaining funding and constructing the facility contribute to the time between action and recovery. In more complex cases (e.g., dealing with nonpoint sources), planning and implementation take even longer due to the time needed to engage landowners and integrate new practices into cropping and land management cycles. After action has been taken, another series of lags may be encountered before the desired change in stressor levels is produced and the desired ecological response is observed (Meals et al., 2010). To evaluate whether an action has achieved the desired outcomes, attention must be sustained across time and information shared among the participants involved in the different steps.

The cooperation of stakeholders has been highlighted as a key factor influencing the success and acceptance of management actions taken under the TMDL program (Hoornbeek et al., 2008; U.S. GAO, 2013). Most successful river restoration projects have significant community involvement and an advisory committee (Bernhardt et al., 2007) with substantial involvement in the initiation and implementation phases. One way of engaging stakeholders is to build simple simulation models that allow for the exploration of the consequences of alternative actions through "what if" scenarios (McLain and Lee, 1996). Stakeholders may suggest innovative ways to mitigate exposure or to better implement the action. For example, the Old Order Mennonite community in the Muddy Creek watershed of Virginia, USA, identified ways to implement agricultural BMPs consistent with their values (U.S. EPA, 2013g). Stakeholder involvement and buy-in are also essential for sustaining improvements because most actions require a continuing effort (e.g., the maintenance of fences to prevent cows from entering streams or the adjustment of fertilizer application rates).

The causal assessment process described in this book provides several opportunities to engage stakeholders. These include reporting observations of undesirable effects that prompted the assessment, presenting conceptual models to elicit comment, and presenting the results of the causal assessment. A written report provides a record documenting the reasons for undertaking actions that may be implemented and sustained by future participants.

21.5 Summary

Identification of the cause is only one step in a series of activities needed to attain environmental improvement. In one common sequence, causal assessments are preceded by the identification of an undesirable ecological effect and are followed by assessments that evaluate different management options, the implementation of a management action, and an assessment of whether that action produced the desired outcome. Strong linkages between the steps can help set expectations and ensure management actions are directed at the causes that are likely to improve biological conditions. The products and process of causal assessment inform the choice and evaluation of management actions. In addition, many of the principles of causal assessment are transferable to the issues of source identification and evaluating why some management actions perform below expectations.

Reliable and practical understanding of ecological processes and causes in specific situations is urgently needed in environmental management because management actions are in and of themselves ecosystem disturbances, sometimes major disturbances. Causal assessment results can help managers select actions that will have the greatest chances of success and prevent the implementation of projects that have little chance of improving biological condition.

This chapter began by describing a typical sequence of assessments of condition, causes, options, and outcomes. We would like to end by suggesting that an important and complementary course of action is identifying high-quality ecosystems and protecting them before they become degraded. The prospects for restoring physical and chemical attributes appear to be bright, albeit expensive. Statistics compiled by U.S. EPA's Non-Point Source Program concluded that between 2005 and 2011, it costs 100 million US dollars per year to remediate problems in just 60 water bodies, or about 1.67 million dollars per watershed (U.S. EPA, 2011b). Most of these stream and river restorations have targeted physical and chemical attributes. It is likely that restoration of biological condition will prove to take longer and cost more. In the face of this reality, it seems prudent to extend the three pillars for managing the loss of wetlands under the U.S. Clean Water Act to all types of biological degradation in ecosystems. These principles are, in priority order, avoid, minimize and, only then, mitigate (ELI, 2007).

Part 3

Case Studies

The last part of the book provides three aquatic case studies and one terrestrial case study of causal assessment. These case studies show how the process can accommodate different variations and different types of evidence. Although the case studies are biased toward our work and interest in streams and rivers, the principles they illustrate can be adapted for other systems and places.

Chapter 22. The Long Creek case in South Portland, Maine, has an international airport to the north, a large shopping mall to the west, and many channel modifications throughout the watershed. It exemplifies a complex case with many interacting causes and highlights comparisons of conditions in Long Creek to those in nearby, less affected stream and to other field and laboratory studies.

Chapter 23. The Clear Fork case study assesses individual tributaries and their effect on the main stem of Clear Fork in the center of West Virginia's coal mining district. It illustrates the systematic use of regional comparison values and community symptomology for assessing causes.

Chapter 24. The Athabasca River case study provides an opportunity to highlight field, laboratory, and mesocosm studies.

Chapter 25. A decline in the endangered San Joaquin kit fox is the focus of the Elk Hills, California, case study. Set in an oil field, the case study demonstrates how an assessment's temporal and spatial frame can influence conclusions and highlights the use of a simulation model. It also demonstrates combining all the evidence into a coherent explanation.

But there is no reason to be limited by these case studies. Refer to Table 1.1 or the CAODIS website for many more examples.

22

Causal Assessment in a Complex Urban Watershed—The Long Creek Case Study*

C. Richard Ziegler and Susan B. Norton

The Long Creek case study illustrates the application of causal assessment in an urban watershed with many sources and stressors. The case highlights the comparison of conditions in Long Creek to those in a nearby but less affected stream, and to stressor–response relationships from other field and laboratory studies.

CONTENTS

---

* This chapter has been adapted from Ziegler et al. (2007a), with permission from the authors.

22.1 Summary

Long Creek is a sandy-bottomed stream in an urbanized region of coastal southern Maine. Watershed monitoring in the 1990s and early 2000s revealed that several stream reaches in Long Creek did not meet state biological standards (MEDEP, 2002); this case study explored the reasons why. A high proportion of impervious surfaces cover the lower part of the Long Creek watershed, so the finding of a degraded macroinvertebrate assemblage may have been anticipated. However, reaches further upstream with lower proportions of impervious surfaces also did not meet state standards. This case study focuses on two sites: LCN .415, located low in the watershed, and LCMn 2.274, located farther upstream.

Not surprisingly, the causal assessment concluded that multiple causes contributed to the degraded macroinvertebrate assemblages. Likely causes at both sites included increased salts, altered flow regime, decreased dissolved oxygen, increased temperature, and decreased large woody debris. Measurements and observations relevant to these candidate causes were at more stressful levels compared with a site in a nearby but less affected stream. Salt levels, oxygen concentrations, and temperatures were at levels associated with effects in other field or laboratory studies.

With so many likely causes in urban streams, it can be just as useful to identify unlikely causes. Two candidate causes were concluded to be unlikely contributors at LCMn 2.274. Stressor–response associations argued against increased autochthony (i.e., in-stream algal and macrophyte production). Specifically, the levels of nutrients did not exceed benchmarks and so was not considered sufficient to have produced the effects. This finding was consistent with the functional feeding group analysis, because the relative abundance of organisms that benefit from in-stream productivity was not greater relative to the comparison site. Increased sediment was concluded to be an unlikely contributor: neither base-flow levels of suspended solids nor substrate habitat scores differed between LCMn 2.274 and the comparison site.

Several of the likely causes in Long Creek interact in ways that may worsen effects. In particular, low dissolved oxygen interacts with low base flow and increased temperature in ways that will produce more severe effects than would be expected if these stressors acted independently. Invertebrates with passive gill respiration (e.g., some mayflies) require flowing water, particularly as levels of dissolved oxygen decrease. Low base-flow velocities and low dissolved oxygen at the affected Long Creek sites may be sufficient to cause problems for these types of invertebrates. Similarly, higher temperatures increase many coldwater organisms' investment in respiratory processes, increasing their susceptibility to low dissolved oxygen. Temperature and dissolved oxygen at Long Creek's affected sites are each at levels associated with adverse effects. Interactions among these two causes and low flow are likely making effects worse.

Although no single dominant cause was identified, the assessment clarified the specific ways that human activities in the Long Creek watershed lead to effects on stream biota. The results have informed ongoing efforts to improve Long Creek, including discharge permitting and implementation of a watershed management plan (MEDEP, 2009; LCWMD, 2014).

22.2 Problem Formulation

The U.S. Clean Water Act (CWA) provides statutory context for the Long Creek case study. The CWA requires states to identify water bodies that do not attain water quality standards and to develop plans to return pollutant-impaired water bodies to attainment. Long Creek's listing on Maine's 1998 CWA Section 303(d) list of impaired water bodies partly triggered this case study. The U.S. EPA chose Long Creek as an example urban watershed for study under CWA funding in early 1999. Maine Department of Environmental Protection (MEDEP) and U.S. EPA personnel partnered to conduct this causal assessment.

22.2.1 The Case

Long Creek is a sandy-bottomed stream in an urbanized region of coastal southern Maine. The assessment of two specific reaches within the Long Creek watershed (represented by sites LCN .415 and LCMn 2.274; Figure 22.1) will be described here. The project team analyzed sites individually because differences in macroinvertebrate assemblages among sites suggested that causes of impairment also may have differed.

Site LCN .415 is located low in the watershed on a northern branch of Long Creek. The watershed contributing to LCN .415 is heavily urbanized: impervious surfaces cover approximately 33% of the 262 acres upstream of the study site. At the time of the study, the watershed included a portion of Portland's airport, portions of two semiconductor manufacturing plants, major roadways, retail development, and a soft drink bottling plant. The site

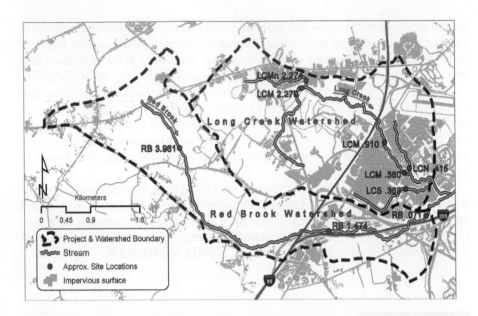

FIGURE 22.1

Impervious surfaces and project site locations. (Map was generated using data obtained from Maine Office of Geographic Information Systems (Adapted from State of Maine. 2014. Maine Office of GIS. http://www.maine.gov/megis/(accessed December 29, 2005)).)

has a sandy substrate and an intact riparian corridor. The benthic macroinvertebrate assemblage observed at LCN .415 includes organisms typical of flowing water, but MEDEP biologists did not observe the sensitive taxa that were present at the comparison site (RB 3.961; see below and Figure 22.1). Biologists found fewer organisms than would be expected and a high percentage of noninsects relative to total macroinvertebrates.

Site LCMn 2.274 is located further up into the Long Creek watershed. It is described as a narrow channel flowing through a "wooded island" refuge, with pond-like habitat and a predominance of fine sediments. Impervious surfaces (largely office parks and roadways) cover approximately 14% of the 427 acres upstream of this site. The dominant macroinvertebrate assemblage observed at LCMn 2.274 reflects a pond-like community tolerant of fine sediment. The absence of passive filter feeders suggests low flow velocity. Over 60% of the organisms found were *Dubiraphia*, an elmid beetle that can cling to vegetation and woody debris and climb out of silt. These beetles have a plastron or physical gill, which may allow them to tolerate low dissolved oxygen levels. The site's dominant mayfly, *Caenis*, is tolerant of low-velocity conditions and high water temperature (U.S. EPA, 2012c).

LCN .415 and LCMn 2.274 were compared to a comparison site in the adjacent Red Brook watershed: RB 3.961. Red Brook has a sinuous sandy-bottomed channel and an intact riparian corridor at this location. The RB 3.961 comparison site was, at the time of observation, dominated by species typical of low gradient, sandy-bottomed streams in Maine and met State biological standards. Several organisms characterized as less tolerant of human disturbance, including the mayfly *Paraleptophlebia*, were observed at this site. The alderfly *Sialis* was observed as the most abundant organism at the site, comprising 13% of organisms collected. No organisms stood out as dominant at this site. Impervious surfaces cover approximately 2% of the 508 acres upstream from this location.

22.2.2 Ecological Effects

For the analysis, the project team focused on three biological statistics (i.e., metrics) that showed conspicuous changes compared with the Red Brook comparison site. The metrics are a subset of those used by the State of Maine to evaluate assemblage quality: number of genera belonging to orders *Ephemeroptera*, *Plecoptera*, and *Trichoptera* (EPT generic richness, abbreviated as "EPT richness" hereafter), and the Hilsenhoff Biotic Index (HBI). EPT richness is often used as an indicator of stream condition (Wallace et al., 1997; see also Bednarek and Hart, 2005). While some individual taxa under the EPT umbrella may be tolerant of particular stressors, EPT are generally more sensitive than other macroinvertebrates to common stressors and often provide a reasonable measure of aquatic environmental quality, that is, greater EPT richness indicates better conditions. Conversely, HBI values often increase as certain aspects of stream condition decline. HBI was originally designed to assess low dissolved oxygen levels caused by organic loading in streams (Hilsenhoff, 1987), but the index may also reflect the presence of other proximate causes. In addition to EPT richness and HBI, the increased proportion of noninsects was evaluated at LCN .415 (proportion of noninsects did not increase at LCMn 2.274, and so was not evaluated there). Table 22.1 summarizes the values of these statistics (see also Box 22.1). Table 22.2 shows the dominant species at the two affected sites and the comparison site.

22.2.3 Candidate Causes

The State of Maine originally listed Long Creek as impaired because of decreased dissolved oxygen and unspecified nonpoint-source pollution. The project team identified seven candidate causes, described below. Conceptual

TABLE 22.1

Specific Biological Statistics for Long Creek Study Sites

Site	EPT Richness	Percent Noninsects	HBI
RB 3.961 (comparison site)	15	7.8	4.2
LCN .415	6	35.6	6.6
LCMn 2.274	7	1.4	6.2

BOX 22.1 LESSONS LEARNED—WHAT ARE OPTIMAL BIOLOGICAL RESPONSE METRICS FOR CAUSAL ASSESSMENT?

Proportions (e.g., percent noninsects) are not optimal measures of effect for a causal assessment because they depend on relative responses of different organic groups. More broadly, however, this raises a bigger question: What are the best response metrics for causal assessment? The project team identified a series of responses along a spectrum of complexity. The full report (Ziegler et al., 2007a) assessed "brook trout," which was the simplest metric used; the team related the response as a binary variable, indicating presence or absence. The next simplest response metric was EPT richness. The percentage of noninsects was less definitive because it is relative to insect abundance; further, it does not identify which specific invertebrates are more or less abundant. The HBI index was the most complicated metric. HBI is calculated by multiplying the abundance of observed organisms by assigned tolerance values, specific to organic pollution, summing the products, and dividing by the total number of individuals.

Specific effects may be of greater benefit to causal assessment than percentages, proportions, and indexes. The more complex the interactions between biological response and candidate cause, the more difficult it is to identify mechanisms by which changes occur and to draw conclusions. Further, calculated effect endpoints may not be distinct; for example, EPT richness and HBI are correlated. Additionally, a biotic index, such as HBI, responds to multiple stressors, not solely stress related to a specific index's focus (i.e., organic pollution for HBI). This is not to say that a simpler endpoint such as presence/absence of one species is going to allow a clear path to diagnostic certainty, but it is a step in the right direction. We recommend causal assessors select response variables that are as specific and unambiguously defined as possible.

models were developed for each candidate cause. A combined, study-wide conceptual model is shown in Figure 22.2.

1. *Increased autochthony* (i.e., increased algal and macrophyte production). Different benthic invertebrates prefer different sources of food. Benthic invertebrate assemblages shift as the proportion of organic matter derived from algae and macrophytes (i.e., in-stream or autochthonous sources) within the stream increases relative to material derived from terrestrial sources like leaf litter (i.e., allochthonous

TABLE 22.2

Dominant Invertebrate Taxa from Rockbag Samples

Class	Order	Family	Genus	HBI[a]	FFG[b]	MOE[c]	% of Total Individuals at Site[d]
Site RB 3.961 (total mean abundance = 120.3)[d]							
Insecta	Megaloptera	Sialidae	*Sialis*	4	Pr	B-Cb-Cg	13
Insecta	Diptera	Chironomidae	*Tanytarsus*	6	C-F,G	Cb,Cg	12
Insecta	Diptera	Chironomidae	*Micropsectra*	7	C-G	Cb,Sp	7
Insecta	Trichoptera	Odontoceridae	*Psilotreta*	0	Sc,C-G	Sp	7
Insecta	Diptera	Chironomidae	*Stempellinella*	5	C-G	Sp	7
Subtotal							46
Site LCN .415 (total mean abundance = 62.7)[d]							
Crustacea	Amphipoda	Hyalellidae	*Hyalella*	8	Sh,G	Sw	20
Insecta	Diptera	Chironomidae	*Procladius*	9	Pr,C-G	Sp	15
Gastropoda	Limnophila	Physidae	*Physella*	8	Sc	Cg,Gl	11
Insecta	Trichoptera	Phryganeidae	*Ptilostomis*	5	Sh,Pr	Cb	7
Insecta	Trichoptera	Limnephilidae	*Limnephilus*	3	Sh,C-G	Cb,Sp, Cg	6
Subtotal							61
Site LCMn 2.274 (total mean abundance = 97)[d]							
Insecta	Coleoptera	Elmidae	*Dubiraphia*	6	C-G,Sc	Cg,Cb	60
Insecta	Ephemeroptera	Caenidae	*Caenis*	7	C-G,Sc	Sp,Cb	9
Insecta	Diptera	Chironomidae	*Microtendipes*	6	C-F,G	Cg	8
Insecta	Diptera	Chironomidae	*Procladius*	9	Pr,C-G	Sp	5
Insecta	Diptera	Chironomidae	*Tanytarsus*	6	C-F,G	Cb,Cg	2
Subtotal							84

[a] Hilsenhoff Biotic Index (HBI) tolerance value.

[b] Functional feeding group (FFG): C = Collector; F = Filterer; G = Gatherer; Pr = Predator; Sc = Scraper; Sh = Shredder (classification based on Merritt and Cummins, 1996, and project team knowledge).

[c] Mode of existence (MOE): B = Burrower; Cb = Climber; Cg = Clinger; Gl = Glider; Sp = Sprawler; Sw = Swimmer (classification based on Merritt and Cummins, 1996, and project team knowledge).

[d] Organisms collected in three rockbags over 32 days. Total mean abundance = total # of individuals from all three rockbags divided by three samples.

sources). Plant production within streams increases when nutrients, light, and other resources required by primary producers are abundant, and physical conditions such as low water velocity favor the establishment and accumulation of algae and macrophytes (Biggs, 2000; Mosisch et al., 2001). These conditions are often simultaneously

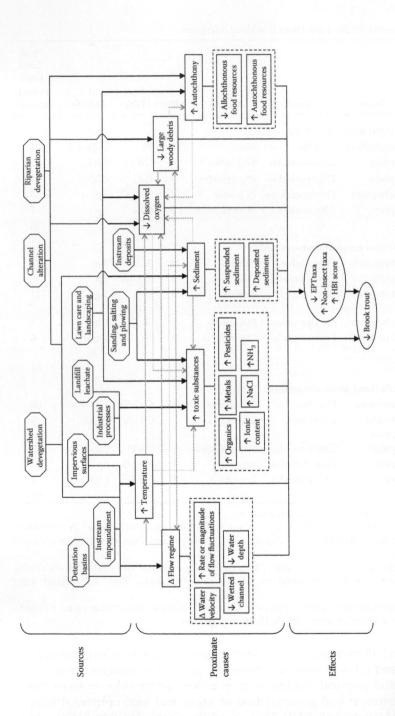

FIGURE 22.2

Conceptual model diagram of candidate causes evaluated in the Long Creek case study. Candidate causes are shown in rectangles. Dark lines depict pathways potentially linking sources, proximate causes, and effects. Light gray lines depict pathways by which candidate causes may interact. Boxes within dotted lines provide more specific descriptions of candidate causes.

associated with reduced terrestrially derived organic matter such as leaf litter and wood (Gregory et al., 1991). In the Long Creek watershed, human activities that could increase in-stream productivity include in-stream impoundments that reduce water velocity, lawn care and landscaping that contribute nutrients, and riparian devegetation that increases light and decreases inputs of leaf litter and wood.

2. *Decreased dissolved oxygen.* Reductions in dissolved oxygen concentration can asphyxiate organisms, ultimately resulting in decreases in sensitive taxa, such as mayflies (Connolly et al., 2004) and stoneflies (Barwick et al., 2004), and increases in tolerant noninsect taxa, such as oligochaetes and pulmonate snails (Peckarsky et al., 1990).

   Dissolved oxygen decreases as a result of several pathways that involve other candidate causes, including increases in water temperature, flow alteration, increased in-stream plant productivity, increased fine sediments, and decreased woody debris. Increases in water temperature decreases dissolved oxygen concentrations because the solubility of oxygen decreases with increasing water temperature. Water turbulence increases aeration, which incorporates atmospheric oxygen into the water column. Thus, factors reducing turbulent flow tend to reduce dissolved oxygen; these factors include decreased large woody debris (Mutz, 2000) and channel alterations that decrease water velocity (Genkai-Kato et al., 2005). Increased sediment deposition covers and clogs interstitial spaces, reducing the flow of oxygenated water into macroinvertebrate habitats (Argent and Flebbe, 1999). Increased plant biomass and/ or productivity decreases dissolved oxygen under low-light conditions when respiration rates exceed photosynthesis. In Long Creek, human activities that could decrease dissolved oxygen include channel alterations and in-stream impoundments that slow flow, lawn care and landscaping that contribute nutrients, riparian devegetation that increases temperatures and decreased inputs of woody debris.

3. *Altered flow regime.* The flow regime in Long Creek has been altered in several ways, including changes in water velocity, decreases in base discharge (or base flow), and increases in storm discharge (or storm flow). Many organisms have preferred flow regimes. High storm flows dislodge organisms or bury their habitat with sediments, low flows result in dry reaches with desiccated organisms. Channel modifications of Long Creek such as in-stream impoundments and channel straightening alter flows and decrease flow heterogeneity. Impervious surfaces in Long Creek's watershed have likely increased the intensity of storm flows and decreased base flows (Ziegler, 2007).

4. *Decreased large woody debris.* Large woody debris provides stable substrate for aquatic organisms and is especially important in low gradient systems with relatively unstable bottom sediments (Benke and Wallace, 2003; Benke et al., 1984; Smock et al., 1989). Large woody debris helps retain macroinvertebrate food resources, such as leaves. Devegetation of riparian areas along Long Creek reduces inputs of woody debris, and impervious surfaces in the watershed intensify storm flows, which subsequently reduce woody debris retention.

5. *Increased sediment.* Suspended and deposited sediments affect aquatic biota by abrading gills, decreasing visibility, and filling habitat. Several studies have shown negative effects of sediment on EPT taxa (e.g., McClelland and Brusven, 1980). In contrast, certain noninsect taxa such as oligochaetes are tolerant of fine sediments (Zweig and Rabeni, 2001). Fine sediments (i.e., sediments <2 mm in diameter) come from eroded soils from the watershed or are mobilized from the banks and bed of the stream itself. Winter road treatments can also contribute sand.

6. *Increased temperature.* Increases in stream temperature lead to thermal stress resulting in increases in warm-water-tolerant taxa and decreases in taxa preferring colder waters, such as stoneflies (Lessard and Hayes, 2003). Stream water temperatures increase through several major pathways. For example, riparian devegetation allows more light to reach the water surface. Decreases in water velocity exacerbate this situation by increasing retention time and thus heat transfer to a given volume of water. Decreased base flow from groundwater inputs increases temperature because groundwater tends to be colder than surface waters during summer months. Finally, increased inputs of heated surface runoff (e.g., from impervious surfaces) raises stream water temperatures (Paul and Meyer, 2001).

7. *Increased toxic substances.* For the Long Creek watershed, increased toxic substances is a category that includes several individual stressors and subgroups, namely increased salts, various metals, and polycyclic aromatic hydrocarbons. Toxicity is expected to vary by chemical and organism, and organisms would be expected to be exposed to multiple toxic substances simultaneously. Point sources that could contribute toxic substances to Long Creek include industrial effluent discharges and storm water overflows. Nonpoint sources include runoff from impervious surfaces (e.g., oil and winter road treatments) and pesticides from lawn care and landscaping. Increased toxic substances were evaluated in terms of acute exposure (short duration linked to storm-flow conditions) and chronic exposure (long duration linked to base-flow conditions) for multiple toxic stressors (e.g., copper, lead, salts).

22.2.4 Data and Causal Assessment Methods

22.2.4.1 Data Sources and Measurements

The project team primarily used water-chemistry and biological-sampling data collected by MEDEP biologists within the Long Creek watershed beginning in the 1990s. The biological sampling protocols are described by Davies and Tsomides (2002). A variety of relevant data were available for most candidate causes at the study sites (see Table 22.3).

TABLE 22.3

Measurements Used in the Long Creek Case Study

Increased autochthony
  Aquatic vegetation
  Canopy shade
  Chlorophyll *a*
  Rapid bioassessment protocol (RBP) habitat score: riparian vegetative zone width
  Water chemistry, 2000 and 2001 storm flows: total phosphorous, ortho-phosphorous, total Kjeldahl nitrogen, nitrite, and nitrate
  Water chemistry, 2000 base flows: total phosphorous, ortho-phosphorous, total Kjeldahl nitrogen, nitrite, and nitrate
Decreased dissolved oxygen
  Canopy shade
  Chlorophyll *a*
  RBP scores: channel alteration and riparian vegetative zone width
  Water chemistry, 2000 and 2001 storm flows: total phosphorous, ortho-phosphorous, total Kjeldahl nitrogen, nitrate, and nitrite
  Water chemistry, 2000 base flows: total phosphorous, ortho-phosphorous, total Kjeldahl nitrogen, nitrate, and nitrite
  Water quality, 2000 base flow: dissolved oxygen
Altered flow regime
  Base-flow discharge
  Base-flow thalweg velocity
  Percent impervious surface
  RBP scores: channel alteration, channel sinuosity, and riparian vegetative zone width
  Storm flow, 1994 event
  Storm flow, 2001 event
Decreased large woody debris (LWD)
  LWD count
  RBP scores: channel alteration, channel sinuosity, and riparian vegetative zone width
Increased sediment
  Chlorophyll *a*
  Muck mud
  Pfankuch score (a measure of channel stability)
  Percent impervious surface
  RBP scores: epifaunal substrate, pool substrate, sediment deposition, channel alteration, channel sinuosity, riparian vegetative zone width, bank vegetation protection, and bank stability
  Sediment size
  Water chemistry, 1994 storm flow: total suspended solids (TSS)

continued

TABLE 22.3 (continued)

Measurements Used in the Long Creek Case Study

Water chemistry, 2000 and 2001 storm flows: TSS
Water chemistry, 2000 base flows: TSS

Increased temperature

Canopy shade
Percent impervious surface
RBP scores: channel alteration, channel sinuosity, and riparian vegetative zone width
Temperature: weekly minimum, maximum, and mean

Increased toxic substances

Sediment chemistry, 1993
Sediment chemistry, 2003
Sediment toxicity, 2003
Water chemistry, 1992 base flow: copper, lead, and zinc
Water chemistry, 1994 storm flow: copper, lead, and zinc
Water chemistry, 2000 and 2001 storm flows: cadmium, copper, lead, nickel, and zinc
Water chemistry, 2000 base flows: cadmium, chloride, copper, lead, and nickel
Water chemistry, 2000 storm flow PAHs
Water chemistry, 2001 storm flow PAHs
Water chemistry, 2003 low flow: aluminum, antimony, arsenic, barium, beryllium, cadmium, calcium, chromium, cobalt, copper, iron, lead, magnesium, manganese, molybdenum, nickel, selenium, silver, thallium, vanadium, and zinc
Water quality, 2000 base flow: specific conductivity and salinity

Note: Storm-flow water chemistry measurements were not taken at site LCMn 2.274, woody debris counts were not made at LCN .415, and flow regime was measured differently at the two sites.

22.2.4.2 *Causal Assessment Method*

The project team applied U.S. EPA's step-by-step Stressor Identification (SI) process to determine Long Creek's causes of biological impairment (U.S. EPA's CADDIS website describes the process; U.S. EPA, 2012d). Full details are provided in Ziegler et al. (2007a,b).

Six types of evidence were evaluated: (1) spatial/temporal co-occurrence, (2) causal pathway, (3) stressor–response associations from the case and nearby comparison sites, (4) laboratory tests of site media, (5) evidence of mechanism or mode of action, and (6) stressor–response associations from the laboratory or other field studies. The evidence was scored using the following system:

+++	Convincingly supports
– – –	Convincingly weakens
++	Strongly supports
– –	Strongly weakens
+	Somewhat supports
–	Somewhat weakens
0	Neither supports nor weakens
NE	No evidence

After each candidate cause was evaluated individually, possible interactions among supported causes were evaluated.

22.3 Evidence

22.3.1 Spatial/Temporal Co-Occurrence

The project team compared data from each affected site to comparison site data for each candidate cause to evaluate spatial/temporal co-occurrence (see Tables 22.4 and 22.5). For example, predawn dissolved oxygen measurements are lower at the Long Creek affected sites than at the comparison site, which resulted in a positive score, supporting spatial/temporal co-occurrence of decreased dissolved oxygen and the adverse effects. The project team compared samples collected on the same day and at similar times when possible; out-of-sync comparisons, for example, cross-year comparisons, were not used. Only data directly representing proximate causes (i.e., the candidate causes) were used as evidence for spatial/temporal co-occurrence. Data representing other steps in the causal pathway and surrogate measurements were considered under other types of evidence.

The project team did not discriminate between small and large measured differences among data for the purpose of scoring spatial/temporal co-occurrence. Even when the difference between the affected sites and the comparison site was small, the project team still considered this supporting evidence for the purpose of scoring.

At site LCN .415, most evidence of co-occurrence was judged as supporting or neutral for all candidate causes. The largest differences between LCN .415 and the Red Brook comparison site were observed for altered flow regime and some specific stressors grouped under toxic substances (e.g., salts). Several other individual toxic substances did not appear to be elevated at site LCN .415 compared with the comparison site, although data were sparse. No data were available to evaluate woody debris.

22.3.2 Causal Pathway

Comparisons between Red Brook (comparison site) and Long Creek site conditions were used to demonstrate evidence of interim steps in each causal pathway connecting human activities (e.g., land use) with proximate causes. For example, interim steps linking human activities with low dissolved oxygen include riparian devegetation that leads to decreased woody debris, thus reducing turbulence and aeration. Riparian devegetation could also lead to decreased dissolved oxygen by decreasing shade, thereby increasing algal photosynthesis and respiration, resulting in dissolved oxygen swings and

TABLE 22.4

Spatial/Temporal Co-Occurrence Evidence and Scores: LCN .415

Candidate Cause	Variable, Units	RB 3.961	LCN .415	Difference	Score	Comments
Increased autochthony	Dominant aquatic vegetation, estimated% of local reach	Diatoms 25%	Diatoms 25%	0%	0	The higher chlorophyll a measurement at LCN .415 was not considered strong enough to merit a positive score, given the similarity in dominant aquatic vegetation
	Chlorophyll a, mg/m²	10.4	15.7	51%	+	
Decreased dissolved oxygen	Dissolved oxygen, mg/L	8.7 [3] (8.0–9.5)	6.3 [3] (5.3–7.8)	−28%	+	
Altered flow regime	Base-flow discharge/watershed area, m³s/ac	0.000051 [2] (0.000050–0.00052)	0.000039 [2] (0.000024–0.000053)	−25%	+	The base-flow data suggest that less groundwater recharge may be occurring at the affected site, although this is based on two samples only. The four storm flow variables indicate that the affected site responds to storm runoff with flashier discharge than the comparison site (also see Ziegler et al., 2007a)
	Storm event peak discharge/watershed area, m³ s/ac	0.00024	0.0094	3752%		
	Storm event volume/watershed area, m³/ha	12.5	84.1	573%		
	Storm event duration, hours	25.4	5.5	−78%		
	Storm event time to peak discharge, hours	9.4	2.3	−76%		
Decreased large woody debris	No LWD data at affected site				NE	

Cause	Variable				Score	Note
Increased sediment	Base-flow TSS, mg/L	<10 [3] (<2–<10)	<10 [3] (3–<10)	≈0%	+	The positive score is based on storm flow TSS and the Rapid Bioassessment Protocol (RBP) sediment deposition score. The project team recognizes that other variables listed indicate similarity between the two sites, and the muck-mud variable indicates better conditions at the affected site. As such, the positive score is borderline and based on ambiguities, this cause could have been given a score of zero. We note that some TSS data did not meet Maine Department of Environmental Protection (MEDEP) quality standards
	Storm flow TSS, mg/L	<10–118 [9]	<10–271 [9]	130%		
	Muck-mud, %	60	40	–33%		
	RBP epifaunal substrate, score, and category	13 suboptimal	13 suboptimal	0%		
	RBP pool substrate, score, and category	10 marginal	10 marginal	0%		
	RBP sediment deposition, score, and category	18 optimal	11 suboptimal	LCN .415 worse than RB 3.961		
Increased temperature	Weekly minimum, °C	12.9 [3] (11.4–14.0)	16.3 [3] (15.4–17.3)	27%	+	
	Weekly maximum, °C	21.1 [3] (20.3–22.1)	22.7 [3] (21.6–24.2)	7%		
	Weekly mean, °C	16.7 [3] (16.1–17.4)	19.2 [3] (18.6–20.0)	15%		

continued

TABLE 22.4 (continued)

Spatial/Temporal Co-Occurrence Evidence and Scores: LCN .415

Increased toxic substances
Water column sampling (units in ppm or mg/L, except specific conductivity µS/cm):

Candidate Cause	Variable, Units	RB 3.961	LCN .415	Difference	Score	Comments
Salts	Base-flow chloride	29 [3] (26–30)	122 [3] (91–141)	324%	+	
	Storm flow chloride	17–57 [9]	15–296 [9]	419%		
	Low flow calcium	6.8	67	885%		
	Low flow magnesium	2.2	17	673%		
	Base-flow specific conductivity, µS/cm	129 [3] (79–155)	745 [3] (659–796)	476%		
Cadmium	Base flow	<0.0005 [3]	<0.0005 [3]	ND	+	The cadmium positive score is considered borderline; due to ambiguity, this could have been scored zero. Only one of nine storm samples at the affected site registered positive for cadmium (0.0007 ppm), and cadmium was not detected in any other measurement
	Low flow	<0.0002	<0.0002	ND		
	Storm flow	<0.0005 [9]	<0.0005–0.0007 [9]	>0%		
Copper	Base flow	<0.002 [3]	<0.002 [3]	ND	+	
	Low flow	Contaminated sample		NA		
	Storm flow	<0.002–0.003 [9]	0.002–0.018 [9]	500%		
Lead	Base flow	<0.003 [3]	<0.003 [3]	ND	+	
	Low flow	<0.0002	<0.0002	ND		

Nickel	Storm flow	<0.003–0.004 [9]	0.003–0.031 [9]	675%	+
	Base flow	<0.004 [3]	<0.004 [3]	ND	
	Low flow	0.00045	0.0032	611%	
	Storm flow	<0.004 [9]	<0.004–0.013 [9]	>0%	
Zinc	Base flow	<0.005 [3]	0.014 [3] (0.013–0.015)	>0%	+
	Low flow	<0.005	0.0064	>0%	
	Storm flow	0.008–0.024 [9]	0.043–0.14 [9]	483%	
Aluminum	Low flow	0.045	0.006	−87%	– –
Antimony	Low flow	<0.0005	<0.0005	ND	0
Arsenic	Low flow	<0.0005	0.00098	>0%	+
Barium	Low flow	0.0054	0.021	289%	+
Beryllium	Low flow	<0.0002	<0.0002	ND	0
Chromium	Low flow	<0.0005	0.0032	>0%	+
Cobalt	Low flow	0.00085	0.0029	241%	+
Iron	Low flow	0.091	0.14	54%	+
Manganese	Low flow	0.025	0.37	1380%	+
Molybdenum	Low flow	<0.0005	0.00092	>0%	+
Selenium	Low flow	<0.001	<0.001	ND	0
Silver	Low flow	<0.0002	<0.0002	ND	0
Thallium	Low flow	<0.0005	<0.0005	ND	0
Vanadium	Low flow	0.0003	0.00082	173%	+

continued

TABLE 22.4 (continued)

Spatial/Temporal Co-Occurrence Evidence and Scores: LCN .415

Candidate Cause	Variable, Units	RB 3.961	LCN .415	Difference	Score	Comments
PAHs water column sampling (ppm or mg/L):						
Acenaphthene	For all PAHs: Storm flow PAH samples are from two events, occurring on 23 October 2000 and 25 September 2001. Values and differences shown at right are separated by "and" to distinguish between the two storm events. PAHs were tested for but not detected for either storm event at the comparison site (surrogate site RB 1.694)		0.0001 and <0.0001	>0% and ND	+	
Acenaphthylene			<0.00005 and <0.0001	ND and ND	0	
Anthracene			0.0002 and <0.0001	>0% and ND	+	
Benzo(a) anthracene			0.0001 and 0.00033	>0% and >0%	+	
Benzo(a) pyrene			0.0001 and 0.00048	>0% and >0%	+	
Benzo(b) fluoranthene			0.0002 and 0.00111	>0% and >0%	+	
Benzo(ghi) perylene			0.0001 and 0.0005	>0% and >0%	+	
Benzo(k) fluoranthene			<0.00005 and 0.00029	ND and >0%	+	
Chrysene			0.0002 and 0.0008	>0% and >0%	+	
Dibenzo(a,h) anthracene			<0.00005 and 0.00011 (note that RL = 0.0002)	ND and ND	0	
Fluoranthene			0.0005 and 0.0016	>0% and >0%	+	

Fluorene	0.0001 and <0.0001	>0% and ND	+
Indeno (1,2,3-cd) pyrene	0.0001 and 0.00056	>0% and >0%	+
Naphthalene	0.0001 and <0.0001	>0% and ND	+
Phenanthrene	0.00025 and 0.00067	>0% and >0%	+
Pyrene	0.0003 and 0.00115	>0% and >0%	+

Sediment sampling (mg/kg): one sediment sample taken on 10 October 2003

Antimony	<10	<10	ND	0	
Arsenic	<20	<20	ND	0	
Barium	43	41	-5%	– – –	Toxicity tests were conducted using these sediment samples; for results and related information, see the stressor–response analysis
Beryllium	<1.0	<1.0	ND	0	
Cadmium	<3.0	<3.0	ND	0	
Chromium	6.3	18	186%	+	
Cobalt	5.5	8.2	49%	+	
Copper	3.2	6.3	97%	+	
Lead	14	13	-7%	– – –	
Nickel	<6.0	10	>0%	+	
Selenium	<10	<10	ND	0	
Silver	<3.0	<3.0	ND	0	
Thallium	<20	<20	ND	0	

continued

TABLE 22.4 (continued)

Spatial/Temporal Co-Occurrence Evidence and Scores: LCN .415

Candidate Cause	Variable, Units	RB 3.961	LCN .415	Difference	Score	Comments
Vanadium		9.2	24	161%	+	
Zinc		32	54	69%	+	

Note: Base and low flow values shown as mean [*n*] (range), where more than one value available, and storm flow values shown as range [*n*]. (Note that a range is provided for base flow only if a toxic substance is detected.)

Difference calculation: ND = not detected; NA = not applicable; Majority of differences are expressed as a percent = [(affected site value – comparison site value)/comparison site value] × 100%; Differences between RBP values are shown as greater or less than the comparison site value based on RBP qualitative condition categories (see further below); Differences between two ranges of values are calculated using the maximum values.

Lower thresholds of essential elements are not considered in this causal assessment.

RBP:

Habitat Parameter	Score and Condition Category
Epifaunal substrate/available cover	0–5 poor, 6–10 marginal, 11–15 suboptimal, 16–20 optimal
Pool substrate characterization	0–5 poor, 6–10 marginal, 11–15 suboptimal, 16–20 optimal
Sediment deposition	0–5 poor, 6–10 marginal, 11–15 suboptimal, 16–20 optimal

Evidence scoring system for spatial/temporal co-occurrence.

+ The effect occurs where or when the candidate cause occurs OR the effect does not occur where or when the candidate cause does not occur.

0 It is uncertain whether the candidate cause and the effect co-occur.

– – The effect does not occur where or when the candidate cause occurs OR the effect occurs where or when the candidate cause does not occur.

R The effect does not occur where and when the candidate cause occurs OR the effect occurs where or when the candidate cause does not occur and the evidence is indisputable.

NE No evidence.

TABLE 22.5

Spatial/Temporal Co-Occurrence Evidence and Scores: LCMn 2.274

Candidate Cause	Variable, Units	RB 3.961	LCMn 2.274	Difference	Score	Comments
Increased autochthony	Dominant aquatic vegetation, % of local reach	Diatoms 25%	Diatoms 20%	−20%	0	The higher diatom observation at the comparison site counteracts the higher chlorophyll *a* measurement at the affected site; uncertainty yielded a score of zero
	Chlorophyll *a*, mg/m²	10.4	17.5	68%		
Decreased dissolved oxygen	Base flow, mg/L	8.7 [3] (8.0–9.5)	5.5 [3] (4.4–6.2)	−37%	+	Dissolved oxygen data were collected approximately 10cm above the stream bottom; therefore, these values may be more applicable to fish habitat than benthic invertebrate habitat
Altered flow regime	Mean thalweg velocity, m/s	0.10	0.03	−73%	+	Flow regime differences between the two sites cannot be characterized by mean velocity alone. Flow heterogeneity adds support for a positive score (also see Ziegler et al., 2007a)
	Base-flow velocity measured at 2m increments along 100m reach in site vicinity	Highly variable longitudinal channel velocity and normally above zero	Low longitudinal velocity variability and often equal to zero	Qualitative support of cause		
Decreased large woody debris	LWD diameter ≥5cm, # of pieces	91	43	−53%	+	
	LWD diameter ≥10cm, # of pieces	39	12	−69%		

continued

TABLE 22.5 (continued)

Spatial/Temporal Co-Occurrence Evidence and Scores: LCMn 2.274

Candidate Cause	Variable, Units	RB 3.961	LCMn 2.274	Difference	Score	Comments
Increased sediment	Base-flow TSS, mg/L	<10 [3] (<2–<10)	<10 [3] (4–<10)	≈0%	—	
	Muck-mud, %	60	40	−33%		
	RBP epifaunal substrate, score, and category	13 suboptimal	12 suboptimal	0%		The difference in muck-mud between the two sites does not provide enough evidence over the other variables, which are essentially equal at both sites; therefore, this was scored minus. Note that some TSS data did not meet MEDEP quality standards
	RBP pool substrate, score and category	10 marginal	8 marginal	0%		
	RBP sediment deposition, score, and category	18 optimal	18 optimal	0%		
Increased temperature	Weekly minimum, °C	13.1	13.2	0%	+	The weekly maximum is higher, and this value may be more important that the others (see stressor–response analysis); therefore, while the minimum and mean values for the two sites are relatively similar, a positive score was still given
	Weekly maximum, °C	20.3	21.8	8%		
	Weekly mean, °C	16.6	16.5	0%		

Increased toxic substances

Water column sampling (units in ppm or mg/L, except specific conductivity µS/cm):

Salts	Base-flow chloride	29 [3] (26–30)	66 [3] (58–73)	128%	+	
	Low flow calcium	6.8	31.5	363%		
	Low flow magnesium	2.2	11	400%		
	Base-flow specific conductivity, µS/cm	129 [3] (79–155)	459 [3] (376–510)	256%		

continued

Cadmium	Base flow	<0.0005 [3]	<0.0005 [3]	ND	0
	Low flow	<0.0002	<0.0002	ND	0
Copper	Base flow	<0.002 [3]	0.0013 [3]	ND	0
	Low flow	contaminated sample	(<0.002–0.002)	NA	
Lead	Base flow	<0.003 [3]	<0.003 [3]	ND	0
	Low flow	<0.0002	<0.0002	ND	0
Nickel	Base flow	<0.004 [3]	<0.004 [3]	ND	+
	Low flow	0.00045	0.0019	322%	
Zinc	Base flow	<0.005	0.0042 [3] (<0.005–0.005)	ND	0
	Low flow	<0.005	<0.005	ND	
Aluminum	Low flow	0.045	0.019	−58%	− −
Antimony		<0.0005	<0.0005	ND	0
Arsenic		<0.0005	0.00235	>0%	+
Barium		0.0054	0.011	104%	+
Beryllium		<0.0002	<0.0002	ND	0
Chromium		<0.0005	0.00205	>0%	+
Cobalt		0.00085	0.000555	−35%	− −
Iron		0.091	0.22	142%	+
Manganese		0.025	0.092	268%	+
Molybdenum		<0.0005	0.000385	ND	0
Selenium		<0.001	<0.001	ND	0
Silver		<0.0002	<0.0002	ND	0

TABLE 22.5 (continued)

Spatial/Temporal Co-Occurrence Evidence and Scores: LCMn 2.274

Candidate Cause	Variable, Units	RB 3.961	LCMn 2.274	Difference	Score	Comments
Thallium		<0.0005	<0.0005	ND	0	
Vanadium		0.0003	0.000695	132%	+	
Sediment sampling (mg/kg): one sediment sample taken on 10/10/2003						
Antimony		<10	<10	ND	0	Toxicity tests were conducted using these sediment samples; for results and related information, see the stressor–response analysis
Arsenic		<20	<20	ND	0	
Barium		43	38	-12%	- -	
Beryllium		<1.0	<1.0	ND	0	
Cadmium		<3.0	<3.0	ND	0	
Chromium		6.3	16.5	162%	+	
Cobalt		5.5	6.95	26%	+	
Copper		3.2	6.25	95%	+	
Lead		14	7.5	-46%	- -	
Nickel		<6.0	10.5	>0%	+	
Selenium		<10	<10	ND	0	
Silver		<3.0	<3.0	ND	0	
Thallium		<20	<20	ND	0	

| Vanadium | 9.2 | 20 | 117% | + |
| Zinc | 32 | 58.5 | 83% | + |

Note: Base and low flow values shown as mean [n] (range), where more than one value available, and storm flow values shown as range [n].
(Note that a range is provided for base flow only if a toxic substance is detected.)

Difference calculation: ND=not detected; NA=not applicable; Majority of differences are expressed as a percent=[(affected site value–comparison site value)/comparison site value]×100%; Differences between RBP values are shown as greater or less than the comparison site value based on RBP qualitative condition categories (see further below); Differences between two ranges of values are calculated using the maximum values.

Lower thresholds of essential elements are not considered in this causal assessment.

RBP:

Habitat Parameter	Score and Condition Category
Epifaunal substrate/available cover	0–5 poor, 6–10 marginal, 11–15 suboptimal, 16–20 optimal
Pool substrate characterization	0–5 poor, 6–10 marginal, 11–15 suboptimal, 16–20 optimal
Sediment deposition	0–5 poor, 6–10 marginal, 11–15 suboptimal, 16–20 optimal

Evidence scoring system for spatial/temporal co-occurrence.

+ The effect occurs where or when the candidate cause occurs OR the effect does not occur where or when the candidate cause does not occur.

0 It is uncertain whether the candidate cause and the candidate cause co-occur.

— The effect does not occur where or when the candidate cause occurs OR the effect occurs where or when the candidate cause does not occur.

R The effect does not occur where and when the candidate cause occurs OR the effect occurs where or when the candidate cause does not occur, and the evidence is indisputable.

NE No evidence.

low predawn measurements. Using detailed conceptual model diagrams as a guide (expanding on the pathways shown in Figure 22.2), the project team found supporting evidence for at least some causal pathway steps across all candidate causes at both affected sites. Because causal pathway evidence did not discriminate among candidate causes, it is not presented here. Full documentation is available in Ziegler et al. (2007a).

22.3.3 Covariation of the Stressor and the Effect from the Affected Sites and Nearby Comparison Sites

The project team developed study-wide scatterplots to assess whether the magnitudes of effects increase or decrease with increasing or decreasing stressor exposure. Figure 22.3 provides example scatterplots, illustrating biological endpoints as a function of dissolved oxygen. The nine sites shown in Figure 22.1 were used for this analysis; these sites included LCN .415, LCMn 2.274, and RB 3.961. Sample size was not sufficient to make judgments about individual sites or stream reaches, nor were data sufficient to use a multivariate modeling approach. Rather, the project team sought only to characterize bivariate, study-wide trends, when possible. The project team interpreted the scatterplots by looking for linear and curvilinear trends in the data. The team supplemented visual interpretation with statistical correlation coefficients.

At least one of the effect endpoints, EPT taxa richness, percent noninsects or HBI, was judged to covary with the following candidate causes: increased autochthony, decreased dissolved oxygen, decreased large woody debris, increased temperature, and increased salts. Specific conductivity and chloride (indicative of increased salts; see also Ziegler et al., 2007b) were the only two variables for which the project team visually interpreted an association for all three biological endpoints of interest. As specific conductivity and

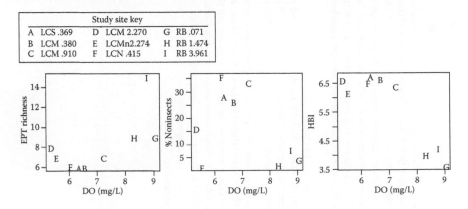

FIGURE 22.3
Example scatter plots, illustrating biological endpoints as a function of mean base flow DO concentration (mg/L). This evidence was judged to support decreased DO as the cause of HBI increases, but was ambiguous for EPT richness and % noninsects.

chloride increase throughout the two study watersheds, EPT taxa richness decreased while percent noninsects and HBI values increased.

22.3.4 Laboratory Tests of Site Media

Sediment samples were used to assess whether sediment toxicity—a subcategory of increased toxic substances—might play a role in the biological impairment. Samples were taken from the Red Brook comparison site, LCN .415 and LCMn 2.274 and tested in the laboratory for toxicity to chironomids (*Chironomus dilutus*) and amphipods (*Hyallela azteca*) (U.S. EPA, 2004a). No statistically significant differences were found in chironomid survivorship among the two affected sites tested, the comparison site and the laboratory control (U.S. EPA, 2004a). Amphipod survival was significantly lower at the Red Brook site and LCN .415 compared to the laboratory control; however, survival at LCMn 2.274 was similar to the control.

Interpretation of these results is not straightforward. The project team used chironomids (*C. dilutus*) and amphipods (*H. azteca*) as surrogates for EPT and noninsects, respectively. However, chironomids are generally thought to be more tolerant of environmental degradation than most EPT species (this was taken into consideration when scoring). In addition, survival of *H. azteca* would be expected to decrease if chemicals in the sediment were toxic, but the proportion of noninsects increased at LCN .415. Finally, laboratory tests were conducted in controlled environments and only tested survivorship, unlike conditions found in the field where more factors often interact and affect organism occurrence. For the above reasons, this category of evidence was judged to be ambiguous, neither clearly supporting nor weakening the case for sediment toxicity.

22.3.5 Evidence of Mechanism or Mode of Action

Our understanding of how each cause produces effects was evaluated for consistency with observed changes in EPT richness, percent noninsects, and HBI. When available, data that reflect mechanisms underlying the biological changes were included in the evaluation.

Mechanistic information argued both for and against increased autochthony. For EPT taxa richness, the project team expected that the food resource changes associated with increased autochthony would increase the abundance of scraping taxa and decrease shredding taxa at affected sites relative to the comparison site. However, the data do not reflect this pattern: the highest relative abundance of scrapers is seen at the comparison site, and the functional feeding group analysis does not show a clear pattern among the sites for percentage of shredders (see Table 22.6). Thus, the project team judged that the argument for increased autochthony decreasing EPT richness was not supported. Increased abundance of snails (noninsects) is often associated with increased autochthony, and data show that snails were

TABLE 22.6

Functional Feeding and Mode of Existence Groups

	Study Site Percent		
Group	RB 3.961	LCN .415	LCMn 2.274
Functional Feeding			
Filterers	18.3	1.1	10.3
Gatherers	18.6	16.5	71.8
Predators	36.6	30.9	13.1
Scrapers	17.2	13.3	1.4
Shredders	7.8	34.6	2.7
Mode of Existence			
Burrower-sprawlers	57.9	38.3	19.2
Swimmers	5.3	22.9	2.1
Clingers	6.4	5.9	70.1
Climbers	27.7	18.6	7.6

found at LCN .415 but not at the comparison site; therefore, the argument for increased autochthony increasing the proportion of noninsects was supported. The argument for autochthony leading to increased HBI was also judged to be supported, as HBI was originally designed to assess low dissolved oxygen caused by organic loading, and excess autochthony can lead to increased organic carbon.

Mechanistic information supported the argument for decreased dissolved oxygen for all endpoints at both sites. Low dissolved oxygen levels can cause asphyxiation in EPT taxa and relative increases in tolerant noninsect taxa. HBI would be expected to increase, as this index was originally designed to assess low dissolved oxygen caused by organic loading.

The argument for altered flow regime was also supported for all three biological endpoints for both sites. The project team focused on lower day-to-day base-flow conditions (a component of hydrologic flashiness) as the specific candidate cause for this type of evidence. Some EPT taxa prefer running water habitats and are found on substrate surfaces in riffles. Conversion of higher water velocity areas into lower flow areas reduces lotic habitat. Certain noninsect taxa (e.g., oligochaetes and snails) are tolerant of lentic conditions; similarly, several taxa with high HBI tolerance values (e.g., some chironomids and oligochaetes) are less reliant on lotic conditions.

Large woody debris provides habitat and cover for EPT taxa, supporting the argument for this cause. Caddisflies were observed using submerged woody debris as habitat in the Long Creek watershed. The mechanisms by which large woody debris would increase the proportion of noninsects or HBI values were unknown.

For increased sediment, the project team expected to observe increases in suspended sediment leading to decreases in abundance of filter-feeding taxa, many of which are trichopterans. The percent of filter-feeding taxa was highest at the comparison site, which supports the argument that increased suspended sediment decreased EPT richness at both affected sites. Noninsect taxa such as oligochaetes often increase in abundance with increasing fine sediments. The project team could not identify a mechanism by which increased HBI values tracked increased sediment levels. However, HBI is a complex endpoint (in the context of causal assessment; see also Box 22.1), and such tracking may happen as an artifact of the taxa found at the affected sites. That is, if one or two taxa with a high HBI tolerance value also respond to increased sediment, then HBI would follow. Zweig and Rabeni (2001) indicate that HBI may be insensitive to increases in deposited sediments and that traits associated with susceptibility to organic enrichment (as related to HBI) are often not related to traits associated with sediment deposition. The project team scored this as ambiguous owing to the uncertain relationship between the HBI endpoint and sediment.

Plecoptera taxa have lower temperature optima than many other groups of stream invertebrates (Galli and Dubose, 1990; Lessard and Hayes, 2003), supporting the argument that increased temperatures could have reduced EPT richness endpoints for both affected sites. The team scored this line of evidence in support of increased temperature as a potential stressor, but with low confidence, as only one stonefly was found at the Red Brook comparison site. The team could not find mechanistic information associating increased temperature with increases in the percentage of noninsects or HBI values.

Increased toxic substances were judged to be a mechanistically plausible cause for decreased EPT richness. The project team assumed EPT richness is likely to decline in the presence of increased toxic substances, based on the sensitivity of mayflies to metals. HBI and noninsects were both scored 0 (zero) because the project team did not find documentation of the relative group- or index-level sensitivities to toxic substances.

22.3.6 Stressor–Response Relationships from the Laboratory and Other Field Studies

Observed levels of stressors were compared with available benchmarks derived from, or found in, the literature. In general, if observed levels were below benchmarks considered to be protective, then the evidence was judged to argue against a given candidate cause. Alternatively, observed levels above benchmark values were judged to support a candidate cause. The biological endpoints and the degree of protectiveness used to derive the benchmark were considered when making comparisons.

22.3.6.1 Autochthony

The benchmarks used to evaluate autochthony and their comparison to site conditions are shown in Table 22.7. For LCN .415, base-flow total phosphorus is in the range where eutrophication might be seen, and the mean was twice that of the regional comparison value. The chlorophyll *a* site observation is approximately one order of magnitude less than benchmark values found in the literature. Total nitrogen levels at the affected sites fall below the level for eutrophication risk, and all nitrogen measures are relatively close to the regional comparison condition. Although phosphorus values argue for the cause, chlorophyll *a* and nitrogen values argue against it. Support for this cause was judged to be ambiguous, because both phosphorus and nitrogen would be expected to cause effects through increased chlorophyll *a*.

At LCMn 2.274, the chlorophyll *a* site observation is approximately one order of magnitude less than benchmark values found in the literature. Total nitrogen and phosphorus site levels fall under the level for eutrophy risk, and all nitrogen and phosphorus measures are relatively close to the regional comparison condition values. These values were considered to weaken the case that autochthony caused decreased EPT richness and increased HBI values.

22.3.6.2 Dissolved Oxygen

The U.S. EPA criterion continuous concentration (CCC) was compared with measured dissolved oxygen values (see Table 22.8). At LCN .415 and LCMn 2.274, the minimum measured dissolved oxygen value (5.3 and 4.4 mg/L, respectively) and the range of measured levels all were less than the U.S. EPA criterion (8.0 mg/L for early life stages in cold water). The project team considered these data supporting evidence for decreased dissolved oxygen reducing EPT taxa and increasing HBI, but was unsure how noninsects would respond to these dissolved oxygen levels.

22.3.6.3 Altered Flow

The project team chose to use impervious surface area as a surrogate measure for altered flow regime in the context of this type of evidence (see Table 22.9). The use of impervious surface area as a surrogate for increased hydrologic flashiness allowed the team to take advantage of endpoint-specific stressor–response data from other studies. At 6% impervious surface area, Morse et al. (2003) reported an abrupt decrease in EPT species with an increase in gastropods. Impervious surfaces were estimated at 32.6 and 14.3% of the watersheds contributing to sites LCN .415 and LCMn 2.274, respectively. Both values are sufficient to expect a decrease in EPT species and increase in gastropods. However, because impervious surfaces are antecedent of many candidate causes, this evidence may also reflect the influence of other candidate causes such as increased temperature and toxic substances.

TABLE 22.7

Stressor–Response Relationships and Site Data Used to Evaluate Increased Autochthony

Variable, Units	Stressor–Response Benchmark Description	Value	RB 3.961	LCN .415	LCMn 2.274
Chlorophyll *a*, mg/m²	Reduced invertebrate diversity (Nordin, 1985)	100	10.4	15.7	17.5
	Eutrophy risk range (U.S. EPA, 2000c, summary document)	100–200			
Base-flow total nitrogen, ppm	U.S. EPA reference for ecoregion XIV, 59, northeastern coastal zone (U.S. EPA, 2000d)	0.57	0.310 [3] (0.280–0.350)	0.617 [3] (0.610–0.620)	0.457 [3] (0.330–0.540)
	Eutrophy risk range (U.S. EPA, 2000c, summary document)	1.5			
Base-flow total Kjeldahl nitrogen, ppm	U.S. EPA reference for ecoregion XIV, 59, northeastern coastal zone (U.S. EPA, 2000d)	0.30	0.167 [3] (0.100–0.200)	0.300 [3] (0.300–0.300)	0.400 [3] (0.300–0.500)
Base-flow nitrate + nitrite, ppm	U.S. EPA reference for ecoregion XIV, 59, northeastern coastal zone (U.S. EPA, 2000d)	0.31	0.143 [3] (0.150–<0.20)	0.317 [3] (0.310–0.320)	<0.057 [3] (0.03–<0.20)
Base-flow total phosphorus, ppm	Deleterious effects on fish communities (Miltner and Rankin, 1998)	0.06	0.009 [3] (0.008–0.010)	0.048 [3] (0.040–0.061)	0.030 [3] (0.024–0.035)
	U.S. EPA reference for ecoregion XIV, 59, northeastern coastal zone (U.S. EPA, 2000d)	0.024			
	Eutrophy risk range (U.S. EPA, 2000c, summary document)	0.035–0.075			

Note: Base and low flow values shown as mean [*n*] (range), where more than one value available, and storm flow values shown as range [*n*].

TABLE 22.8

Stressor–Response Relationships and Site Data Used to Evaluate Decreased
Dissolved Oxygen

Variable, Units	Stressor–Response Benchmark			RB 3.961	LCN .415	LCMn 2.274
	Description	Value				
Minimum dissolved oxygen, mg/L	U.S. EPA (1986b) cold freshwater aquatic life criteria	8.0		8.0 [3] (8.0–9.5)	5.3 [3] (5.3–7.8)	4.4 [3] (4.4–6.2)
	30-day LC$_{50}$ values for four different EPT organisms (Nebeker, 1972)	4.4–5.0 [4]				

Note: Base and low flow values shown as mean [n] (range), where more than one value available, and storm flow values shown as range [n].

22.3.6.4 Large Woody Debris

No quantitative benchmarks were found that relate decreased large woody
debris to aquatic invertebrates. However, Benke et al. (1984) reported that
aquatic invertebrate productivity was 3–4 times higher for submerged
wooden substrates or snags than for sandy or muddy benthic habitats. In
addition, debris dam abundance has been positively correlated with macroinvertebrate abundance and relative abundance of shredders (Smock et al.,
1989; applies to EPT taxa). The project team considered this supporting evidence for the observed effects on EPT taxa at LCMn 2.274 (LWD was not
measured at LCN .415). It is not known how HBI values or percent noninsects
would respond to site-specific levels of woody debris.

TABLE 22.9

Stressor–Response Relationships and Site Data Used to Evaluate Altered
Flow Regime

Variable, Units	Stressor–Response Benchmark		RB 3.961	LCN .415	LCMn 2.274
	Description	Value			
Impervious surface area,%	Abrupt decline in taxonomic richness, specifically EPT, and an increase in non-insects, specifically gastropods (Morse et al., 2003)	6	2.1	32.6	14.3
	Shift to tolerant species (Maxted, 1996)	10–15			

Note: % impervious surface area also may be a surrogate for other proximate causes, such as increased temperature and toxic substances.

22.3.6.5 Sediment

No published stressor–response associations were found that relate bedded sediments to macroinvertebrate endpoints in sandy streams. However, benchmarks are available that relate TSS to aquatic invertebrate mortality (see Table 22.10). Storm-flow TSS measurements were compared with 24-h duration exposures. Exposure to 53–92 mg/L TSS caused decreased invertebrate populations (Gammon, 1970, as cited in Newcombe and MacDonald, 1991). Storm-flow measurements of TSS at site LCN .415 ranged from below detection to 271 mg/L. Because of the variability in the measurements, the evidence was judged ambiguous for EPT taxa richness. It is not known how HBI values or percent noninsects would respond to these TSS levels (also see discussion on sediment and HBI in Section 22.3.5: Evidence of mechanism or mode of action and Box 22.1).

22.3.6.6 Temperature

A literature review by Galli and Dubose (1990) indicates that the optimum temperature for EPT species is generally <17°C (see Table 22.11). Temperatures corresponding to 50% mortality (LC_{50}) for the mayfly species *Baetis rhodani*, *Baetis tenax*, and *Caenis* sp. are 21.1, 21.3, and 26.7°C, respectively (Galli and Dubose, 1990).

TABLE 22.10

Stressor–Response Relationships and Site Data Used to Evaluate Increased Suspended Sediment

Variable, Units	Stressor–Response Benchmark		RB 3.961	LCN .415	LCMn 2.274
	Description	Value			
Base-flow TSS, mg/L	Exposure causing 40–60% aquatic invertebrate mortality and severe habitat degradation at greater than 1000 h duration (similar to base-flow condition) (mean [*n*] (range), from literature review, Newcombe and MacDonald, 1991)	33 [4 studies] (8–77)	<10 [3] (<2–<10)	<10 [3] (3–<10)	<10 [3] (4–<10)
Storm flow TSS, mg/L	Exposure causes decreased invertebrate population at approximately 24 h duration (similar to storm event) (Gammon, 1970, as cited in Newcombe and MacDonald, 1991)	53–92	<10–118 [9]	<10–271 [9]	NE

Note: Base and low flow values shown as mean [*n*] (range), where more than one value available, and storm flow values shown as range [*n*].

TABLE 22.11

Stressor–Response Relationships and Site Data Used to Evaluate Increased
Temperature

Variable, Units	Stressor–Response Benchmark		RB 3.961	LCN .415	LCMn 2.274
	Description	Value			
Temperature, weekly maximum, °C	Severe stress to most cold-water organisms (literature review by Galli and Dubose, 1990)	21	21.1 [3] (20.3 – 22.1)	22.7 [3] (21.6 – 24.2)	21.8
	Generalized optimum for EPT (literature review by Galli and Dubose, 1990)	<17			
	50% mortality for *B. rhodani, B. tenax,* and *Caenis* sp. (Ephemeroptera), respectively (literature review by Galli and Dubose, 1990)	21.1, 21.3, and 26.7			

Note: Base and low flow values shown as mean [*n*] (range), where more than one value available, and storm flow values shown as range [*n*].

At LCN .415, the mean weekly maximum temperature exceeds most—and the range exceeds all—stressor–response benchmark values, except the LC_{50} for *Caenis* sp. The stressor–response evidence supports the case for increased temperature decreasing EPT richness. It is unclear how percent noninsects and the HBI might respond to the site's temperatures.

At LCMn 2.274, the mean weekly maximum temperature exceeds most of the stressor–response benchmark values listed. The site's second-most dominant organism, *Caenis* sp., is tolerant of high temperatures. In contrast, Caenidae were not found at the comparison site. However, this evidence is weakened somewhat by the fact that comparison site temperatures also were equal to, or exceeded, temperature benchmarks except for the *Caenis* value.

22.3.6.7 Toxic Substances

Concentrations of toxic substances measured during base-flow and two storm-flow events (only near site LCN .415) were compared with available benchmark values and/or species sensitivity distributions developed specifically for this project (see Tables 22.12 and 22.13). Benchmarks were

TABLE 22.12

Stressor–Response Relationships and Site Data Used to Evaluate Increased Toxic Substances

Variable (ppm or mg/L)	Flow	Stressor–Response Benchmark		RB 3.961	LCN .415	LCMn 2.274
		Description	Value			
Salts, water column sampling						
Calcium	Low	Not available		6.8	67	31.5
Chloride	Base	U.S. EPA CCC	230	29 [3] (26–30)	122 [3] (91–141)	66 [3] (58–73)
	Storm	U.S. EPA criterion maximum concentration (CMC)	860	17–57 [9]	15–296 [9]	NE
Magnesium	Low	Not available		2.2	17	11
Salinity	Base	Range of 48-h LC_{50} values for various salt combinations tested on *Ceriodaphnia* (Mount et al., 1997)	250–5700	67 [3] (0–100)	367 [3] (300–400)	200 [3]
Specific conductivity, µS/cm	Base	EPT effects based on interpretation of statewide data from Maine (Davies et al., unpublished), Florida (Florida Department of Environmental Protection, 2005), and Kentucky (Pond, 2004), see also Ziegler et al., 2007b)	100–200	129 [3] (79–155)	745 [3] (659–796)	459 [3] (376–510)
		Presence of amphipods in place of EPT might be indicative of higher conductivity, as amphipods are more tolerant (Kefford et al., 2003)	NA			
Elements, water column sampling						
Aluminum	Low	U.S. EPA CCC	0.087	0.045	0.006	0.019

continued

TABLE 22.12 (continued)

Stressor–Response Relationships and Site Data Used to Evaluate Increased Toxic Substances

Variable (ppm or mg/L)	Flow	Stressor–Response Benchmark		RB 3.961	LCN .415	LCMn 2.274
		Description	Value			
Arsenic	Low	Invertebrate SSD, LC_{50} for 10% of species	0.66	<0.0005	0.00098	0.00235
		U.S. EPA CCC	0.15			
Cadmium	Base	U.S. EPA CCC	0.00025	<0.0005 [3]	<0.0005 [3]	<0.0005 [3]
	Low	Invertebrate SSD, LC_{50} for 10% of species	0.032	<0.0002	<0.0002	<0.0002
	Storm	U.S. EPA CMC	0.002	<0.0005 [9]	<0.0005–0.0007 [9]	NE
Chromium	Low	Invertebrate SSD, LC_{50} for 10% of species	0.17	<0.0005	0.0032	0.00205
Copper	Base	Invertebrate SSD, LC_{50} for 10% of species	0.008	<0.002 [3]	<0.002 [3]	0.0013 [3] (<0.002–0.002)
		U.S. EPA CCC	0.009			
	Storm	Invertebrate SSD, LC_{50} for 10% of species	0.013	<0.002–0.003 [9]	0.002–0.018 [9]	NE
		U.S. EPA CMC	0.013			
Iron	Low	U.S. EPA CCC	1.0	0.091	0.14	0.22
Lead	Base	U.S. EPA CCC	0.0025	<0.003 [3]	<0.003 [3]	<0.003 [3]
	Low			<0.0002	<0.0002	<0.0002
	Storm	U.S. EPA CMC	0.065	<0.003–0.004 [9]	0.003–0.031 [9]	NE
Nickel	Base	Invertebrate SSD, LC_{50} for 10% of species	0.61	<0.004 [3]	<0.004 [3]	<0.004 [3]

Metal	Flow	Benchmark	Benchmark value			
	Base	Chordate SSD, LC$_{50}$ for 10% of species	2.9	0.0045	0.0032	0.0019
	Base	U.S. EPA CCC	0.052			
	Low	*Use base-flow values immediately above*				
	Storm	Invertebrate SSD, LC$_{50}$ for 10% of species	1.9	<0.004 [9]	<0.004–0.013 [9]	NE
	Storm	Chordate SSD, LC$_{50}$ for 10% of species	6.2			
	Storm	U.S. EPA CMC	0.47			
Selenium	Low	U.S. EPA CCC	0.005	<0.001	<0.001	<0.001
Zinc	Base	Invertebrate SSD, LC$_{50}$ for 10% of species	0.087	<0.005 [3]	0.014 [3] (0.013–0.015)	0.0042 [3] (<0.005–0.005)
	Low	U.S. EPA CCC	0.12	<0.005	0.0064	<0.005
	Low	*Use base-flow values immediately above*				
	Storm	Invertebrate SSD, LC$_{50}$ for 10% of species	0.45	0.008–0.024 [9]	0.043–0.14 [9]	NE
	Storm	U.S. EPA CMC	0.12			
Antimony	Low	NA		<0.0005	<0.0005	<0.0005
Barium	Low	NA		0.0054	0.021	0.011
Beryllium	Low	NA		<0.0032	<0.0002	<0.0002
Cobalt	Low	NA		0.00085	0.0029	0.000555
Manganese	Low	NA		0.025	0.37	0.092
Molybdenum	Low	NA		<0.0005	0.00092	0.000385
Silver	Low	NA		<0.0002	<0.0002	<0.0002
Thallium	Low	NA		<0.0005	<0.0005	<0.0005
Vanadium	Low	NA		0.0003	0.00082	0.000695

Note: All values displayed in ppm or mg/L, unless otherwise noted. Base and low flow values shown as mean [n] (range), where more than one value available, and storm flow values shown as range [n]. Note that a range is provided for base flow only if a toxic substance is detected.

U.S. EPA CCC and CMC from U.S. EPA, 2004b.

NE = no evidence.

NA: no stressor-response benchmarks were available

TABLE 22.13

Stressor–Response Evidence and Scores for Increased PAHs at LCN 415

PAH (Units are µg/mL)	Storm, 10/18/00 1.1" over 21 hours			Storm, 9/25/01 1.7" over 24 hours			U.S. EPA, 2000e[a]	U.S. EPA, 1986a (Goldbook)	Eisler, 1987	ECOTOX[b]	Sub-Score	British Columbia, 1993	Sub-Score	Canada, 2013 (Fresh Water Aquatic Life)	Sub-Score	Score
	LCN .585	RB 1.694	RL	LCN .585	RB 1.694	RL										
Acenaphthene	0.10	ND	0.05	ND	ND	0.1	Human consumption values only	≤1700 and 520 acute and chronic freshwater, respectively	Not listed	1280 (tox)	–	6 chronic	–	5.8	–	–
Acenaphthylene	ND	ND	0.05	ND	ND	0.1	Listed, but no benchmarks	Not listed	Not listed	Not listed	NE	Not listed	NE	No data	NE	NE
Anthracene	0.20	ND	0.05	ND	ND	0.1	Human consumption values only	Not listed	Info not pertinent	95 (tox)	–	4 chronic and 0.1 phototox	+	0.012	+	0
Benzo(a) anthracene	0.10	ND	0.05	0.33	ND	0.1	Human consumption values only	Not listed	LC87 (6 month) bluegill at 1000	Not listed	–	0.1 chronic and phototox	+	0.018	+	0
Benzo(a) pyrene	0.10	ND	0.05	0.48	ND	0.1	Human consumption values only	Not listed	LC50 (96 h) sandworm at >1000	Not listed	–	0.01 chronic	+	0.015	+	0
Benzo(b) fluoranthene	0.20	ND	0.05	1.11	ND	0.1	Human consumption values only	Not listed	Info not pertinent, but listed many times	Not listed	NE	Not listed, but pending ref to 0.01 (NPS, 1997)	+	Not listed	+	0
Benzo(ghi) perylene	0.10	ND	0.05	0.5	ND	0.2	Listed, but no benchmarks	Not listed	LC50 (96 h) sandworm at >1000	Not listed	–	Not listed	NE	Not listed	NE	0

Benzo(k)fluoranthene	0.20	ND	0.05	0.29	ND	0.1	Human consumption values only	Not listed	Not listed	Not listed	NE	Not listed	Not listed, but pending ref to 0.01 (NPS, 1997)	+	0
Chrysene	0.05	ND	0.05	0.8	ND	0.1	Human consumption values only	Not listed	LC_{50} (96 h) sandworm at >1000	Not listed	–	Insufficient data	Insufficient data	NE	0
Dibenzo (a,h) anthracene	ND	ND	0.05	0.11	ND	0.2	Human consumption values only	Not listed	Not listed	Not listed	–	Not listed	Not listed	NE	0
Fluoranthene	0.50	ND	0.05	1.6	ND	0.1	Human consumption values only	≤3980 acute freshwater, and ≤40 and 16 acute and chronic, respectively, saltwater	LC_{50} (96 h) sandworm 500	38 (feeding) – 200 (tox)	–	4 chronic and 0.2 phototox	0.04	+	0
Fluorene	0.10	ND	0.05	ND	ND	0.1	Human consumption values only	Not listed	LC_{50}s (various orgs, 96 h) 320–5600	Not listed	–	12 chronic	3.0	–	– –
Indeno (1,2,3-cd) pyrene	0.10	ND	0.05	0.56	ND	0.2	Human consumption values only	Not listed	Not listed	Not listed	NE	Not listed	Not listed	NE	NE
Naphthalene	0.10	ND	0.05	ND	ND	0.1	Listed, but no benchmarks	≤2300 and 620 acute and chronic freshwater, respectively	LC_{50} (10 day) copepod at 50, and LC_{50}s (various orgs, 24–96 h) 920–150,000	1700 (behavior) ~5700 (physiology)	–	1 chronic	1.1	–	– –

continued

TABLE 22.13 (continued)

Stressor–Response Evidence and Scores for Increased PAHs at LCN .415

PAH (Units are μg/mL)	Storm, 10/18/00 1.1" over 21 hours			Storm, 9/25/01 1.7" over 24 hours			U.S. EPA, 2000e[a]	U.S. EPA, 1986a (Goldbook)	Eisler, 1987	ECOTOX[b]	Sub-Score	British Columbia, 1993	Canada, 2013 (Fresh Water Aquatic Life)	Sub-Score	Score
	LCN .585	RB 1.694	RL	LCN .585	RB 1.694	RL									
Phenanthrene	0.25	ND	0.05	0.67	ND	0.1	Listed, but no values	Not listed	LC$_{50}$ (24 h) grass shrimp at 370, and LC$_{50}$ (96 h) sandworm 600	340 (tox)	−	0.3 chronic	0.4	+	0
Pyrene	0.30	ND	0.05	1.15	ND	0.1	Human consumption values only	Not listed		1020 (tox)	−	0.02 phototox	0.025	+	0

Note: Evidence scoring system for plausible effect given stressor–response relationships

++ *The observed relationship between exposure and effects in the case agrees quantitatively with stressor–response relationships in controlled laboratory experiments or from other field studies.*

+ The observed relationship between exposure and effects in the case agrees qualitatively with stressor–response relationships in controlled laboratory experiments or from other field studies.

0 The agreement between the observed relationship between exposure and effects in the case and stressor–response relationships in controlled laboratory experiments or from other field studies is ambiguous.

− The observed relationship between exposure and effects in the case does not agree with stressor–response relationships in controlled laboratory experiments or from other field studies.

−− The observed relationship between exposure and effects in the case does not qualitatively agree with stressor–response relationships in controlled laboratory experiments or from other field studies or the quantitative differences are very large.

NE no evidence.

a U.S. EPA (2000c) lists only human health consumption criteria for some PAHs.
b ECOTOX database (U.S. EPA, 2014c).

unavailable for antimony, barium, beryllium, cobalt, manganese, molybdenum, silver, thallium, and vanadium.

Salts: At both sites, levels of specific conductivity were within a range in which effects have been observed by MEDEP personnel and corroborated by observations from Florida and Kentucky (FDEP, 2005; Pond, 2004). Further supporting evidence comes from the most dominant species at LCN .415, a salt-tolerant amphipod (Kefford et al., 2003). Increased abundances of salt-tolerant amphipods and isopods were not observed at site LCMn 2.274. At both sites, chloride concentrations fell below U.S. EPA water quality criteria (CCC and CMC). Salinity values were within the range of LC_{50} values for *Ceriodaphnia dubia* at site LCN .415, but below that range at site LCMn 2.274. The project team judged this evidence to argue for salts causing decreased EPT taxa at both sites and increased noninsects at site LCN .415. The implications of this evidence for HBI results are unclear (also see discussion on sediment and HBI in Section 22.3.5: Evidence of mechanism or mode of action, and Box 22.1).

Other toxic substances: Concentrations of all other measured toxic substances were below benchmark values with a few exceptions. The base-flow reporting limits for cadmium and lead at both sites are greater than corresponding U.S. EPA chronic criteria (CCC); therefore, while cadmium and lead were not detected in base-flow samples, they could still exceed CCC values. At LCN .415, one of nine storm samples at the affected site registered positive for cadmium (0.0007 ppm). Copper, in one of nine storm events, exceeded the invertebrate SSD 10% threshold (Figure 22.4) and the EPA acute criterion (CMC), and another one of nine equaled those two criteria; this adds supporting evidence to the EPT endpoint, but evidence for noninsects and HBI is unclear. Storm flow measurements were not taken at site LCMn 2.274.

22.4 Conclusions

The CADDIS system of scoring was used to summarize the results obtained from the different pieces of evidence. After each candidate cause was evaluated individually, possible interactions among supported causes were evaluated.

22.4.1 Likely Causes

Evidence supported the involvement of most of the candidate causes evaluated (listed here in no particular order):

- Decreased dissolved oxygen
- Altered flow regime
- Decreased large woody debris

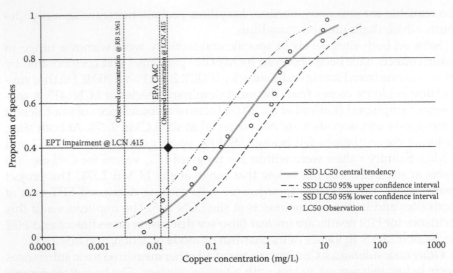

FIGURE 22.4
Species sensitivity distribution plot comparing laboratory LC$_{50}$ values for freshwater invertebrates (points) to storm-flow copper concentrations at site LCN .415. The proportional decrease
in EPT taxa richness (i.e., 40% lower than the comparison site) intersects with observed storm
flow copper concentrations at a point (the solid diamond) outside the lower confidence interval
of the SSD. This evidence indicates that copper concentrations are insufficient alone to account
for the reduction in EPT taxa. Test results were obtained from the ECOTOX database (U.S. EPA,
2014c) and selected for site-appropriate water hardness, pH, and temperature. U.S. EPA's acute
criterion (CMC) also shown.

- Increased temperature
- Increased salts

These conclusions are presented by site and biological endpoint in Table
22.14. The summary of scores for each type of evidence is shown in Tables
22.15 and 22.16, for LCN .415 and LCMn 2.274, respectively. The project team
ranked probable causes in order of importance for each site and endpoint.
Evidence scores (i.e., ++ , +, 0, and –) weighed heavily in the team's consideration for ordering the causes, and additionally, it is within this table that
the project team employed a degree of professional judgment, based on the
entire analysis and all available information.

22.4.2 Unsupported and Uncertain Causes

A few of the candidate causes were concluded to be unlikely contributors.
Stressor–response associations argued against increased autochthony at
LCMn 2.274; specifically, the level of nutrients did not exceed benchmarks,
and so was not considered sufficient to have produced the effects. This finding was consistent with the functional feeding group analysis, which did not

TABLE 22.14

Likely Causes of Effects

Affected Site	Biological Effect		
	Decreased EPT Generic Richness	Increased% Non-Insect Taxa Individuals, Relative to Total Macroinvertebrate Abundance	Increased Hilsenhoff Biotic Index (HBI) Score
Long Creek, northern branch (LCN .415)	Increased salts, Altered flow regime, Decreased dissolved oxygen, Increased temperature, Decreased large woody debris	Increased salts, Altered flow regime	Decreased dissolved oxygen, Altered flow regime
Long Creek, northern main branch (LCMn 2.274)	Decreased dissolved oxygen Increased temperature Increased salts, Decreased large woody debris, Altered flow regime	Net assessed: % non-insect taxa were not more abundant relative to the Red Brook comparison site (RB 3.961)	Decreased dissolved oxygen Altered flow regime

Note: Likely causes are listed in order, from highest to lowest importance as judged by project team, within each cell.

support the increased autochthony hypothesis, because the relative abundance of scrapers and filterers did not increase relative to the comparison site. Increased sediment was concluded to be an unlikely contributor: neither base-flow levels of suspended solids nor substrate habitat scores differed between LCMn 2.274 and the comparison site. In addition, there was little evidence that toxic substances—other than salts—caused the decreased EPT and increased HBI at LCMn 2.274.

22.4.3 Interactions

Streams in urban watersheds are often subject to multiple, interacting causes of adverse biological effects. The direct combined effects of multiple stressors on organisms may be independent (effects additive), concentration additive, or synergistic—resulting in a greater than additive effect. In addition, stressors may have indirect combined effects through interactions in the causal pathway.

For example, decreased base flow and water depth, two common manifestations of altered flow regime, may directly reduce suitable habitat for some organisms. Decreased base flow may also reduce turbulence, decreasing dissolved oxygen. Decreased water depth may increase water temperature, because the temperature of shallow water rises more quickly than that of deep water, and increase metabolic rates in organisms. Subsequently, higher metabolic rates may increase demand for dissolved oxygen, while decreased turbulence decreases availability of dissolved oxygen. Dissolved oxygen is

TABLE 22.15

Summary of Evidence Scores: LCN .415

Biological Endpoint Candidate cause	Types of Evidence That Use Data From The Case			Types of Evidence That Use Data From Elsewhere		
	Spatial/Temporal Co-Occurrence	Covariation of the Stressor and Effect	Causal Pathway	Evidence of Mechanism or Mode of Action	Stressor–Response Relationships From Laboratory or Other Field Studies	Body of Evidence
EPT richness						
Increased autochthony	0	0	+	0	0	0
Decreased dissolved oxygen	+	0	+	+	+	++
Altered flow regime	+	NE	+	+	+	++
Decreased large woody debris	NE	+	+	+	NE	+
Increased sediment	+	0	+	+	0	+
Increased temperature	+	+	+	+	+	+
Increased salts	+	+	+	+	+	++
% non-insects						
Increased autochthony	0	0	+	+	0	0
Decreased dissolved oxygen	+	0	+	+	0	+
Altered flow regime	+	NE	+	+	+	++
Decreased large woody debris	NE	0	+	+	NE	0
Increased sediment	+	0	+	+	0	+
Increased temperature	+	0	+	0	0	+
Increased salts	+	+	+	0	+	++
HBI						
Increased autochthony	0	+	+	+	0	+
Decreased dissolved oxygen	+	+	+	+	+	++
Altered flow regime	+	NE	+	+	0	+
Decreased large woody debris	NE	+	+	+	NE	+
Increased sediment	+	0	+	0	0	+
Increased temperature	+	+	+	0	0	+
Increased salts	+	+	+	0	0	+

Note: NE = No evidence.

TABLE 22.16

Summary of Evidence Scores: LCMn 2.274

	Types of Evidence That Use Data From The Case			Types of Evidence That Use Data From Elsewhere		
Biological Endpoint[a] Candidate cause	Spatial/Temporal Co-Occurrence	Covariation of the Stressor and Effect	Causal Pathway	Evidence of Mechanism or Mode of Action	Stressor–Response Relationships From Laboratory or Other Field Studies	Body of Evidence
EPT richness[a]						
Increased autochthony	0	0	+	0	–	–
Decreased dissolved oxygen	+	0	+	+	+	+ +
Altered flow regime	+	NE	+	+	+	+
Decreased large woody debris	+	+	+	+	+	+ +
Increased sediment	–	0	+	+	0	–
Increased temperature	+	+	+	+	+	+ +
Increased salts	+	+	+	+	+	+ +
HBI						
Increased autochthony	0	+	+	+	–	–
Decreased dissolved oxygen	+	+	+	+	+	+ +
Altered flow regime	+	NE	+	+	0	+
Decreased large woody debris	+	+	+	0	NE	+
Increased sediment	–	0	+	0	0	–
Increased temperature	+	+	+	0	0	+
Increased salts	+	+	+	0	0	+

Note: NE = No evidence.

[a] The% non-insects biological endpoint was not assessed at site LCMn 2.274. See text for more information.

less soluble at both higher temperatures and higher salinity levels, and when salinity climbs above favorable levels, sensitive species will spend more metabolic activity on ionic regulation, thereby limiting energy normally dedicated to other processes.

In Long Creek, low dissolved oxygen may be interacting with low base flow and increased temperature in ways that produce more severe effects than would be expected if these stressors acted independently. In the context of dam modifications and flow management, Bednarek and Hart (2005) report that "the combined influence of flow and DO could be non-additive, which would further complicate efforts to evaluate their individual effects."

As dissolved oxygen decreases, invertebrates with passive gill respiration (e.g., mayflies such as *Baetis* and *Rhithrogena*) require flowing water to move oxygen over their gills (Jaag and Ambühl, 1964). Low base-flow velocities and decreased dissolved oxygen at the affected Long Creek sites may be sufficient to cause problems for these types of invertebrates. Similarly, higher temperatures increase many coldwater organisms' investment in respiratory processes, increasing their susceptibility to low dissolved oxygen. Allan (1995) describes this interaction and makes specific mention of caddisflies being susceptible. Both Long Creek sites have decreased dissolved oxygen and temperatures described by Allan as stressful. It is likely that interactions among these two causes and low flow is making effects more severe.

22.5 Management Implications

The results of the causal assessment are being used to inform ongoing watershed-level activity, including discharge permitting and evidence-based sustainability efforts (e.g., MEDEP, 2009; LCWMD, 2014).

Urban development can lead to many sources of stress for aquatic systems. Two sources, in particular, altered channels (e.g., straightening) and impervious surfaces produce stressors that likely cause of biological impairment in Long Creek. About 33% of LCN .415 watershed is covered by impervious surfaces including a portion of Portland's airport, portions of two semiconductor manufacturing plants, major roadways, extensive retail development, and a soft drink bottling plant. The LCMn 2.274 site includes "pond-like habitat," likely resulting from both historically altered channels and impervious surfaces (largely office parks and roadways) covering approximately 14% of the contributing watershed. Decreasing the watershed's percent impervious cover (e.g., through building and/or landscaping design changes and best management practices) would increase base flow and thereby potentially reduce the effects caused jointly by low dissolved oxygen and low flow velocity. Additionally, decreased storm-flow velocities may increase retention of woody debris, providing habitat for invertebrates and increasing aeration.

23

Clear Fork Watershed Case Study: The Value of State Monitoring Programs

Lei Zheng, Jeroen Gerritsen, and Susan M. Cormier

This case study illustrates an approach for watershed-wide causal assessment that has several advantages for making analysis efficient, consistent, and defensible. Models of common causes of adverse biological effects were developed using West Virginia's monitoring data, so multiple streams could be easily assessed relative to regional expectations. The case study also illustrates the use of four causal characteristics, antecedence, co-occurrence, sufficiency, and alteration, in providing a useful framework for both reaching causal conclusions and communicating them.

CONTENTS

23.1 Summary

The Clear Fork of the Coal River, West Virginia, USA, and its tributaries are impaired based on the state's aquatic life index. Where practical, candidate causes were defined in terms of the sources as well as the proximate causes to support future management action. Evidence was derived for four causal characteristics: antecedence, co-occurrence, sufficiency, and alteration. First, unlikely candidate causes were eliminated from further consideration based on lack of antecedents (primarily lack of sources) and lack of spatial co-occurrence compared to state-wide reference values. For the remaining candidate causes, evidence of sufficiency and alteration was scored for each cause. Stressor–response threshold values from state-wide data analyses were used to assess whether a candidate cause occurred at a sufficient intensity to cause biological effects. A nonmetric multidimensional scaling model of community composition was used to assess similarity of a site's community composition with those of communities with known causes. Field notes were also consulted for other evidence or explanations of any inconsistencies. Likely causes were identified as those demonstrating the four evaluated causal characteristics. A candidate cause was rejected by demonstrating the lack of one or more causal characteristics.

Likely causes differed among the tributaries in the watershed and the contribution of all these causes was evident in the Clear Fork mainstem, which exhibited some resiliency due to dilution and different geophysical attributes. Causes identified in the various reaches and tributaries included metal contamination and acidification from mine drainage, aluminum toxicity in association with low pH, increased bicarbonate, sulfate and chloride salts from surface coal mining, sediment deposition, organic enrichment from sewage and from algal productivity enhanced by nutrients, and low dissolved oxygen.

23.2 Problem Formulation

23.2.1 Geography and Geology

Clear Fork and its tributaries drain a rugged ecoregion that is primarily forested with surface and underground coal mining higher in the mountains and towns and small-scale livestock farms scattered along the flood plains (Woods et al., 1999). The Clear Fork watershed is entirely within the Cumberland Mountains (Level IV, Ecoregion 69d) (Woods et al., 1996) (see Figure 23.1). The crests of the mountains are 366–1097 m above sea level with steep slopes terminating in narrow valleys 107–168 m below the crests. The bedrock is sandstone, siltstone, shale, and coal. Mostly mixed mesophytic forests grow on well-drained soils of low to moderate fertility (Woods et al., 1999).

23.2.2 Ecological Effects

23.2.2.1 Measure of Biological Effect: Multimetric Benthic Invertebrate Index

The biological condition in Clear Fork did not meet West Virginia's narrative water quality criteria for aquatic life. A multimetric index based on

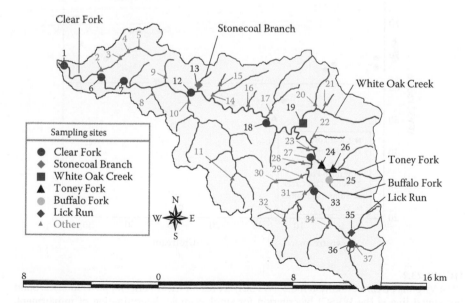

FIGURE 23.1
Clear Fork Watershed Showing Numbered Sampling Sites and Several Tributaries (West Virginia Department of Environmental Protection). Clear Fork flows northwesterly from the head waters near Clear Fork sampling site 36 to sampling site 1 near its confluence with the Little Coal River.

> **BOX 23.1   A SPECIFIC BIOLOGICAL EFFECT WAS
> NOT IDENTIFIED IN THIS CASE STUDY**
>
> As discussed in Box 22.1 for Long Creek, an aggregated measurement
> like a biotic index makes it difficult to distinguish biological effects
> among the several affected sites and may lead to a conclusion that the
> same effect and cause occurs at each location. To avoid this potential
> misconception, it is better to identify the biological effect more specifi-
> cally. In the Clear Fork case study, the lack of specificity in defining an
> effect at the outset was compensated for by multivariate analyses link-
> ing genera with specific causes.

family-level taxonomic identification was used to assess condition. The
maximum West Virginia Stream Condition Index (WVSCI) score is 100, with
impaired waters having a score <60.6 based on a one-time measurement and
<68 based on multiple samples (see also Box 23.1).

The WVSCI scores for Clear Fork ranged from about 55 to about 75 (see
Figure 23.2). Although both the headwaters and the mouth of Clear Fork

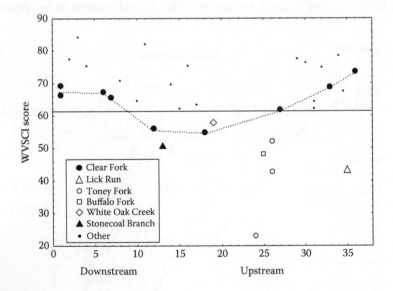

FIGURE 23.2
West Virginia Stream Condition Index (WVSCI) scores for Clear Fork and its tributaries. Solid
horizontal line is the WVSCI biocriterion for single-sample determination of impairment.
Symbols indicate sampled streams. Large symbols are sites assessed in this chapter; small
symbols indicate sites not assessed in this chapter. Locations are in order relative to the Clear
Fork mainstem from the river mouth (left) to furthest upstream location (site #36) on the right.
Tributary sampling location is the location of the confluence with Clear Fork. Dotted line connects Clear Fork sites for easier viewing.

scored above 68.0, mid-reach sections of the stream scored below 60.6. Several tributary streams had WVSCI scores in the "fair" range (60.6–68.0). The five other tributaries (Lick Run, Toney Fork, Buffalo Fork, White Oak Creek, and Stonecoal Branch) were selected for causal assessments because they scored below 60.6 and were distributed along the mainstem and therefore could affect the mainstem of Clear Fork. They are also highlighted in this chapter to demonstrate how causal assessments can discriminate among a variety of stressors.

23.2.3 Potential Sources and Candidate Causes

Coal, oil, and natural gas extraction have contributed to the degradation of streams in the Clear Fork watershed. Coal mining, primarily in the form of mountaintop removal, is the major industry in this region. This method of surface mining is commonly used to unearth coal seams once considered too thin to be mined. As a byproduct of this process, excess rock (overburden) is placed into headwater valleys, creating valley fills. These large landscape alterations significantly affect stream hydrology and morphology, and otherwise stress the receiving streams (U.S. EPA, 2011a). Residential land-use also affects streams in the Clear Fork watershed. Residential inputs include organic and nutrient enrichments via discharges from improperly sewered homes and failed septic systems. Agriculture also occurs throughout the watershed but has declined since the 1950s.

The candidate causes of an altered benthic macroinvertebrate assemblage as seen in the affected tributaries of Clear Fork are listed below and shown in Figure 23.3. High pH from ammonia and surface mining are not shown in the conceptual model because pH >9 and ammonia levels >0.5 mg/L did not occur.

1. Low pH/high pH from acid mine drainage, acid mine drainage treatment with ammonia, and surface mining
2. Dissolved metals causing direct toxicity or formation of flocs and embedded substrates
3. Dissolved minerals (dominated by calcium, magnesium, bicarbonate, and sulfate salts) from coal mining
4. Suspended sediment from erosion
5. Altered habitat by sedimentation, algal growth, or stream channel alterations
6. Increased temperature from forest removal
7. Low dissolved oxygen from organic enrichment
8. Organic enrichment, including organic matter or algae that alter the quantity and quality of food
9. Ammonia toxicity from human and animal waste

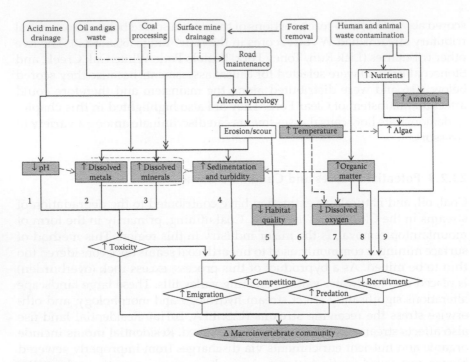

FIGURE 23.3
Conceptual model of sources and stressors in the Coal River Watershed, WV. Sources are listed in upper oblongs; stressors appear as rectangles with grey rectangles being proximate stressors; mode of action is in diamonds; and the effect is in the gray oval. Interacting stressors are indicated by dashed arrows. Candidate causes are numbered (1) through (9).

23.2.3.1 Acidity, Metals, and Minerals

Mining of high-sulfur coal exposes iron sulfide minerals, such as pyrite (FeS_2), to oxidation by chemosynthetic bacteria, forming sulfuric acid. Acidic water is toxic on its own, but it also dissolves minerals and metals and increases their solubility in water. In addition, Central Appalachia is exposed to atmospheric acidic deposition from coal-burning power plants and other sources. Consequently, poor geological buffering capacity and sources of acid from both atmospheric acidic deposition and acid mine drainage (AMD) often cause acidification of streams (DeNicola and Stapleton, 2002).

The metals most associated with acid mine drainage are aluminum, iron, and manganese. They are all toxic to invertebrates and fish when present in acidic waters as dissolved ions. As acid drainage mixes with uncontaminated water and the pH rises, the metals precipitate. Soluble iron, aluminum, and manganese do not occur above pH levels of 4, 5, and 8, respectively. Flocs and other precipitates are abrasive and smother invertebrates where they occur at sufficient levels (Diz, 1997).

Present-day coal mine operations are required to treat AMD by adding alkaline materials or anhydrous ammonia. The treatment neutralizes acidity and causes potentially toxic metals to precipitate from solution, but leaves soluble salts in solution. The salts generally are less toxic than the metals that are precipitated but can be present at concentrations high enough to adversely affect aquatic biota. In addition, surface mining produces large quantities of unweathered, crushed rock that demineralizes with contact with water from precipitation raising specific conductivity in some streams to 10 – 1000× background for the region. Treated and untreated mine drainage is typically very high in bicarbonate, sulfate, calcium, and magnesium ions. Organisms that are adapted to low-conductivity waters may be particularly vulnerable to increased levels of these ions (Koel and Peterka, 1995; Cormier et al., 2013a, b; U.S. EPA, 2011c).

23.2.3.2 Excess Sediment and Habitat Alteration

Fine sediment is a common stream pollutant that results from agriculture, logging, mining, road construction, and urbanization (Henley et al., 2000; Walsh et al., 2005; Waters, 1995). Studies of adverse effects of increased sediment on aquatic life in streams (see summaries in Cormier et al., 2008; Waters, 1995) have shown that aquatic invertebrates decrease in abundance and that the benthic macroinvertebrate assemblage changes taxonomically (Wood and Armitage, 1997). Deposited sediments reduce the amount of habitat available to benthic invertebrates by filling interstitial spaces between boulder, cobble, and gravel substrates. Many stream-dwelling aquatic animals deposit their eggs in gravel or on cobble substrates. When substrates are buried under fine sediment, egg mortality increases due to reduced availability of DO. Suspended sediments reduce water transparency and lower rates of primary production by aquatic plants which are food for animals (Relyea et al., 2000; Vannote et al., 1980). Suspended particles damage and clog the delicate gill structures of aquatic organisms (Wood and Armitage, 1997). Stressed by excess sediment, aquatic invertebrates emigrate by drifting which removes them from the system and puts them at greater risk from predation (Shaw and Richardson, 2001).

Bank stability, one of the habitat assessment measurements used by the West Virginia Department of Environmental Protection (WVDEP), is an estimate of the erosion potential of a streambank (Barbour et al., 1999). It is affected by armoring, bank vegetation cover, excessive stream energy, and the long-term stability of the stream valley. Increased bank erosion may lead to extensive habitat degradation, including embeddedness, scour, habitat instability, and reduced habitat availability for both fish and macroinvertebrates (Allan and Castillo, 2007; Cummins, 1974; Hynes, 1970).

Measurements of vegetative bank protection primarily indicate whether vegetation cover is sufficient to stabilize the stream bank and shade streams.

In Central Appalachia, industrial development and residential use cause increased flow variability which causes erosion during high flows and drying during low flows. In contrast, surface and underground mining change the annual flow regimes from highly variable to fairly stable. Although such a flow regime augments stream flow during low-flow conditions and provides additional habitat for some species, flow stabilization may be particularly harmful to organisms that are well adapted to variable flow and related thermal regimes.

23.2.3.3 Temperature

Water temperatures greater than 30°C are uncommon in the region. Increased temperatures in the region tend to be associated with decreased shading, which occurs on mine sites, at sediment ponds, and on residential and farmed lands. Increased temperature increases the metabolism and the food requirements of aquatic animals. It also reduces the solubility of oxygen.

23.2.3.4 Organic Enrichment, Nutrients, Ammonia, and Low Dissolved Oxygen

Organic enrichment, nutrients, and ammonia change the food base from biofilms on leaf litter to organic waste and algae by direct input of organic matter or as nutrients for algal growth. Sources of organic matter and nutrients to Clear Fork include sewage discharges, animal wastes, runoff from fertilized fields and lawns (WVDEP, 2006), and atmospheric deposition of nitrogen. Some homes in the Clear Fork watershed still lack satisfactory sewer systems, and even properly designed septic systems may fail, releasing organic wastes.

In addition to promoting algal growth, ammonia is toxic at sufficient concentrations. The relative amounts of NH_4^+ and unionized NH_3, is pH-dependent (Strumm and Morgan, 1996). Higher pH values drive the reaction towards greater concentrations of the unionized (more toxic) form (U.S. EPA, 2002). In eutrophic waters, relatively low concentrations of ammonia may episodically increase to toxic levels due to elevated water temperatures and photosynthesis-driven increases in pH. During warm and sunny periods, photosynthesis reduces carbon dioxide (CO_2) and bicarbonate ion concentrations, which increases pH (Wetzel, 2001). Alkaline pH contributes to the formation of unionized ammonia. When this occurs, toxicity increases, at least until levels of dissolved CO_2 are re-established as photosynthesis declines. In the dark, ammonium is nitrified by bacteria which consume oxygen in the process (e.g., Rysgaard et al., 1994). The addition of bicarbonates from mining also raises pH.

Direct organic enrichment and algal growth not only alters the food base, and therefore, the types of organisms that can exploit it, but also increases decomposition, which lowers the concentrations of DO. If sufficient light is available, nutrient enrichment increases algal growth. Greater levels of algae result in greater production of oxygen during periods of photosynthesis and greater consumption of oxygen during periods dominated by respiration. Thus, DO declines to levels that are stressful to biota at night, when photosynthesis ceases but respiration continues (e.g., Hynes, 1960, 1970).

23.2.3.5 Causes Not Considered

Pesticides, oil, and other hydrocarbons were not assessed because no measurements were available. Pathways associated with high pH were not assessed because pH >9 was not reported in the watershed.

23.2.4 Methods

23.2.4.1 Physical, Biological, and Chemical Data from the Study Sites and State-Wide

All data used in this assessment were collected by the West Virginia Department of Environmental Protection. Statistical analyses were performed by Tetra Tech, Inc. using state-wide data from West Virginia's Water Analysis Database (WABbase; WVDEP, 2008a). Some analyses used data from selected ecoregions within the state and are noted with the analyses. The WABbase contains data from a mixed sampling program that collects measurements from long-term monitoring stations, targeted sites within watersheds, randomly selected sample sites (Smithson, 2007), and sites chosen to further define impaired stream segments and plan ways to reduce inputs. Most sites were sampled once during an annual sampling period, but the case study sites were sampled monthly for water quality. The data set contains water quality, habitat, watershed characteristics, macroinvertebrate data (both raw data and calculated metrics), and geographic location (WVDEP, 2008a). The WABbase includes assignment of reference status using a tiered approach. Analyses involved 107 reference sites drawn from the Level 1 reference status (WVDEP, 2008b) which selects reference sites that "are thought to represent the characteristics of stream reaches that are least disturbed by human activities and are used to define attainable chemical, biological, and habitat conditions for a region" (WVDEP, 2013).

Macroinvertebrate data are based on collections from a total area of 1 m² in a 100-m reach at each site (WVDEP, 2008b).

WVSCI scores are calculated from the family composition and relative abundances of benthic macroinvertebrates using the average of six standardized metrics (Tetra Tech, Inc., 2000): (1) total Taxa Richness (the number of distinct taxa); (2) total EPT (number of taxa within the orders Ephemeroptera [mayflies], Plecoptera [stoneflies], or Trichoptera [caddisflies]); (3) % EPT (percentage of individuals that are in the orders *Ephemeroptera, Plecoptera,* or *Trichoptera*); (4) % Chironomidae (percentage of individuals in the family that includes true midges); (5) two dominant taxa (the cumulative percentage of individuals within the two numerically dominant taxa); and (6) the Hilsenhoff Family Biotic Index (HBI) (Plafkin et al.,1989). Determination of biological impairment using the WVSCI is based on dissimilarity (fifth quantile) from 107 high-quality reference sites in the WABbase.

23.2.4.2 Pollutant Sources and Source Tracking

GIS-based reports developed by WVDEP were used to identify possible pollutant sources, including permitted outlets (mining and nonmining), permitted mining areas, valley fills, abandoned mine lands, oil and gas wells, history of managed forest land, water quality sampling locations, roads, weather stations, U.S. GS gauging stations, towns, streams defined in the NHD, subwatershed delineation, and land use.

Potential sources of stressors in subwatersheds were checked by WVDEP by walking all NHD-designated stream reaches listed as impaired and documenting and photographing potential sources of pollution (point sources, nonpoint sources, and general riparian condition and activities). This "ground-truthing" located undocumented abandoned mine drains and other discharges, livestock, and unmapped streams and roads.

23.2.4.3 Measurements of Candidate Causes

Proximate causes and antecedents are listed with associated measurements and thresholds in Table 23.1. The concentrations of three metals were assessed: aluminum, iron, and manganese (as dissolved or total in mg/L). The total amount of dissolved minerals (i.e., major ions) was measured as specific conductivity. Sulfate and chloride were also measured because they are related to distinct sources of high specific conductivity. Various measurements of suspended and deposited sediment and habitat quality are listed in Table 23.1. Because the time of day that DO was sampled was not always clear and DO has a diurnal cycle, these measurements were interpreted with caution. Because nutrients were not sampled regularly in the Clear Fork stream system, and phosphorus detection limits (most commonly 0.10 mg/L) were too high to determine background or reference phosphorus concentration, nutrient levels could not be directly assessed within the causal pathway leading to habitat or food-base alteration. Biological oxygen demand was only rarely measured but was used when available. Fecal coliform levels were sampled regularly throughout the watershed and used with caution as a surrogate for organic enrichment.

23.2.4.4 Causal Assessment Process

Likely and contributing causes were identified using the method described in this book with some modification (see Figure 23.4). Evidence was analyzed in a step-wise fashion followed by a weight-of-evidence analysis that considered all the evidence. The State of West Virginia preferred to use categories rather than plus and minus scores to weight evidence of sufficiency (see Table 23.1). Weakening evidence is equivalent to a minus, weakly plausible is equivalent to ambiguous, plausible is similar to a single plus sign, and substantial and sustained are similar to two plus signs. A scoring table developed for this book is provided in the conclusions for illustrative purposes.

TABLE 23.1

General Scoring Table for Evaluating Co-occurrence and Sufficiency

Candidate Cause	Stressor Measures	Measured in Lick Run			Thresholds from State-Wide Data Set				
		Min	Median	Max	Weakening Evidence (−) Min < Ref	Weakly Plausible (0) Min > Thresh. Med < Thresh.	Plausible S–R (+) Med >Thresh.	Substantial Effects (++) Med >Thresh.	Sustained Effects (++) Min >Thresh.
1. Acidity/ alkalinity	pH	7.63	8.01	8.54	Min > 6.5	Min < 6.5	Med ≤ 4	Med >Thresh.	Max < 4
	pH	7.63	8.01	8.54	Max < 9	Max > 9	Med > 9		Min > 9
2. Metal toxicity	Al (dissolved)	0.020	0.050	0.200	Max < 0.18	Max > 0.18	Med > 0.2	Med > 0.4	Min > 0.4
	Fe (total)	0.090	0.470	19.700	Max < 0.8	Indeter. no S–R			
	Mn (total)	0.070	0.320	2.110	Max < 0.05	Min > 0.05	Med > 0.05		Min > 0.05
3. Salt	Specific conductivity (µS/cm)	263	475	804	Max < 180	Max > 180	Med > 180	Med > 300	Min > 300
	SO$_4$	70	106	223	Max < 43	Max > 43	Med > 43		
	Chloride		ND		Max < 10	Min > 10	Med > 10	Med > 17	Min > 17
4. Sedimentation and turbidity	TSS	3	7	528	Max < 7	Indeter. no S–R			
	% fines (SSC)		35		Max ≤ 30%	Max > 30%	Med > 30%		Min > 30%
	RBP embeddedness		5		≥13		<13	<9	
	RBP sediment		2		≥11		<11	<8	
	RBP bank stability		12		≥13		<13	<12	
5. Habitat quality	RBP total score		85		≥147	<147	<140	<130	
	RBP: channel alteration		7		≥16		<10		

continued

TABLE 23.1 (continued)

General Scoring Table for Evaluating Co-occurrence and Sufficiency

Candidate Cause	Stressor Measures	Measured in Lick Run			Thresholds from State-Wide Data Set				
					Weakening Evidence (−)	Weakly Plausible (0)	Plausible S–R (+)	Substantial Effects (++)	Sustained Effects (++)
		Min	Median	Max	Min < Ref	Min > Thresh. Med < Thresh.	Med > Thresh.	Med > Thresh.	Min > Thresh.
5. Habitat (continued)	RBP: cover		10		≥15	<15 ≥10	<10	No subst. effects	
	RBP: riparian vegetation		6		≥14		<10	No subst. effects	
6. Temperature (direct)	Temperature	3.95	13.03	24.94	Max < 30.6				
7. Low DO	DO	8.53	10.93	13.20	Min ≥ 5.0		Min < 5.0	Min < 4.0	
8. Organic enrichment	Fecal coliform	2	42	950	Max < 250	Max > 250	Med > 250	Med > 500	Min > 500
	Algae obs.	Algae low to high			Few observed		Moderate	High	
	NO₃		ND		Max < 0.6	Indeter. no S–R			
	TKN		ND		Max < 1.7				
	TP		ND		Max < 0.04	Indeter. no S–R			
9. Ammonia toxicity	mg/L		ND		Max ≤ 0.5		Max > 0.5		

Note: Thresholds are listed in the five right-hand columns. Data for a specific location is entered in columns 3–5, in this example, for Lick Run. Gray highlighted cells in threshold columns indicate score for Lick Run determined by comparing measurement (columns 3–5) with thresholds. Scores used to illustrate scoring are shown in parentheses after text threshold descriptors. ND, no data; Max, maximum; Min, minimum; Indeter., Indeterminate; RBP, rapid bioassessment protocol; SSC, suspended sediment concentration; DO, dissolved oxygen; S–R, stressor–response; Thresh., threshold; subst., substantial.

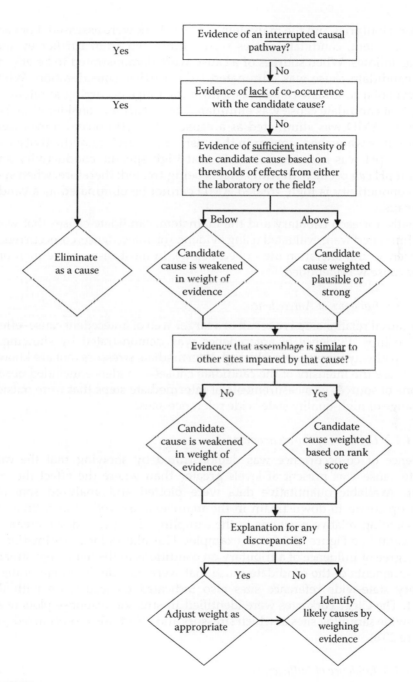

FIGURE 23.4
The process used to identify causes in this case study.

Four tributaries and the mainstem of Clear Fork were assessed. For each subassessment, candidate causes were eliminated from further evaluation as follows. When sources of a cause were demonstrated to be absent, that candidate cause was eliminated from further consideration. When the level of a candidate cause was within the range occurring at reference sites, that candidate cause was eliminated from further consideration. For example, AMD was eliminated as a cause when: (1) no coal mines were present in a watershed, now or in the past, or (2) specific conductivity was low and pH was near-neutral. Note that high specific conductivity and neutral pH *can* occur with AMD that is being treated; therefore, when specific conductivity is high, treated AMD cannot be eliminated as a candidate cause.

Finally, for each tributary and the mainstem, candidate causes that were not eliminated were evaluated using evidence of antecedence, co-occurrence, sufficiency, and alteration (described below). In most cases, more than one likely cause was identified.

23.2.4.4.1 *Evidence of Antecedence*

Each causal relationship is a result of a larger web of antecedent cause–effect relationships. Evidence of antecedence was demonstrated by showing a causal pathway from sources through intermediate stressors that are known to increase the intensity of the candidate cause(s). Evidence included observations of sources or measurements of intermediate steps that were outside the range of high-quality state-wide reference sites.

23.2.4.4.2 *Evidence of Co-occurrence*

Evidence of co-occurrence was demonstrated by showing that the candidate cause was present at levels greater than where the effect did not occur. Available quantitative data were plotted and analyzed spatially from upstream to downstream in the mainstem, as well as in tributaries, by assigning relative positions to the sampling sites (from downstream to upstream) (see Figure 23.5 for an example). This allows for an estimation of the degree of influence of a tributary on conditions in the mainstem stream. Measurements of the candidate cause that were outside the range of high-quality state-wide reference sites also indicated co-occurrence with the effect. The Clear Fork sites were identified in stressor–response plots of all state-wide sampling sites at which stressors and effects also co-occurred (see Figure 23.6).

23.2.4.4.3 *Evidence of Sufficiency*

Stressor–response (S–R) relationships were generated from each quantified stressor and the WVSCI score using state-wide data. Thresholds of response and nonresponse were estimated by (1) graphical analyses of scatter plots of biological indicator values and measured stressors values (see Figure 23.6)

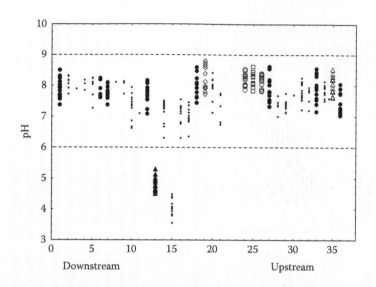

FIGURE 23.5

Geographic-order scatterplot for pH used to assess co-occurrence. X-axis indicates order of sites as shown in Figure 23.2. Graph illustrates one example of how observations for mainstem and tributaries were depicted for monthly sampling in 2003. The mainstem of Clear Fork is indicated by large solid circles; data from Clear Fork tributaries are represented by other large symbols; see Figure 23.2 for tributary names. Horizontal lines at pH 9 and 6 demarcate the range considered unlikely to cause an effect. These types of plots were prepared for each biological, physical, and chemical measurement endpoint. Based on this graphic, low pH was not considered further as a candidate cause for those sites with pH consistently within the range of 6–9. Low pH co-occurs with low WVSCI scores at Stonecoal Branch (solid triangles) which is assessed in this chapter. Dow Fork, at site 15, is acidic, but was not assessed because it had a WVSCI score of 62 thus unimpaired by state biological standards.

and (2) several statistical techniques for deriving thresholds of response (Gerritsen et al., 2010).

In scatterplots for each stressor, three regions of "plausibility" of a biological response were defined (see Figure 23.7):

1. When the concentration or intensity of the stressor was similar to that found in state-wide reference sites, then that stressor was an implausible candidate cause or unlikely to cause an adverse effect which weakens the case for that candidate cause (below a response threshold).

2. Intermediate concentrations of the stressor may have effects on the biota (above a response threshold), but alone, the level of the candidate cause may not be sufficient to result in substantial biological change on a regular basis (response detectable, but below substantial change threshold). Such candidate causes may cause effects in combination with other stressors.

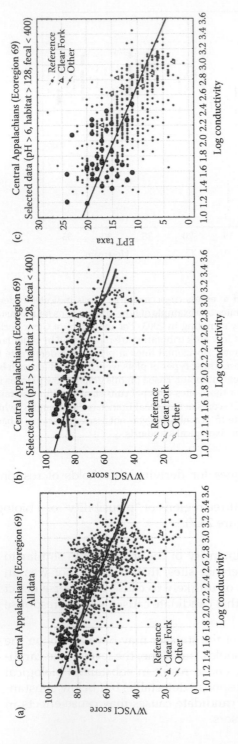

FIGURE 23.6

Scatterplots of WVSCI and Specific Conductivity (scale is log μS/cm). Central Appalachian (Ecoregion 69) reference sites and Clear Fork sites are identified. The bold, red curve is a LOWESS-fitted line, and straight black lines show linear regressions. A smaller data set is shown here to better resolve the sampling points. Thresholds used in the assessment used the larger state-wide data set. (a) Full data set and b) selected data set to evaluate the effect of potential confounders. Note threshold in locally weighted estimation (LOWESS) estimate at specific conductivity near 60 μS/cm (log specific conductivity 1.8), and crossing WVSCI = 71 at specific conductivity near 250 μS/cm (log specific conductivity 2.4). (b) Same as (a), but data selected to remove other stressors such that all sites had a pH > 6, habitat score >128, and fecal coliform <400 colonies. LOWESS estimated threshold is the same, but specific conductivity at WVSCI = 71 increased to approximately 400 μS/cm (log specific conductivity 2.6). c) Same as b (selected data) but showing a sensitive indicator that is a component of WVSCI, the number of taxa that are mayflies, stoneflies, or caddisflies (EPT taxa). Linear regression only is shown because LOWESS showed no improvement over linear; hence, the response is linear with no distinct threshold. Note that this is not a wedge-shaped scatter plot. Reference 95th quantile of specific conductivity is 180 μS/cm (log specific conductivity 2.25), and regression line crosses the 5th quantile of EPT taxa at 250 μS/cm.

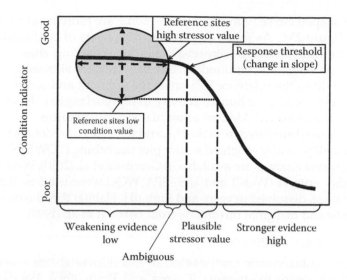

FIGURE 23.7
Diagram showing interpretation of stressor–response relationship. The data are plotted, and a LOWESS regression reveals an S-shaped association between the stressor and the response or condition indicator. The ellipse with broad crossed arrows indicates the 90% distribution of both stressor and response indicator values in state-wide reference sites. The solid vertical line shows the 95th quantile of stressor value in reference sites, and the horizontal dotted line shows the fifth quantile of biological indicator or response values in the reference sites. The curve reveals a response threshold (vertical dashed line) where the mean response value begins to decline as the stressor increases (breakpoint). The dot-dash line indicates the point where mean predicted response value (from LOWESS or linear regression) is equal to the fifth quantile of biological condition (i.e., where the mean condition is substantially less than the reference condition). Relative strength of evidence is indicated by the brackets below the *x*-axis, weakening evidence (implausible), plausible evidence supports, and stronger evidence strongly supports.

3. At high concentrations of the stressor, the biota are clearly different from reference in the state-wide data set and therefore the concentration is deemed sufficient to cause substantial changes in the specific location, thus strengthening the case for that candidate cause (above substantial response threshold). These candidate causes were identified as causing substantial or sustained effects depending on intensity.

These three thresholds were scored weakening, plausible, and substantial, respectively.

We estimated these thresholds from two information sources: (1) the distribution of the stressors in state-wide reference sites to estimate the range of the stressor with no effect, or almost no effect, on biological response and (2) an S-shaped response curve based on the selected stressor gradient, showing an initial decline in condition (the response threshold) at the shoulder of the curve, and also showing the point where the mean response declines below the fifth quantile of reference condition (see Figure 23.7). Not all responses show an S-shaped curve with a shoulder—some are more nearly a straight-line as in Figure 23.6. As much as possible, the effects of confounding and collinear multiple stressors were evaluated by removing sites with high levels of these potential confounders. Several stressors showed a response "shoulder" within the reference 95% envelope (e.g., salt and sedimentation). Given the extent of historic human activity in this coal region, these could be in part due to undetected AMD or other historic disturbances.

Thresholds used are listed in Table 23.1, and the analytical details for conditional probability, locally weighted scatter plot smoothing (LOWESS) analysis, and change point analysis are available in Gerritsen et al. (2010). West Virginia water quality criteria (WQC) and U.S. EPA WQC were used as thresholds for temperature, dissolved oxygen, and high pH. Habitat thresholds used the fifth quantile and marginal thresholds from Barbour et al. (1999).

23.2.4.4.4 Evidence of Alteration

Changes in the taxonomic composition of macroinvertebrate assemblages occur with exposure to stressors (Cairns and Pratt, 1993). We developed empirical models to predict the stressors most likely to have caused symptomatic change in assemblage composition using a subset of the WV WABbase data set from Ecoregion 67 and 69 from 1999 to 2007. Stressor interactions were not examined because the interaction terms are many, and although the data set is large, even moderate collinearity among the stressors severely reduces the ability to detect interactions. This was not the case for pH and aluminum toxicity, where the interaction could be confidently characterized (Gerritsen et al., 2010).

A "dirty" reference approach was used to define groups of sites affected by single stressors in comparison to high-quality reference sites. The approach uses the macroinvertebrate assemblage composition at sites with known causes to develop a classification model based on assemblage pattern. The macroinvertebrate assemblages at the affected sites were then compared to the model results to identify the group that they most closely resemble. Five "dirty" reference groups were identified and consisted of sites primarily affected by a single stressor category: acidic mine drainage (AMD; characterized by high specific conductivity and low pH); acidic deposition (characterized by low specific conductivity and low pH); excessive sedimentation; high nutrients and organic enrichment (using fecal coliform as a surrogate measure of wastewater and livestock runoff); and increased salt primarily from surface mining (using sulfate concentration as a surrogate measure, and excluding AMD). In addition, a "clean" reference group of sites was identified based on low levels of all measured stressors. Nonmetric multidimensional scaling (NMS) and permutation procedures were used to examine the separation of the "dirty" reference groups from each other and from the "clean" reference group based on the biological assemblages observed among the groups.

The results indicated that the centroids of the "dirty" reference groups were significantly different from the "clean" reference group ($p < 0.0001$).

Of the "dirty" reference groups, the AMD and acidic deposition groups were significantly different from the other three "dirty" reference groups ($p < 0.001$). The other three "dirty" reference groups, though overlapping in ordination space to some extent, were also significantly different from one another ($p < 0.05$). Overall, each of the five "dirty" reference models were significantly different from one another ($p < 0.001$), suggesting that differences among stressors led to consistent differences among macroinvertebrate assemblages.

Independent biological samples known to be impaired by a single stressor were used to test the performance of these diagnostic models. Most of the "clean" test sites (80%) were correctly identified as unimpaired, with 10% considered unclassified. None of the "dirty" test sites was classified as "clean." In addition, the sites in the AMD group were either correctly classified as impaired by AMD (87.5%) or not classified (12.5%). The majority of the high specific conductivity test sites (75%) were correctly identified as affected by salt measured as specific conductivity. The "dirty" reference models also identified most of the fecal test group (organic enrichment) (78%) as fecal impaired, although 22% of the fecal test sites (organic enrichment) were misclassified as sediment affected. Some of the sediment test sites (37.5%) were also misclassified as affected by fecal contamination (see Figure 23.8). The model predictions during model validation were correct more often than not, so the Bray–Curtis similarity index was used to assess similarity of the biological assemblages at sites in Clear Fork and the tributaries with sites influenced by one stressor (Bray and Curtis, 1957; Legendre and Legendre, 2012).

For each Clear Fork and Tributary site, multiple stressors were ranked according to the measured similarity to each reference group. The relative similarity and the variation explained by each model were accounted for in the final ranking of the predicted stressors for each impaired site. The majority of test results indicated that the model agreed with the stressor conclusions based on the physical and chemical data collected at each site. Discrepancies between the model predictions from the "dirty" reference models from field observations weakened the candidate cause, but in some cases had no effect on the weight of evidence when there was evidence of episodic exposures, or when the model was unable to discriminate certain stressors, such as nutrients and sedimentation. The distances (in ordination space) between affected site and candidate cause centroids were used to weight evidence of similarity between the assemblage composition at the site and other sites with that candidate cause (see Figure 23.9 for a graphical visualization).

To evaluate the similarity of a site to a reference group (clean or dirty), the abundance-based similarity index was calculated. The composite sample based on the average relative abundance of taxonomic composition of each "reference" group was selected as the centroid of that reference group. The abundance-based similarity index (Bray–Curtis similarity index) calculates the minimum similarity between a study site and the

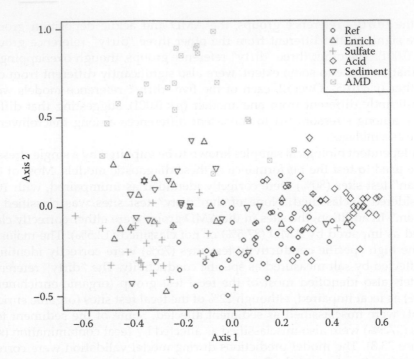

FIGURE 23.8
Discrimination of sites affected by single stressors: acidic mine drainage (AMD); acidic deposition (acid); excessive sedimentation and habitat (sediment); high nutrients and organic enrichment (Enrich; using fecal coliform as a surrogate measure of wastewater and livestock runoff); and increased salt concentration (sulfate; using sulfate concentration as a surrogate measure for a mixture of salts). In addition, a "clean" reference group of sites (Ref) was identified using WVDEP's selection procedures.

centroids (mean relative abundance) of all the reference groups. After similarity indexes were calculated between a site and each of the six reference groups, these were compared to the centroid of the reference condition and the "dirty" reference sets. The "dirty" reference model yields an ordinal ranking of stressors. The study site was classified to a stressor group with the highest ranking (Table 23.2).

23.3 Evidence and Conclusions

The evidence for Lick Run, Toney Fork and its tributary Buffalo Fork, White Oak Creek, Stonecoal Branch, and Clear Fork are described below. Table 23.1 shows the thresholds used for all sites and the measurements only for Lick Run. Tables for all streams are available from Gerritsen et al. (2010). Table 23.2 shows the Bray–Curtis values for each tributary.

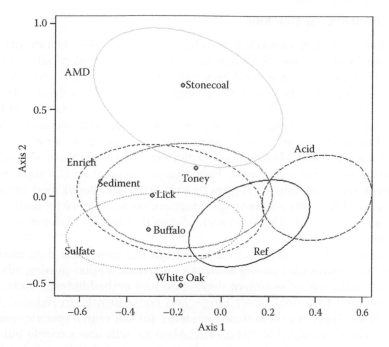

FIGURE 23.9

The distance of the affected sites coordinates from the candidate cause centroids illustrates the relative discrimination among candidates based on assemblage composition at the site. The ovals are the 75% confidence interval around the centroid for each candidate cause.

TABLE 23.2

Tributaries to Clear Fork: Bray–Curtis Similarity Index Values and Percentage Similarity (Parenthesis) among Several Categories of Impaired Streams

Stream Name	Acidity[a]	Acidity with Metals[b]	Enrichment	Sediment	Salt[c]	Reference
Stonecoal	0.14 (2%)	0.59 (38%)	0.23 (2%)	0.17 (2%)	0.21 (2%)	0.15 (1%)
White Oak	0.24 (2%)	0.21 (2%)	0.27 (2%)	0.29 (2%)	0.35 (5%)	0.28 (1%)
Toney	0.33 (2%)	0.26 (2%)	0.36 (6%)	0.41 (17%)	0.29 (2%)	0.25 (1%)
Buffalo	0.25 (2%)	0.26 (2%)	0.35 (6%)	0.50 (26%)	0.46 (8%)	0.32 (1%)
Lick	0.29 (2%)	0.44 (18%)	0.53 (26%)	0.61 (64%)	0.55 (18%)	0.36 (1%)

Note: Higher values indicate greater similarity of the biological assemblage with sites affected by acid deposition, acid mine drainage, organic enrichment, sediment, salt, or unaffected reference sites. Percentages refer to "dirty" reference samples in question; a value of 2% indicates that the sample is farther from the centroid than 98% of the model's "dirty" reference group, i.e., it relatively far away.
[a] Acid deposition.
[b] Acid mine drainage (AMD).
[c] Sulfate used as measurement.

23.3.1 Evidence for Lick Run

Antecedence. Lick Run is a second-order tributary near the headwaters of Clear Fork (Site 35; see Figure 23.1). It is severely impaired biologically (WVSCI score = 44). Lick Run has several permitted mining discharges and a current mining area on its northern watershed ridge and consists mostly of reclaimed mine land throughout. Forests were clear-cut and gravel roads were built in advance of surface mining. There was no residential land use and no livestock. These sources could contribute to any of the candidate causes.

Co-occurrence. Field observations indicated severe sediment deposition supporting sediment as a cause. Poor riparian vegetation throughout the catchment and moderate algal growth supported an altered food resource as the cause. There was weakening evidence of co-occurrence for metals and acidity associated with AMD and acidic deposition and for high temperature (see Table 23.1).

Sufficiency. The S–R evidence derived from the state-wide data analysis (see Table 23.1) provided strong evidence for sedimentation causing adverse effects as indicated by sediment deposition and embeddedness metrics in the substantial threshold effect range, and consequently, in reduced total habitat score. There was also strong evidence for salt with a specific conductivity maxima measured at 804 µS/cm. Algal growth was variable but was recorded as substantial. The S–R analysis suggested that iron is not toxic to benthic macroinvertebrates (see Gerritsen et al., 2010), but plausible albeit weak evidence was present for manganese toxicity (data indicated only weak evidence for magnesium toxicity at the highest observed concentrations). There was also plausible but weak evidence for the candidate cause of organic enrichment measured by fecal coliform.

Alteration. The "dirty" reference model for the Lick Run sample indicated sediment as the strongest stressor, followed by salt and organic enrichment (see Table 23.2).

23.3.2 Conclusions for Lick Run

Overall, the evidence for excess sediment deposition was strong for Lick Run. Secondary stressors in Lick Run include salt effects and algal growth causing a food source shift for the benthic macroinvertebrates, but the effects of food alteration are masked to some extent by the sedimentation. The likely sources of salts are reclaimed mine lands and current mining activity. Algal growth is most likely stimulated because the open tree canopy allows increased light penetration into this headwater stream. There are no known anthropogenic nutrient sources in the Lick Run Watershed aside from the mining activity.

The principal cause of biological impairment of Lick Run appears to be sediment deposition most likely from abandoned mine lands, and riparian disturbance along the stream corridor, both of which also contribute to degraded aquatic habitat. Salt, likely from abandoned mine lands and current mining activity, is also a likely cause.

23.3.3 Evidence for Toney Fork and Buffalo Fork

Antecedence. Toney Fork and its tributary Buffalo Fork are two of the most impaired tributaries to Clear Fork. The southern half of the Toney Fork Watershed, which includes Buffalo Fork, is an active mining area with numerous permitted mining discharges. The West Virginia pollutant source database recorded three permitted valley fills in upper Toney Fork and five in Buffalo Fork. Field observations indicated a moderate amount of houses and lawns, and some cattle and poultry. These sources could contribute any of the candidate causes.

Co-occurrence. On the sampling days of biological monitoring, evidence weakened the case for excess sediment, high temperature, AMD, and acidic deposition based on thresholds in Table 23.1. However, field observations suggested intermittent sediment deposition and removal, including "fine black sludge" of small coal particles. These observations were made monthly at each of the three sites in Toney and Buffalo and covered the entire length of the reach sampled by the field crews. Quantitative sediment measurements made at the time of macroinvertebrate sampling were confined to the 100-m sampling reach at the time of sampling only. The numerous qualitative observations were judged to be more reliable indications of the potential for sediment co-occurring at the affected site than the one-time quantitative sediment measurements, because sediment deposits shift with changing hydrology. Field observations also indicated a moderate level of algal abundance, no sewage odors, and no observations of domestic sewage pipes.

Sufficiency. The S–R evidence derived from the state-wide data analysis suggested strong evidence for salt (measured as specific conductivity—median 1206, maximum 1650) which exceeded the substantial effects threshold; moderate evidence for organic enrichment in Toney Fork, but weak in Buffalo Fork; and weak evidence for iron (maximum 2.890 mg/L) and manganese (maximum 0.290 mg/L).

Alteration. The community similarity from the "dirty" reference model indicated that the macroinvertebrate assemblage was most similar to communities strongly affected by sediment (Table 23.2). Organic enrichment was a moderate secondary stressor in Toney Fork and third in Buffalo Fork based on the Bray–Curtis similarity. Salt was a secondary stressor in Buffalo Fork.

23.3.4 Conclusions for Toney Fork and Buffalo Fork

The principal stressors causing the impairment are salinity and sedimentation. The chronic source of salts is surface mining and the mine effluent treatment ponds. Sedimentation, which may be variable and intermittent as coal sludge, is a likely cause affecting habitat quality and exerting toxic effects. The potential sources of the sediments include current mining operations, abandoned mine lands and tailings piles, valley fills, roads and tracks, and residential activities and construction. The third contributing stressor is moderate alteration of food resources potentially due to enrichment from septic systems, lawns, and livestock.

The principal cause of impairment appears to be excess salt exacerbated by intermittent coal sludge contamination. Mining activities and mine effluent ponds are present within these watersheds and are known sources of dissolved ions.

23.3.5 Evidence for White Oak Creek

Antecedence. White Oak Creek has two tributaries, Left Fork and Road Fork. The WV Pollutant Source database recorded mining in the watershed, including several small valley fills in the headwaters of White Oak Creek and two large valley fills in the headwaters of Left Fork. More than 75 dwellings occur in the creek valley, near the stream channel. These are antecedents of all candidate causes except acid mine drainage.

Co-occurrence. Observational data collected at the White Oak Creek indicated that the food base was altered by inputs of organic matter and periphyton which co-occurred with the effect. Evidence included moderate to high periphyton scores, noticeable sewage odors, fecal coliform concentrations exceeding the substantial threshold during base flow, and evidence of agricultural runoff. Visual source-tracking by walking the entire stream also indicated sedimentation stress in parts of the catchment, but not at the location of biological sampling. There was evidence against high temperature, AMD, and acidic deposition.

Sufficiency. The S–R evidence derived from the state-wide data analysis indicated moderate-to-strong evidence for an altered food resource from waste and algal growth. Organic enrichment was indicated by episodically high fecal coliform concentrations exceeding the plausible threshold for effects during base-flow periods and high amounts of algae were observed in the substantial threshold range. Dissolved oxygen ranged from 7.69 to 12.91 which is indicative of a strong algal production-decay cycle somewhat supporting low DO as a cause. There was also plausible evidence for salt (median 469 µS/cm, max 750 µS/cm). There was plausible but weak evidence for manganese (median 0.03 mg/L, max 0.88 mg/L) and for habitat degradation (RBP 137).

Alteration. The "dirty reference" model did not identify any stressor as being stronger or more likely than others (Table 23.2). This result indicates that the biological assemblage, while impaired, was not more similar to any one of the single-stress communities than to any other, suggesting that the stream was subject to multiple, cumulative causes.

23.3.6 Conclusions for White Oak Creek

Organic enrichment was the principal stressor in White Oak Creek and appeared to be from inadequately treated domestic sewage. Salt (measured as specific conductivity) was identified as a secondary stressor in White Oak Creek. Surface mining in the headwater reaches is the most likely source of dissolved ions. Sedimentation and moderately degraded habitat were identified as tertiary stressors.

The principal cause is most likely organic and nutrient enrichment and appears to be from inadequately treated domestic sewage.

23.3.7 Evidence for Stonecoal Branch

Antecedence. Stonecoal Branch is a small tributary of Clear Fork that is mostly forested with small dirt roads along the stream channel and a mountain top mine/valley fill permit in the headwaters. In addition to the permitted mining activity, there are extensive, unremediated AMD throughout the catchment. No residences or agriculture occur within its catchment. The WVSCI score in Stonecoal Branch (50.7) was well below the biological impairment threshold. Evidence of these sources support all candidate causes except acid deposition and organic enrichment.

Co-occurrence. Acid mine drainage is an extreme case of metal contamination that results in visible evidence. When acid mine drainage mixes with the higher pH water of a receiving stream, the metal hydroxides (ferric hydroxide, aluminum hydroxide, manganese oxide) precipitate and form flocs that coat the streambed. These precipitates of "yellow boy" were observed in Stonecoal Branch.

Embeddedness and sediment RBP scores were less than most reference sites, thus supporting sediment as a cause. At times, fecal coliform was high (maximum of 5600 colony-forming units/L) supporting organic enrichment as a cause; however, the median fecal coliform of 3.5 colony-forming units was less than reference sites weakening the case for organic enrichment altering the food base. There was also evidence against excessively high temperatures and acidic atmospheric deposition.

Sufficiency. The S–R evidence derived from the state-wide data analysis confirmed strong evidence for AMD impairment (mean pH = 4.8; mean dissolved aluminum = 3.7 mg/L, in the plausible and substantial threshold range, respectively). State-wide, 80% of sites in the pH range of 4–6 had WVSCI scores <71 where Al concentration exceeded 1 mg/L (Gerritsen et al., 2010). There was also strong evidence for salt (mean specific conductivity = 499 μS/cm exceeding the substantial stress threshold); however, AMD is always associated with high specific conductivity and sulfate. There was plausible evidence for manganese (minimum 0.9 mg/L). There were substantial effects of embeddedness due to concretions of yellow boy, and AMD was also reflected in an overall median RBP score of 104. The median fecal coliform was well below the effects threshold of 250 colony-forming units and algal growth was low, therefore organic enrichment was not regularly sufficient to alter the food base.

The minimum embeddedness (RBP score of 3) and the minimum sediment (RBP score of 2) were less than the substantial effects thresholds of <9 and <8, respectively. Lower scores indicate greater level of stressor.

Alteration. The "dirty" reference model indicated a strong AMD signature (Table 23.2).

23.3.8 Conclusions for Stonecoal Branch

AMD was the primary stressor in Stonecoal Branch. Indicators of severe AMD noted in the field included cementing of substrate particles by iron hydroxides (yellow boy), and the presence of aluminum hydroxide flocs. Stonecoal Branch is impaired by acid mine drainage and likely by sediment from abandoned mine lands and dirt roads.

23.3.9 Evidence for Clear Fork

Antecedence. The mainstem of Clear Fork receives waters from its biologically affected tributaries described above, as well as several other tributaries that exceed West Virginia water quality standards, but were not listed as biologically impaired. Clear Fork's biological condition is good in the headwaters (Site 36; see Figure 23.2), then declines below Lick Run and other tributaries becoming impaired from Site 27 to Site 12. Below Site 12, the stream condition recovers to "marginal" status to the mouth. Influent tributaries carrying stressors above S–R thresholds are shown in Table 23.3.

TABLE 23.3

Clear Fork Mainstem

Clear Fork Site	Upstream Tributary (Sites)	Stressors in Tributary
33	Lick Run (35)	Sediment, salt, iron, manganese
27[a]	Workman Creek (31, 32)	Manganese, salt
	McDowell Branch (29, 30)	Fecal coliform (organic enrichment)
18[a]	Toney, Buffalo Forks (24–26)	Sediment and salt
	White Oak Creek (19)	Fecal coliform (organic enrichment), some iron
12[a]	Long Branch	Some dissolved aluminum from AMD in Dow Fork (tributary to Long Branch), salt
	Stonecoal Branch	AMD, iron, manganese
7 to mouth	Sycamore Creek	Slight iron, fecal coliform (organic enrichment)

Note: Each mainstem site is matched to contributing tributaries described in this chapter, as well as other tributaries with measured stressors above stressor–response thresholds.

[a] Impaired.

In addition to the primary tributaries, a few mining areas drain directly into Clear Fork. Numerous oil and gas wells are located along the mainstem and along tributaries to the south. Residences and roads occur throughout the stream valley and floodplain, mostly near the stream. There are sources of all the candidate causes.

Co-occurrence. Field records indicate sedimentation in the upper and lower thirds of the mainstem. The gradient of the middle third is too high for sediments to deposit and remain (high hydraulic power). Moderate levels of algae, both periphytic diatoms and soft (filamentous) forms, were observed in the lower portion of the mainstem. Raw sewage and sewage odor was noted several times during sampling in the lower mainstem. Scouring was evident in some locations in Clear Fork indicative of an altered habitat. Measurements did not support low pH (range 7.02–8.61) or high temperature (max 28.73°C). Specific conductivity was much greater than at state-wide reference sites.

Sufficiency. The S–R evidence derived from the state-wide data analysis indicate moderately plausible evidence for organic enrichment as measured by fecal coliform; plausible evidence for salt causing impairment (median 464 µS/cm, max 974 µS/cm); plausible evidence for excess sediment in the upper and lower thirds of the mainstem; and weak evidence for manganese.

Alteration. The "dirty" reference model did not identify any single stressor as the greatest cause of impairment throughout the mainstem.

23.3.10 Conclusions for Clear Fork

Clear Fork, as the receiving stream of the above tributaries, is an example of a stream affected by multiple stressors and multiple, cumulative causes (organic/nutrient enrichment from untreated domestic wastewater, excess sedimentation, and residual metals and salt from mining). No single stressor is overwhelming by itself, and the condition ranges from unimpaired in the upper third to moderately impaired in the middle third, and recovering to marginally impaired in the lower third. The upper third of Clear Fork is affected by scour and suspended sediment, but nevertheless remains above WV's biological threshold. In the middle third, organic enrichment from sewage most likely has the strongest effect, although suspended sediment during high flows and residual metals toxicity from tributaries may be contributing factors. The lower third is most likely affected by sedimentation, poor habitat, greater concentrations of salt, and moderate enrichment causing algal growth. The biota in the lower third of Clear Fork is in fair condition, but the apparent multiple, cumulative or combined stressors in this receiving stream have prevented recovery to good condition.

Likely causes were different among the Clear Fork tributaries and included high levels of dissolved salt measured as specific conductivity, metal contamination and acidification from coal mine drainage, aluminum toxicity in association with low pH, sediment deposition, organic enrichment from direct releases and from algal productivity enhanced by nutrients, and conditions of low DO. In the Clear Fork itself, the combination of all these inputs was evident in the mainstem, which exhibited some resiliency due to dilution and different geophysical attributes.

23.4 Commentary on Scoring

While preparing this chapter, the case study was revisited and scores were assigned as shown in Table 23.1. This type of table is very handy because the thresholds indicating a difference from the state-wide reference sites and thresholds for sufficiency remain the same, making it easy to fill in data from a new site and simply highlight the thresholds that were exceeded. For illustrative purposes, we have made a simpler weighting table (see Table 23.4) with assigned plus and minus scores to show the patterns and summary from the pieces of evidence that may be easier for communicating the findings. We used the data from Lick Run, Table 23.1 to assign scores.

Most candidate causes had potential sources and received a single plus or minus as evidence of antecedence as indicated in the narrative. The level of the stressor at the site was compared with the state-wide reference threshold: when it was less, evidence of co-occurrence was scored with a single minus and when it was greater, it was assigned a single plus. The level of the stressor at the site was compared with the state-wide thresholds that are likely to cause effects, and evidence of sufficiency was thus scored zero for weakly plausible, one plus for plausible, and two plusses for plausible substantial or sustained effects (see Table 23.1). Evidence of alteration was scored on the basis of the results of the Bray–Curtis similarity index. The model yields an ordinal ranking of stressors. The highest ranking was scored three plusses, next highest two plusses, and next one plus. The model did not have the ability to predict habitat, temperature, dissolved oxygen, or ammonia as causes, so these candidate causes were scored as NE (see Table 23.4).

After each piece of evidence was scored (see Table 23.4), the pieces of evidence for different measurements were weighed to assign an overall finding for the characteristics of each candidate cause (see Table 23.5). A minus for co-occurrence in most cases led to a finding of unlikely. Exceeding a sufficiency threshold usually resulted in a likely cause. Other observations were sometimes more influential. For example, although sediment measurements were low on days of biological monitoring, heavy layers of coal sludge were noted in the field notes in Toney Fork and Buffalo Run; therefore, sedimentation was judged a likely cause. When a higher stressor–response threshold was exceeded and the biological community composition was consistent with that candidate cause, the finding was usually very likely. Weak evidence of a specific alteration had two interpretations: the candidate cause was unlikely or several candidate causes were active and interfered with the

TABLE 23.4

Scores for Evidence of Causal Characteristics based on Specific Stressor
Measurements for Lick Run

		Evidence of Causal Characteristic			
		Ante-cedence	Co-occurrence	Sufficiency	Alteration
Candidate Cause	Stressor Measurement	Obser-vation of Sources	Level at Impaired Site Compared to State-wide Reference	Impaired Site Compared to State-wide Thresholds	Similarity of Community Composition
1. Acidity/ alkalinity	pH low	+	–	–	– –
	pH high	+	–	–	
2. Metal toxicity	Al (dissolved)	+	0	0	+
	Fe (total iron)	+	+	0	
	Mn (total manganese)	+	+	+	
3. Salt	Specific conductivity	+	+	+ +	+ +
	Sulfate	+	+	+	
	Chloride	+	+	NE	
4. Sedimen-tation and turbidity	RBP embeddedness	+	+	+ +	+ + +
	RBP sediment	+	+	+ +	
	% fines (SSC)	+	+	+	
	RBP bank stability	+	+	+	
	TSS	+	+	0	
5. Habitat quality	RBP total score	+	+	+ +	NE
	RBP: channel alteration	+	+	+	
	RBP: cover	+	+	0	
	RBP: riparian vegetation	+	+	+	
6. Temperature	Temperature	+	–	–	NE
7. Low DO	DO	–	–	–	NE
8. Organic enrichment	Algae observed	+	+	+ +	+
	Fecal coliform	+	+	+ +	
9. Ammonia toxicity	Ammonia	+	NE	NE	NE

TABLE 23.5

Summary Scores for Nine Candidate Causes Assessed in Lick Run

Candidate Cause	Conclusion	Antecedence	Co-occurrence	Sufficiency	Alteration
1. Acidity/alkalinity	Unlikely	+	−	−	− −
2. Metal toxicity	Ambiguous	+	+	0	+
3. Salt	Very likely	+	+	+ +	+ +
4. Sedimentation and turbidity	Very likely	+	+	+ +	+ + +
5. Habitat quality	Likely	+	+	+	NE
6. Temperature	Unlikely	+	−	−	NE
7. Low DO	Unlikely	−	−	−	NE
8. Organic enrichment	Very likely	+	+	+ +	+
9. Ammonia toxicity	No evidence	+	NE	NE	NE

Note: Likely causes that are highlighted in gray have consistent supporting evidence (synthesized from Table 23.4) for four causal characteristics.

model prediction. Professional judgment was used to make that determination and was reflected in the narrative.

The decision to organize data and score in a single table or to use separate tables is a matter of preference. An important factor to consider is consistency in weighing evidence among causes and from case to case. Another important factor is organizing the data to reveal patterns and to communicate results as described in Chapters 19 and 20.

23.5 Discussion

General causation is well established for many environmental causes, but reliable and relevant general causation models* that can be routinely used to assess site-specific causation are uncommon. This case study provides an exception. State-wide models related to common causes of adverse biological effects were developed using monitoring data so that multiple streams could be easily assessed with a spreadsheet. Site data were compared with the state-wide background estimates to assess whether an antecedent or proximate cause co-occurred at a location. Similarly, site data were compared with state-wide stressor–response thresholds to assess whether a candidate causal relationship exhibited the characteristic of sufficiency. Finally, a causal rela-

---

* By general causation model, we do not imply that the models used in this case study are applicable outside their data set parameters, but rather that they describe the capability of a cause to produce an effect (i.e., general causation) rather than an instance of specific causation.

tionship was assessed for the characteristic of alteration. Invertebrate assemblage composition at each site was compared with the typical assemblages observed across Ecoregions 67 and 69 for common causal relationships, such as taxa abundances altered by aluminum toxicity. The ability to derive evidence based on inference from models of general causation provides an efficient way to develop evidence of several causal characteristics and many candidate causes and intermediate steps.

By highlighting the causal characteristics, this case not only shows what causes are best supported by the evidence, but also shows that the most likely causes display several characteristics of causal relationships. The causal characteristics draw attention to the causal reasoning behind the findings of the case.

hardship is assessed for the characteristics of illustration that whole assemblage composition at each site was compared with the typical assemblages observed across the regions 67 and 68 for common causal relationships such as taxa abundances altered by aluminum toxicity. The ability to derive evidence based on inference from models of general causation provides an efficient way to develop evidence for several characteristics and many candidate causes and intermediate steps.

By highlighting the causal characteristics this case not only shows what causes are best supported by the evidence, but also shows that the most likely causes display several characteristics of causal relationships. The causal characteristics draw attention to the causal reasoning behind the findings of the case.

24

Northern River Basins Study and the Athabasca River: The Value of Experimental Approaches in a Weight-of-Evidence Assessment

Alexa C. Alexander, Patricia A. Chambers, Robert B. Brua, and Joseph M. Culp

This case study illustrates how field, laboratory, and mesocosm studies were used to clarify the relative roles of nutrients and toxicants in altering biota downstream from municipal sewage and pulp mill discharges.

CONTENTS

24.1 Summary

Point-source inputs from pulp mills and municipal sewage contribute both toxicants and nutrients to the Athabasca River. Knowing that nutrients and contaminants can change the structure of aquatic communities by removing sensitive species from the system, investigators initially expected to observe reduced abundance and diversity. Instead, field surveys showed increased benthic algal and invertebrate abundance downstream from outfalls, with abundance decreasing to background levels with increasing distance downstream from point sources. The stimulatory effect of sewage and pulp-mill effluents on benthic communities was confirmed through laboratory and mesocosm studies and continued field monitoring.

The overall conclusion was that added nutrients were important drivers of the observed changes in the benthic community, to the extent of potentially masking negative effects of any toxic constituents. Increased abundances of algae and improved condition of sculpins were routinely detected immediately downstream from sewage and pulp-mill outfalls. Mesocosm tests demonstrated that nutrient enrichment resulted in a plateau beyond which additional nutrients were no longer eliciting a response. These studies indicated that when a threshold for a nutrient is exceeded, different essential nutrients become limiting (i.e., nitrogen instead of phosphorus). Finally, the abundance of some invertebrate species decreased with further distance downstream from point sources. Laboratory tests confirmed that the decline could not be attributed to toxic responses to pulp-mill effluent. Rather, alleviation of nutrient enrichment may contribute to decreased abundance and increased diversity of some populations downstream.

The Northern River Basins Study (NRBS) produced more than 150 technical and synthesis reports. The NRBS is particularly relevant today because the Athabasca River and its tributaries drain Canada's largest economic development project, the Canadian oil sands. These early studies helped establish a base line for evaluating future effects from oil sand development. The NRBS influenced not only how the Athabasca River and its tributaries are managed, but has also laid the foundation for a multidisciplinary framework that modified Canadian environmental regulations. The NRBS provided a clear example of how a multistakeholder approach can successfully design and implement an integrated environmental assessment program.

24.2 Problem Formulation

24.2.1 The Case

The following case study focuses on 720 km of the Athabasca River where the effects of contaminants and nutrients from pulp mill and sewage effluents were investigated as part of the Northern River Basins Study (NRBS; see also Section 24.4). The Athabasca River Basin spans 157,000 km^2 and flows 1300 km northeast across northern Alberta. It is a tributary of the Mackenzie River Basin, Canada's largest and most northern river system, ultimately discharging to the Arctic Ocean (MacLock et al., 1996; Cohen, 1997). Concerns about water quality in the Athabasca River began in the 1950s with the construction of the first pulp mill near Hinton (see Figure 24.1). Before the advent of clarifiers or secondary treatment, the impact of even a single pulp-mill discharge was relatively severe. Low under-ice concentrations of dissolved oxygen and high turbidity were routinely observed downstream from the Hinton mill during the first 20 years of its operation (Chambers et al., 1997, 2000).

Several lines of evidence suggested that the Athabasca River from Hinton to Whitecourt was exposed to environmental stressors including polychlorinated biphenyls (PCBs), dioxins, furans, and metals that were higher than in other river reaches (e.g., Wrona et al., 2000). At the same time, nutrients from effluents had stimulated algal growth to levels of aesthetic concern. At the time of the study, little was known about the impact of effluent on northern rivers. Residents of the basins, predominantly First Nations, were concerned

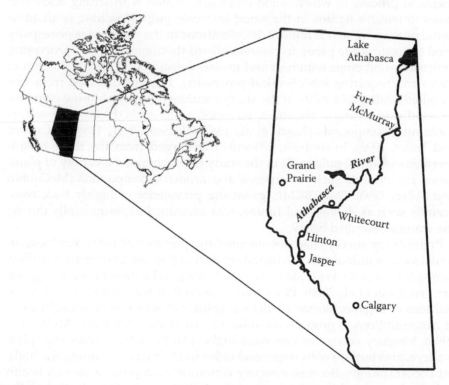

FIGURE 24.1
The Athabasca River in Alberta, Canada. The Athabasca River flows 1300 km northeast from Jasper National Park to Lake Athabasca and is a tributary of the Mackenzie River Basin. The Mackenzie is Canada's largest and most northern river system ultimately discharging into the Arctic Ocean. Between Hinton and Fort McMurray, there were five pulp mills on the Athabasca River or a tributary in the 1990s. Each municipality also discharged municipal waste into the Athabasca main stem. During the period of the study, forestry-related land clearing was expanding as was investment in oil extraction that is currently pervasive in the region.

that people and wildlife were being exposed to harmful chemicals in wastewater and effluents that were being discharged into waterways and accumulating as persistent contaminants in fish.

24.2.2 Candidate Causes and Sources

The prevailing view at the time of the NRBS (i.e., the 1990s) was that the primary concern with the outfalls was toxicity produced by contaminants in the effluents. The NRBS expanded the list of candidate causes to include nutrient enrichment impacts, as well as nutrient-contaminant interactions.

Five pulp mills discharged effluent into the Athabasca River or a tributary during the 1990s. The types of contaminants entering the system were dependent on the processing systems of each mill: bleached kraft mill effluent (BKME) or chemi-thermomechanical mill effluent. Bleached kraft is a chemical process in which wood chips are treated with strong acids and bases to remove lignins in the wood to create pulp. Bleaching is an additional processing step to remove discolorations in the pulp and is principally used to create white paper. In contrast, chemi-thermomechanical processing pretreats wood chips with heat and pressure, facilitating the breakdown of fibers and requiring less chemical processing. Effluent collected from two bleached kraft paper mills in the study contained organochlorine contaminants, dioxins, furans, phenolics, terpenes, resin acids, PAHs, and sulfur-containing compounds (Kuehl et al., 1987; Clement et al., 1989; McCubbin and Folke, 1993). In contrast, effluents discharged from the three chemi-thermomechanical pulp mills in the study area contained a variety of plant-based compounds such as terpenes and aromatic compounds (McCubbin and Folke, 1993). Only BKME, given the prevalence of highly toxic compounds such as dioxins and furans, was examined experimentally during the studies described below.

Preliminary surveys had documented the presence of persistent bioaccumulative contaminants in fish including Burbot (*Lota lota*), northern pike (*Esox lucius*), longnose sucker (*Catostomus catostomus*), and flathead chub (*Platygobio gracilis*) (Cash et al., 2000). PCBs were observed in fish tissue, as were body burdens of organochlorine pesticides, neither of which were thought to be at concentrations to pose a human-health risk (Pastershank and Muir, 1995, 1996). Mercury concentrations were highest in predatory fishes (e.g., pike, walleye, and burbot) with larger and older individuals containing the highest concentrations. Because mercury contamination poses a serious health risk for humans, this component of the NRBS was further studied in the Northern Rivers Ecosystem Initiative (NREI).

Between the Athabasca headwaters and Fort McMurray, four municipalities discharged sewage continuously into the River at the time of the study, while another four municipalities had continuous sewage discharge to tributaries (see Figure 24.1). The Town of Hinton discharged its sewage with the Hinton pulp mill. The mill discharged minimally treated effluent directly into the river until the 1960s (Chambers et al., 2000). Primary clarifiers to remove solids and aerating lagoons were installed between 1967 and 1975. The main types of contaminants found in municipal effluents were dioxins, furans, chlorophenols, volatile organic compounds, and PCBs. However, with the advent of secondary treatment, the major issue surrounding municipal wastewater discharge was nutrient loading, particularly that of nitrogen and phosphorus.

24.2.3 Ecological Effects: Benthic Invertebrates

Although the NRBS had a broad scope encompassing bioaccumulative chemical and human health concerns, this chapter describes the studies used to explain the observed patterns in benthic communities. In field surveys, the abundance of benthic invertebrates was consistently greater downstream from sewage and pulp-mill inputs. The greater downstream abundance was due to an increase in relatively few taxa: midges (chironomidae), mayflies (baetidae), stoneflies, caddisflies (hydropsychidae), and oligochaetes. Consequently, greater abundance was not always accompanied by corresponding increases in taxa richness or diversity. Interestingly, abundance or number of species generally considered to be sensitive to anthropogenic stressors, such as the EPT orders of insects (mayflies, stoneflies, and caddisflies), increased rather than decreased downstream from outfalls. In addition, insect abundance did not increase linearly with increased downstream inputs: addition of more phosphorus from a pulp-mill effluent downstream from a sewage treatment facility did not result in an additive increase in invertebrate abundance. Further downstream from outfalls, the abundance of some benthic macroinvertebrates declined.

24.3 Evidence

Evidence was developed using a tiered (i.e., stepwise) approach that included field observations, laboratory tests, and mesocosm (artificial stream) studies. Throughout, different research groups working with Environment Canada shared the same sites so that observations of sediment and water quality, nutrients and algal biomass, benthic fish, and in situ experiments were all co-located, thus enabling the generation of a large body of evidence for scientists and managers to weigh for decision-making.

24.3.1 Field Observational Studies

Field studies were designed to contrast conditions above and below multiple discharges and were based on previous survey results (see Carey et al., 2001) and predicted characteristics of pulp and paper-mill effluents of relevance in northern rivers (Chambers et al., 2000; McCubbin and Folke, 1993). The studies showed that contaminants and nutrients and benthic algal biomass all increased downstream from outfalls, co-occurring with the somewhat lower diversity but increased abundance of macroinvertebrates. Studies of benthic fish showed enzymatic responses expected from contaminant exposure as well as increased size expected from increased nutrients.

Sediment and water quality. Sediments were collected and analyzed using multispectral gas chromatography–mass spectrometry. Sediment contaminants in sites sampled upstream of development were found to be at low concentrations compared to affected aquatic ecosystems in surrounding regions (Wrona et al., 1996). Near point source inputs, aluminum levels in sediments, as well as concentrations of dioxins, furans, PCBs, and mercury, exceeded Canadian or provincial guidelines (e.g., Canadian guidelines for the protection of aquatic life; CCME, 2013). Generally, concentrations of contaminants in receiving waters were orders of magnitude less than those found in sediments (Wrona et al., 2000) and were below Canadian and provincial guidelines. These water quality results led to the conclusion that ambient river water was an unsuitable predictor of contamination in sediments and biota, and influenced monitoring in more recent assessments (e.g., NREI).

Nutrients and algal biomass. Total phosphorus and total nitrogen were measured monthly with additional sampling occurring during the lower flow periods of late summer to winter when enrichment effects of nutrients from municipal wastewater and pulp-mill effluent were expected to be more pronounced (Chambers et al., 2000).

Phosphorus and nitrogen were found to increase downstream from each pulp mill and sewage outfall, as did periphyton biomass. During these intensive sampling periods, periphyton was collected by scraping known areas on rocks collected from several riffles at multiple sites along the river to evaluate biomass (expressed as chlorophyll *a* content). Algal biomass (see Figure 24.2) increased immediately downstream from every effluent source, never exceeding a threshold value of <500 mg/m^2 chlorophyll *a* and reverting to background concentrations within 30 km of an outfall. During the late fall, the effects of the waste and pulp-mill effluents were exaggerated due to lower flows.

Benthic fish. The spoonhead sculpin (*Cottus ricei*) was monitored as a sentinel species for evaluating the effects of BKME at the Hinton mill (Gibbons et al., 1998). Spoonhead sculpin are small bodied, territorial fishes whose abundance in the basin made them ideal for evaluating spatial patterns in the Athabasca River. The activity of the detoxification enzyme (7-ethoxyresorufin *O*-deethylase [EROD]) was 2.5 times higher in fishes downstream from the pulp mill, confirming exposure that persisted for 48 km downstream. Sculpin were also older, larger, and heavier in study sites exposed to the BKME plume.

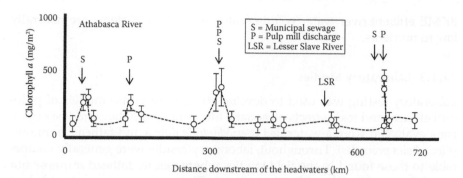

FIGURE 24.2
Periphyton biomass (expressed as chlorophyll *a* concentration) measured in the fall of 1994 for the Athabasca River (after Chambers et al., 2000, with permission). Periphyton biomass increased downstream from each outfall, as did concentrations of phosphorus and nitrogen. Periphyton did not increase continuously with distance downstream but rather reached a threshold value in the presence of added nutrients (chlorophyll *a* ≤ 500 mg/m²).

24.3.2 In Situ Experiments

Strategies to manage nutrient enrichment differ depending on which nutrient (e.g., nitrogen or phosphorus) is in excess. In situ experiments were conducted to evaluate the type and degree of nutrient limitation in the Athabasca River and its tributaries using nutrient diffusing substrates (Chambers et al., 2000; Scrimgeour and Chambers, 2000). The substrate was composed of agar prepared with either or both of the nutrients of interest (phosphorus, nitrogen, or nitrogen plus phosphorus) or as an unamended control. The treated agar was then placed in clay flower pots (with replicates of each nutrient treatment present at each site) that were attached to the river bed for 2–4 weeks to be passively colonized by periphyton. At the end of the exposure period, the colonized surface was collected on each of 10 replicate substrates. Treatments where the abundance of periphyton increased (e.g., on P-amended agar) indicated which nutrient was limiting (e.g., phosphorus).

Experiments conducted using nutrient diffusing substrata showed that low periphyton biomass was maintained due to insufficient phosphorus in the upper reaches of the Athabasca River (Scrimgeour and Chambers, 2000), a condition commonly associated with unimpaired aquatic systems. Likewise, examination of periphyton biomass on substrata deployed at multiples sites showed that the Athabasca River was typically phosphorus-limited upstream of municipal wastewater and pulp-mill discharges, whereas phosphorus limitation was not observed downstream from outfalls. Reaches far downstream from outfalls were typically nitrogen or both N + P limited. These findings showed that even in large rivers, when phosphorus limitation was alleviated, other nutrients, specifically nitrogen, may become limiting. In contrast to responses measured in temperate systems, northern rivers can display strong enrichment effects even at low effluent amendments (<4%

BKME effluent:river water volume/volume) as these systems are naturally low in nutrients.

24.3.3 Laboratory Studies

Laboratory testing was used to develop stressor–response models of pulp-mill effluent and toxic effects on aquatic life. Among the standard test organisms evaluated were a water flea (*Ceriodaphnia dubia*) and fathead minnows (*Pimephales promelas*). Throughout, laboratory results were generally comparable to those found in the field studies where neither fathead minnow nor water fleas were affected by the effluent. The toxicity testing provided an overview of concentrations where direct toxic effects were likely to occur (see guidance for methods; Environment Canada, 1992a, b, c). A gradient of concentrations was chosen to mirror those found in the field and provide data on the types and magnitude of effects that were likely to occur. Grab samples were taken of undiluted effluent and diluted in the laboratory to a percentage of the effluent concentration (% effluent). Acute (median lethal concentration, LC_{50}) and sublethal (no-observable-effect concentration, NOEC; lowest-observable-effect concentration, LOEC; inhibition concentration expressed as the concentration at which a 25% reduction in growth/reproduction was observed, (IC_{25})) responses were measured.

Sublethal and lethal responses in the laboratory were only found in effluent dilutions that far exceeded concentrations measured in the field. Seven-day, static renewal tests conducted on water fleas showed that pulp-mill effluent was not acutely toxic: all LC_{50} values were greater than 100%. Sublethal responses were observed over a wide range in effluent dilution: 6.25–50% effluent for NOEC; 9.38–100% effluent for LOEC values, and 8.68–73.4% effluent for IC_{25}, all of which were higher than typical effluent concentrations in the Athabasca River at complete mixing (1–4%, Culp et al., 2000d). Similar tests conducted on early life stages of fathead minnows showed no significant effect of the pulp-mill effluent on survival or growth; effect concentrations were greater than 100% of the effluent concentration in all cases (IC_{25}; LC_{50}; NOEC). Responses in the alga *Selenastrum capricornutum* were more variable (mean $IC_{25} = 22\%$) and were further examined in subsequent mesocosm studies.

The laboratory test results weakened the prevailing view at the time of the study that the primary effect of the pulp mill effluent was as a toxic stressor.

24.3.4 Mesocosm Studies

Several mesocosm studies were conducted to resolve interactions among nutrients and BKME. One of these studies examined the interactive effects of different nutrients (nitrogen and phosphorus) on periphyton communities and determined phosphorus concentrations that maximized algal biomass (Chambers et al., 2000). A subsequent study compared the effect of nitrogen and phosphorus loading with responses measured in 1% BKME and field responses of benthos measured near a BKME diffuser (Culp et al., 2000d).

Interactive effects of different nutrients: algae. Artificial stream experiments were conducted to determine the concentration of phosphate that would result in peak periphyton biomass. Benthic algae were allowed to colonize tile or styrofoam substrates that had been placed in flow-through channels supplied with natural river water from the upper Athabasca amended with 0, 1, 10, or 25 µg/L phosphorus as PO_4^{3-}. While field results suggested that upstream reaches were typically constrained by lack of phosphorus, the mesocosm work confirmed that periphyton from upstream reaches of the Athabasca River were P limited. In all trials, increases in phosphate corresponded to an increase in periphyton biomass up to an asymptote (see Figure 24.3). The largest increases in peak biomass were measured between 2 and 5 µg/L soluble reactive phosphorus (SRP) concentrations which coincided with those measured downstream from outfalls. In each of the three trials, the peak biomass responded in a curvilinear trend with mean ambient SRP.

Interactive effects of nutrients and contaminants: invertebrates. To isolate the effects of nutrients from those of a complex effluent (BKME) containing both nutrients and contaminants, periphyton and benthic macroinvertebrate assemblages were collected and inoculated into replicate artificial streams supplied with upper Athabasca River water either untreated or amended with 1% BKME or a mixture of nitrogen and phosphorus. Benthic chlorophyll *a* as well as the abundance of benthic invertebrate families showed a similar increase in streams with added nutrients as in the 1% BKME. Furthermore, these increases were comparable to those found downstream from pulp-mill

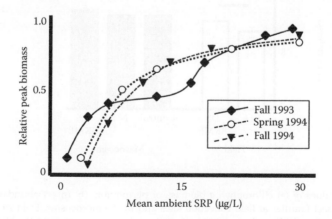

FIGURE 24.3

Response of periphyton grown in artificial streams in fall 1993, spring 1994, or fall 1994 to addition of 0, 1, 10, or 25 µg/L soluble reactive phosphorus (SRP), expressed as relative peak biomass (peak biomass for a treatment normalized to the maximum observed among all treatments). In repeated trials, increased SRP caused biomass to increase to an asymptotic value. (Adapted from Chambers, P. A. et al., 2000. *J Aquat Ecosyst Stress Recov* 8 (1):53–66, with permission.)

outfalls, supporting the hypothesis that the predominant effect of pulp-mill effluent on stream communities in the Athabasca River was one of nutrient enrichment (see Figure 24.4).

These experiments demonstrated that nutrients in the presence of BKME chemical contaminants are sufficient alone to cause increased algal growth and invertebrate abundance and showed that chemical contaminants do not modify the response. At the time of the study, BKME was widely thought to

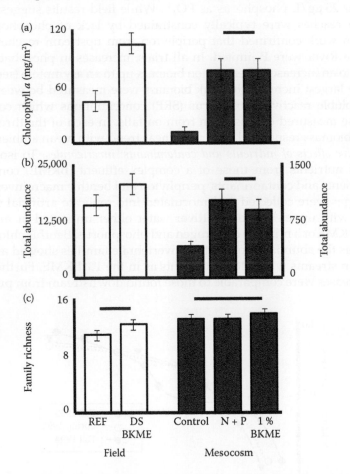

FIGURE 24.4
Mean (±SE) values of (a) chlorophyll *a* biomass of periphyton, (b) insect abundance, and (c) number of insect families at field collection sites and in the mesocosms. Field sites included a reference location (REF) and an effluent-affected reach downstream from the bleached kraft mill effluent (DS BKME). Mesocosm treatments included a control, a nitrogen and phosphorus (N + P) treatment, and a 1% concentration of combined bleach kraft and sewage effluent (1% BKME). Abundance for field samples represents insect density (no./m²), while mesocosm abundance includes the total number of insects per mesocosm. Horizontal bars connect means that are not significantly different ($p \geq 0.05$). (Adapted from Culp, J. M. et al., 2000b. *J Aquat Ecosyst Stress Recov* 7 (2):167–176, with permission.)

be toxic. Thus, the trend of increasing primary and secondary biomass, as opposed to decreasing, was an important finding.

24.4 Conclusions

The above studies all supported the hypothesis that nutrients in pulp-mill efflu-ents and sewage inputs caused the observed enrichment patterns in benthic communities. Increased abundances of algae and invertebrates and improved condition of sculpins were routinely detected immediately downstream from sewage and pulp-mill outfalls. The field study results demonstrated evidence of enrichment below outfalls including increased nitrogen and phosphorus concentrations, periphyton biomass, and improved condition of sculpins. The responses to enrichment did not appear to be additive. Rather, below every sewage and pulp-mill outfall, periphyton biomass increased to a threshold level (250–400 mg/m²) regardless of whether the discharge was sewage or sewage in combination with pulp-mill effluent. Therefore, wastewater and pulp-mill effluent together were no worse than the action of either effluent individually.

The observation of a threshold effect was supported by mesocosm tests that demonstrated a plateau beyond which additional nutrients were no lon-ger eliciting an enrichment response. They confirmed that when a threshold for a nutrient is exceeded, different essential nutrients may become limiting (i.e., instead of phosphates, nitrogen, or possibly silica) as evidenced in the water quality results (e.g., Figures 24.2 and 24.3).

Although concentrations of toxic chemicals in sediments were higher in sediments downstream from outfalls and were higher than national and provincial guidelines, in laboratory studies, only effluent concentrations far greater than those found in the field produced toxic effects in laboratory test species. Finally, the abundance of some invertebrates decreased with further distance downstream from point sources, as both contaminant concentra-tions and nutrient concentrations decreased. It has since been suggested that lower taxa richness downstream from outfalls caused by the eutrophic con-ditions (e.g., increased primary production) may contribute to downstream population declines of sensitive taxa.

24.5 Discussion: The Northern River Basins Study

The assessment described in this chapter was only one of many studies con-ducted as part of the NRBS. The NRBS set out to identify and quantify the impact of multiple complex stressors (e.g., nutrient additions, contaminants, and flow changes) in three large rivers: the Athabasca, Peace, and Slave in northwestern Canada. The study was initiated in 1992 when the ecological understanding of northern and arctic rivers was still in its infancy compared to temperate and tropical systems (Culp et al., 2000c). Among the challenges, the Athabasca, Peace, and Slave Rivers all flow from more southerly, and human-influenced, landscapes and can easily be ice-covered for 6 months of the year (Culp et al., 2000d). As such, developments in these river ecosystems are now recognized as having broad impacts on the biological, chemical, and physical characteristics of proximate as well as downstream, often arctic, habitats. Also at the time of the original studies, relatively little attention had been paid to the impact of nutrient enrichment and its potential masking of contaminant effects, on aquatic biota in waters receiving complex effluents such as pulp mill and sewage effluent (Bothwell, 1992; Dubé et al., 1997).

The NRBS was a multiyear, multidisciplinary investigation to determine anthropogenic impacts on six ecosystem components (hydrology, contaminants, nutrients and dissolved oxygen, drinking water, food chain, and other uses) and the interaction among these components as identified through traditional knowledge and quantitative modeling (Northern River Basins Study Board, 1996).

These studies were among the first to describe the impacts of effluent mixtures in northern aquatic ecosystems. In particular, they highlight the important influence of naturally low levels of nutrients on organism responses in boreal and arctic systems. Management of complex effluents with the potential for multiple (and cumulative) impacts benefit from the use of integrated observational and experimental approaches in order to understand cause and effect relationships.

25

Applying CADDIS to a Terrestrial Case: San Joaquin Kit Foxes on an Oil Field

Glenn W. Suter II and Thomas P. O'Farrell

This terrestrial case study illustrates the use of both spatial and temporal patterns and a simulation model to derive evidence. Conclusions were formed by combining causes, which resulted in a refined conceptual model.

CONTENTS

25.1 Summary

San Joaquin kit foxes on the Naval Petroleum Reserve Number 1 (NPR-1) were observed to decline in abundance during the period 1980–1985. NPR-1 is located on the Elk Hills, on the western edge of the San Joaquin Valley, west of Bakersfield, California. It is an oil field that was held in reserve for the Navy until 1976 when Congress ordered that it be developed to produce at the maximum efficient rate.

The San Joaquin kit fox (*Vulpes macrotis mutica*) is an endangered subspecies. The minimum estimate of kit fox abundance in the NPR-1 study area in summer, based on capture–recapture estimates, declined from a high of 153 in 1981 to a low of 10 in 1991 (Harris et al., 1987; U.S. DOE, 1993). The kit fox decline in the 1980–1986 period suggested that the population was being negatively affected by petroleum development activities. This case study describes the first formal causal assessment of the 1980–1986 decline. Six candidate causes for the decline were considered: (1) prey abundance, (2) habitat alteration, (3) predation, (4) toxic chemicals, (5) vehicular activity, and (6) disease. In addition, for each of the first two candidate causes, two causal pathways were considered: from disturbance and from climate.

The available evidence indicates that the cause of the kit fox decline in the early-to-mid 1980s was increased predation by coyotes. Predation by coyotes was the major cause of death in kit foxes, and a demographic analysis showed that the decline was due to high mortality, particularly of young foxes, with little influence from low fecundity or high emigration. The mechanism (i.e., mortality) was shared with vehicular accidents. However, the mortality rate for accidents was much lower than for predation, so accidents contributed but were not sufficient to account for the decline. After a coyote control program began, coyote abundance declined and the kit fox population stabilized.

The availability and utilization of lagomorph prey (Candidate Cause 1) were strongly related to kit fox abundance, but clinical symptoms of poor condition or starvation were not observed in trapped animals or during necropsies. Prey availability can affect fecundity and females on developed areas produced fewer pups, but the demographic analysis indicated that variance in kit fox fecundity did not significantly contribute to variance in kit fox abundance. Hence, prey availability does not appear to be a significant proximate cause. However, it may be a contributing factor in other sources of mortality. That is, fewer large prey and greater use of small prey implies more time spent hunting and greater exposure to coyotes and vehicles.

Disease (Candidate Cause 6) was eliminated as a contributor. Very few of the trapped foxes were observed to be diseased, little evidence of disease was found during necropsies, and neither serological nor hematological analyses showed evidence of an epizootic that would account for the decline. Disease has caused population declines in other places, but this supporting evidence would be relevant only if there is some positive evidence from the site.

The evidence for habitat alteration (Candidate Cause 2) was ambiguous. The area devegetated is known, but the quality of habitat provided by the vegetated and devegetated areas and the effects of human activities on habitat utility for kit foxes are unknown. The fact that emigration from the developed areas exceeded emigration from the undeveloped areas suggests that habitat quality was lower in developed areas.

The evidence for environmental contaminants (Candidate Cause 4) was inconsistent and complex. Contaminants from oil development were present and potential routes of exposure were identified. Elemental analyses of kit fox fur found that foxes from developed NPR-1 were not highly exposed on average, and only two chemicals, arsenic and barium, were elevated in the fur from most foxes from developed NPR-1 relative to reference sites. Arsenic levels in three foxes reached levels that indicate acute toxicity in humans, but those foxes appeared healthy when captured and their longevity was not apparently reduced. Barium is much less toxic, and although fur levels on NPR-1 were high, they overlapped with fur from reference sites. One fox died after becoming coated in oil. In sum, there was no evidence that toxic exposures could account for the high mortality rates that caused the decline.

Other analyses have attributed variation in kit fox abundance to climate, but precipitation was not particularly low during the decline. In particular, two very good precipitation years occurred in the midst of the decline without influencing the decline in fox abundance. In addition, climatic differences cannot account for the differences between sites. This analysis focuses on a particular localized decline rather than larger-scale and longer-term dynamics addressed by other analyses. In causal assessment, spatial and temporal scales are critical.

25.2 Problem Formulation

In the late 1970s and early 1980s, the Elk Hills, California, were converted from a petroleum reserve into an active oil field through the drilling of more than 1000 production wells and the construction and operation of support facilities. Quantitative monitoring began in 1980 and documented a precipitous decline in the abundance of the endangered San Joaquin kit fox from 1981 to 1986. Concern for the fox population led to intensive demographic studies and studies targeting disease, toxicity, prey abundance, and predation as candidate causes. However, no causal assessment was performed at that time to determine the cause of the decline. The question was recently revisited as a demonstration of the applicability of the CADDIS to a terrestrial case involving a wildlife population. A full presentation of that assessment has been published by Suter and O'Farrell (2008). This chapter summarizes the most important evidence and inferences from that assessment.

25.2.1 The Case

25.2.1.1 The Affected Site

NPR-1 is located about 30 miles southwest of Bakersfield, Kern County, California (see Figure 25.1). It encompasses 47,245 acres, including most of the low foothills of the Temblor Range known as the Elk Hills, which extend southeastward into the San Joaquin Valley. The topography consists of gently rounded slopes with narrow divides, highly dissected draws and dry stream channels in the higher elevations, and gently rolling hills along the perimeter.

Analysis of causal relationships on NPR-1 is limited by the confounding of topography (uplands and lowlands) and degree of development (developed and undeveloped). Almost all of the petroleum developments on NPR-1 were located in the central uplands, and the majority of the lowlands were undeveloped. There were insufficient areas of either developed lowlands or undeveloped uplands to distinguish those factors. Also, the area was not pristine prior to recent disturbances. Oil development has affected the site to some degree since the early twentieth century. Measurements of conditions prior to petroleum developments or before 1974 were unavailable. The analysis is also limited because kit foxes are highly mobile and can have home ranges that include both developed and undeveloped habitats. Hence, it was necessary to consider spatially disjoined reference sites.

25.2.1.2 Comparison Sites

This causal analysis uses three comparison sites. One is a nearby oil field and the other two have no oil development (see Table 25.1).

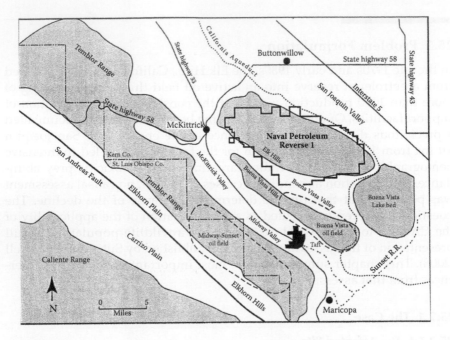

FIGURE 25.1

A map showing the location of NPR-1 on the Elk Hills and its immediate context including the Buena Vista oil field (NPR-2), a developed reference area, and the Carrizo Plain, an undeveloped reference area. (Adapted from U.S. Department of Energy (U.S. DOE). 1993. Supplemental Environmental Impact Statement, Petroleum production at maximum efficient rate, Naval Petroleum Reserve No. 1 (Elk Hills), Kern County, California. Tupman, CA, U. S. Department of Energy. DOE/EIS 0158.)

The Buena Vista Oil Field (NPR-2) occupies the Buena Vista Hills, south of NPR-1. It has lower elevations but is ecologically similar to NPR-1. However, the oil resources there were developed earlier than those on NPR-1, and although some oil development activities occurred in the 1980s, there was no increase in production during the time period of interest. Kit fox demographic studies on NPR-2 began in 1983. The population was apparently stable until 1988, but declined thereafter (see Figure 25.2). It is likely that some movement of foxes between the two reserves occurs, but the fact that the decline on NPR-1 in the early-to-mid-1980s was not mirrored on NPR-2 suggests that they are not a single population.

The Carrizo Plain lies south of NPR-1 and NPR-2, beyond the Temblor Range in San Louis Obispo County (see Figure 25.1). It is primarily grassland, supporting some cattle grazing. It has no oil production.

Camp Roberts (not shown in Figure 25.1) is a California Army National Guard training site in San Louis Obispo and Monterey Counties between the Salinas River floodplain and the Santa Lucia Mountains. It encompasses 172 km² of primarily rolling hills with grassland, oak woodland, and chaparral. It has no oil production.

TABLE 25.1

Sites Considered in the Causal Analysis of the Kit Fox Decline

Site	Development	Kit Foxes	Status
Elk Hills/ NPR-1 developed	Oil field with active drilling, facility construction, and oil production during the kit fox decline	Primary site of the decline	The affected site
Elk Hills/ NPR-1 undeveloped	Low density of oil development with pipe lines and other support facilities	The decline was later and less intense	Near field comparison: low oil activity
Buena Vista Hills/NPR-2	Heavily developed prior to the period of concern. Little oil production or active development	No apparent decline in the period of concern	Near field comparison: high disturbance and contamination, low development activity
Carrizo Plain	No oil development Cattle grazing	No apparent decline in the period of concern	Far field comparison: no oil activity
Camp Roberts	No oil development Military training activities	No apparent decline in the period of concern	Far field comparison: no oil activity

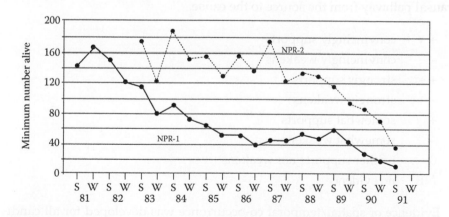

FIGURE 25.2
Minimum numbers of kit foxes on NPR-1 and NPR-2 from summer (S) and winter (W) surveys. The minimum population is the sum of the individuals trapped during each trapping session, plus the number of untrapped foxes that were known to be alive because they were trapped in a previous and a subsequent session. (Adapted from U.S. Department of Energy (U.S. DOE). 1993. Supplemental Environmental Impact Statement, Petroleum production at maximum efficient rate, Naval Petroleum Reserve No. 1 (Elk Hills), Kern County, California. Tupman, CA, U. S. Department of Energy. DOE/EIS 0158.)

25.2.1.3 *The Ecological Effect: Kit Fox Decline*

The minimum number of San Joaquin kit foxes in the NPR-1 study area, based on capture–recapture estimates, declined by approximately 75% from 1981 to 1986 (Harris et al., 1987; U.S. DOE, 1993) (see Figure 25.2). It appeared that the population was being negatively affected by petroleum development activities, prompting a biological opinion by the U.S. Fish and Wildlife Service in 1987 calling for studies to address toxicity as a cause. The population slowly increased from 1986 to 1989, but it declined again from 1989 to 1991 (see Figure 25.2). The period 1987–1991 is of interest primarily in terms of helping to understand the 1981–1986 decline.

25.2.2 Candidate Causes and Sources

Six candidate causes were evaluated (see Table 25.2) and numbered for ease of reference. The first two candidate causes, reduced prey abundance and habitat alteration, were subdivided further to distinguish sources: oil development, climate change, and increased competition from other predators.

25.2.3 Methods

Evidence was developed and organized using the CADDIS types of evidence (see Table 4.2).

The following system was used to score each type of evidence. Where evidence permitted, ranks were developed for the candidate cause and for each causal pathway from the source to the cause.

+ + +   convincingly supports

− − −   convincingly weakens

+ +   strongly supports

− −   strongly weakens

+   somewhat supports

−   somewhat weakens

0   neither supports nor weakens

NE   no evidence

Evidence of spatial/temporal co-occurrence was developed for all candidate causes by comparing conditions from the case, that is, the kit fox population on the Elk Hills (NPR-1), to the different comparison sites (see Table 25.1). Different comparisons were used depending on the data available and the candidate cause being evaluated. Three spatial comparisons were possible (1) The NPR-1 site was divided into developed and undeveloped areas, which allows for comparison of areas in which foxes were directly exposed to drilling, construction, and other development activities during the surge

TABLE 25.2

Candidate Causes

Candidate Causes and Source Subcategories	Notes
1. Reduced prey abundance from: a. Disturbance during oil development b. Climatic effects (especially reduced precipitation) c. Competition from coyotes	Includes changes in the relative abundance of prey species, particularly declines in lagomorphs (black-tailed jackrabbit and desert cottontail) relative to small rodents (primarily kangaroo rats and pocket mice)
2. Habitat alteration from: a. Disturbance during oil development b. Climatic effects (especially reduced precipitation)	Includes the abandonment of the site by foxes seeking more acceptable habitat or to reduced reproductive success due to fewer adequate denning sites. In addition to physical disturbance of the soil and vegetation, human activities may cause stress, disruption of hunting, and increased energy expenditure. Activities close to whelping and pupping dens might disturb vixens and cause them to neglect or even abandon their litters. Habitat alteration may be cumulative (e.g., the total area devegetated by development) or immediate (e.g., the effects of active construction and drilling activities on the willingness of foxes to use an area)
3. Predators	The increased abundance of coyotes results in increased killing of foxes. Oil development may make the site more attractive to coyotes by increasing road kills and food waste to be scavenged, and until the control program began, by protecting coyotes from hunters
4. Toxic chemicals	Toxic effects on the foxes due to exposure to chemicals associated with oil development. The two principal sources were spills of oil or chemicals used in production activities or waste ponds that contained produced water (water pumped up with the petroleum)
5. Vehicular activity	Kit fox mortality due to being struck by vehicles or injured by equipment during oil production. Increased oil production increased vehicle traffic and construction activities that may bury foxes in their dens
6. Disease	Diseases may have been endemic or may have been brought to the site by coyotes or by humans from their pets

in oil production and areas where there was very little development activity. (2) Comparisons could be made between NPR-1 as a whole and NPR-2. This is a comparison of an actively developing oil field and one that is highly developed but where little new development was occurring. (3) Comparisons can be made between the developed and undeveloped portions of both oil fields combined (NPR-1 and NPR-2). This comparison incorporates residual contamination and the loss of habitat due to oil development, but not the effects of active development.

Although no baseline period is available to allow for comparison of the development period with a predevelopment period, temporal comparisons are possible. In the period under investigation (1981–1986), the NPR-1 kit fox population declined rather precipitously but the NPR-2 population was relatively stable at a high level (see Figure 25.2). Hence, we are interested in what happened in the early-to-mid-1980s on NPR-1 that did not occur on NPR-2.

25.3 Evidence

25.3.1 Prey Abundance

25.3.1.1 Prey Abundance: Spatial/Temporal Co-Occurrence

Prey abundance was judged to co-occur with the kit fox decline if prey abundance was low where and when the kit fox decline occurred. Two comparisons were possible.

25.3.1.1.1 Developed versus Undeveloped

Lagomorphs, initially the primary prey of kit foxes on Elk Hills, declined in both developed and undeveloped habitats (1980–1984 based on road surveys; see Figure 25.3), but the decline was much greater (5.3×) in the developed area where they were more abundant than in the undeveloped area (1.9×) (see Table 25.3). Kangaroo rat abundances did not show a trend, but

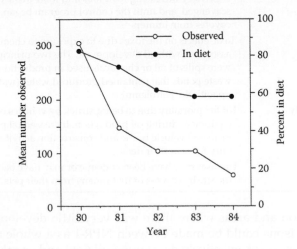

FIGURE 25.3
Number of lagomorphs observed on NPR-1 and their percentage in the diets of San Joaquin kit foxes from 1980 to 1984. (Adapted from U.S. Department of Energy (U.S. DOE). 1993. Supplemental Environmental Impact Statement, Petroleum production at maximum efficient rate, Naval Petroleum Reserve No. 1 (Elk Hills), Kern County, California. Tupman, CA, U. S. Department of Energy. DOE/EIS 0158.)

TABLE 25.3

Relative Abundance of Lagomorphs (Number Observed) and Kangaroo Rats (Trapping Success) in June–November Counts in Two Habitats on Elk Hills, California, 1980–1984

	Undeveloped			Developed		
		Kangaroo Rats			Kangaroo Rats	
Year	Number of Lagomorphs Observed	Trapping Effort (Trap-Nights)	Trapping Success (%)	Number of Lagomorphs Observed	Trapping Effort (Trap-Nights)	Trapping Success (%)
1980	139	250	41.6	850	675	5.0
1981	103	900	41.7	630	1598	2.1
1982	115	900	34.3	246	1800	2.4
1983	89	899	38.9	282	2399	2.0
1984	71	300	57.3	160	900	7.6

Source: Scrivner, J. H., T. P. O'Farrell, and T. T. Kato. 1987. *Diet of the San Joaquin kit fox, Vulpes macrotis mutica, on Naval Petroleum Reserve #1, Kern County, California, 1980–1984.* Santa Barbara, CA, EG&G Energy Measurements, Inc. U.S. Department of Energy Topical Report No. 10282-2168.

trapping success was much higher in the undeveloped area (see Table 25.3). Hence, a decline in the principal prey co-occurred in space with disturbance and with the most rapidly declining component of the kit fox population, which supports prey abundance as a cause (1 = ++). This evidence supports prey abundance through the disturbance pathway as a cause (1a = +). It does not support the climatic pathway because the climate did not differ between areas of NPR-1. However, that evidence is weak because local weather or soil moisture data were not available (1b = –).

25.3.1.1.2 NPR-1 versus NPR-2

Transect surveys from 1983 to 1991 showed a consistent decline in jack rabbits for NPR-1 as a whole (see Figures 25.3 and 25.4). On NPR-2, lagomorph densities did not decline until 1987 but then declined until 1991 (U.S. DOE, 1993) (see Figure 25.5). That is consistent with the delay in onset of kit fox decline on NPR-2 relative to NPR-1 (see Figure 25.2). Hence, the declines in kit foxes on both NPR sites co-occurred with declines in lagomorph prey, which supports prey abundance as a cause (1 = ++), but the declines were not contemporaneous. This evidence supports prey abundance through the disturbance pathway (1a = +). It does not support the climatic pathway, but without local weather or soil moisture data the evidence is weak (1b = –).

25.3.1.2 Prey Abundance: Temporal Sequence

Since the decline in both lagomorphs and foxes appears to have been underway at the beginning of the time series, it is not possible to determine

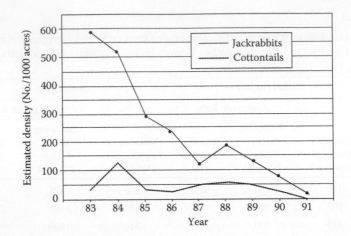

FIGURE 25.4
Estimated density of lagomorphs on NPR-1 from 1983 to 1991. (Adapted from U.S. Department of Energy (U.S. DOE). 1993. Supplemental Environmental Impact Statement, Petroleum production at maximum efficient rate, Naval Petroleum Reserve No. 1 (Elk Hills), Kern County, California. Tupman, CA, U. S. Department of Energy. DOE/EIS 0158.)

whether a decline in prey began before the decline in foxes. Temporal sequence might also be derived from a time series, if there were a consistent lag between a decline in abundance of prey and a decline in kit foxes. However, the steady decline in both predators and prey during the period of concern precludes the identification of a clear temporal sequence (see Figures 25.2 and 25.3). As a result, the correlations of lagomorph and fox abundance are not consistently better with a 1 year time lag than without (see Section 25.3.1.4). The temporal sequence is undefined (1 = 0).

25.3.1.3 Prey Abundance: Evidence of Exposure, Mechanism or Mode of Action

During the period of decline, the proportion of fecal samples from NPR-1 containing fur of lagomorphs decreased and kangaroo rats, usually the secondary prey, increased in developed and undeveloped areas (see Table 25.4). This indicates changes in prey utilization that are consistent with a decline in preferred prey and switching to secondary prey in all areas. This evidence is clear and consistent with declines in prey abundance as a cause (1 = ++). Since it occurred in developed and undeveloped areas, it is consistent with the climate pathway, but not disturbance (1a = − and 1b = +).

25.3.1.4 Prey Abundance: Covariation of the Stressor and the Effect

Kit fox abundance was linearly related to lagomorph abundance in the previous year on developed NPR-1 during 1981–1985 based on road surveys ($r^2 = 0.68$) and less well related to lagomorph abundance in the same year

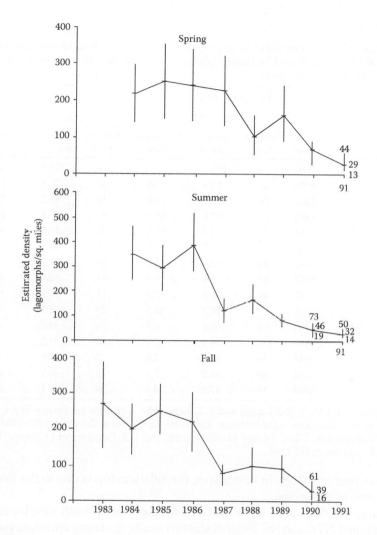

FIGURE 25.5
Lagomorph density estimates by season on NPR-2. (Adapted from U.S. Department of Energy (U.S. DOE). 1993. Supplemental Environmental Impact Statement, Petroleum production at maximum efficient rate, Naval Petroleum Reserve No. 1 (Elk Hills), Kern County, California. Tupman, CA, U.S. Department of Energy. DOE/EIS 0158.)

($r^2 = 0.31$). Kit fox abundance was even better related to lagomorph abundance in the same year on undeveloped NPR-1 during 1981–1985 based on road surveys ($r^2 = 0.98$) and less well related to lagomorph abundance in the previous year ($r^2 = 0.66$).

Kit fox abundance was highly linearly correlated with jack rabbit abundance in the same year on NPR-1 during 1983–1991, based on transect surveys ($r^2 = 0.89$) and less well correlated with jack rabbit abundance in the

TABLE 25.4

Frequency of Occurrence (%) of Lagomorphs and Kangaroo Rats in the Scats of San Joaquin Kit Foxes Collected in Three Habitats and Two Time Periods between 1980 and 1984, Elk Hills, California

| | | Frequency of Occurrence (%) | | | | | |
| | | Dec.–May | | | Jun.–Nov. | | |
Habitat	Year	Sample Size	Lago-morphs	Kangaroo Rats	Sample Size	Lago-morphs	Kangaroo Rats
Undeveloped flat	1980	5	100.0	0.0	4	100.0	0.0
	1981	26	84.6	3.8	60	78.3	13.3
	1982	76	77.6	19.7	31	45.2	32.3
	1983	64	39.1	45.3	33	27.3	42.4
	1984	22	36.4	50.0	32	43.8	37.5
Undeveloped hilly	1980	17	100.0	5.9	5	60.0	0.0
	1981	49	87.8	14.3	21	81.0	0.0
	1982	51	78.4	9.8	15	33.3	20.0
	1983	49	61.2	20.4	13	76.9	15.4
	1984	12	58.3	33.3	29	51.7	20.7
Developed hilly	1980	5	100.0	0.0	24	91.7	4.2
	1981	122	88.5	4.9	108	85.2	0.9
	1982	176	82.7	2.8	78	67.9	5.1
	1983	69	81.2	7.2	22	40.9	18.2
	1984	17	47.1	29.4	28	57.1	17.9

Source: Scrivner, J. H., T. P. O'Farrell, and T. T. Kato. 1987. *Diet of the San Joaquin kit fox, Vulpes macrotis mutica, on Naval Petroleum Reserve #1, Kern County, California, 1980–1984.* Santa Barbara, CA, EG&G Energy Measurements, Inc. U.S. Department of Energy Topical Report No. 10282-2168.

previous year ($r^2 = 0.56$). In both cases, the relationship is due to the first and last 2 years of the series.

In sum, the decline of foxes and of lagomorphs on both developed and undeveloped NPR-1 in the 1980s results in multiple strong correlations from two different lagomorph surveys, with or without a time lag. This result is consistent with loss of prey as a cause (1 = ++), and because the correlations occurred in both developed and undeveloped areas, with the climate pathway (1b = +) but not disturbance (1a = –).

25.3.1.5 Prey Abundance: Causal Pathway

25.3.1.5.1 Disturbance

The primary pathway for disturbance is from oil development activities to reduced vegetation, reduced prey, and reduced kit foxes. The creation of well pads and other construction activity inevitably destroyed vegetation, thereby reducing food and cover for prey organisms. The lagomorph and kit fox declines were greatest in the developed areas of NPR-1. Hence, all steps

in the causal pathway were present which qualitatively supports the disturbance pathway (1a = +).

25.3.1.5.2 Climate

The primary causal pathway for climate is from reduced precipitation
to reduced vegetation, reduced prey, and reduced kit foxes. During the
1981–1986 period of kit fox decline and the three preceding years, effective
precipitation measured in Bakersfield was above average in five years and
below average in three (see Figure 25.6). In particular, the good and extremely
good precipitation in 1982 and 1983 had no apparent effect on the ongoing
kit fox decline (see Figure 25.2). This is contrary to other studies that found
a relationship when they included both NPR sites and a longer time period
(Cypher et al., 2000). That discrepancy suggests that something was negating
the expected precipitation effects on NPR-1 in the period of concern. This
evidence weakens climate (1b = –).

Kit fox abundance on NPR-1 was not correlated with precipitation in the
same year, the prior year, 2 years previously, or 3 years previously. This was
true for both the period of decline (1981–1986) and for the entire study period
(1981–1990). (The time lags account for the time required for vegetation and
prey to respond to precipitation.) This evidence weakens climate (1b = –).

In addition, if precipitation was the source of the kit fox decline, one
would expect to see the same pattern of decline on NPR-2. However, kit fox

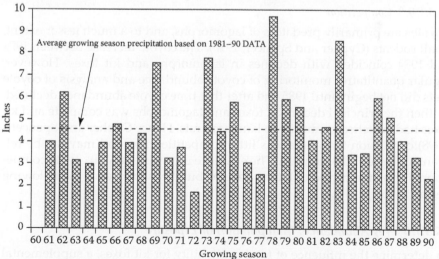

Growing season defined as precipitation for Jan.–Mar. of current year plus Oct.–Dec. of previous year

FIGURE 25.6
Average growing season precipitation for Bakersfield, California. (Adapted from U.S.
Department of Energy (U.S. DOE). 1993. Supplemental Environmental Impact Statement,
Petroleum production at maximum efficient rate, Naval Petroleum Reserve No. 1 (Elk Hills),
Kern County, California. Tupman, CA, U. S. Department of Energy. DOE/EIS 0158.)

abundance on NPR-2 was stable during 1983–1987 and declined thereafter (see Figure 25.2). (Note that the apparent fluctuations in Figure 25.2 are seasonal rather than annual.) This evidence weakens climate (1b = –).

The relationship of total lagomorph counts (mostly jack rabbits) from road surveys on NPR-1 in 1980–1984 to precipitation in the same year and the previous year was analyzed by linear regression. With one exception, correlations were extremely low for both developed and undeveloped areas. In undeveloped areas, lagomorph abundance was negatively correlated with precipitation in the previous year, which is contrary to expectations. Jack rabbit abundance by transect survey was weakly positively correlated with precipitation in the same year ($r^2 = 0.30$) or in the previous year ($r^2 = 0.32$) during 1983–1990. This evidence weakens climate (1b = –).

Based on visual inspection, vegetation production on NPR-1 declined between 1988 and 1991 (U.S. DOE, 1993). Also, at an undisturbed 32-acre site on NPR-1, annual dry matter production declined from 1596 pounds/acre in 1988, to 644 pounds/acre in 1989 and to 85 pounds/acre in 1990. This corresponds to a period of steady decline in precipitation (see Figure 25.6). This evidence is consistent with precipitation as a cause of reduced plant production, but it does not relate to the principal period of kit fox decline when precipitation was higher (1981–1986). This evidence is ambiguous (1b = 0).

In sum, the evidence for the causal pathway from climate to vegetation, lagomorph prey, and kit fox is negative (overall 1b = –).

25.3.1.5.3 Competition

Coyotes are primarily predators of lagomorphs, and to a much lesser extent, small rodents (Cypher and Spencer, 1998). The coyote increase between 1979 and 1984 coincided with declines in lagomorphs and kit foxes. However, regular quantitative monitoring of coyote abundance and analysis of coyote diets did not begin until 1985 and after that time coyote abundance declined. By then the principal decline of foxes and lagomorphs was complete and kit foxes had switched primarily to kangaroo rats. Hence Cypher and Spencer's (1998) conclusion that there was little competition for food may not be relevant to the period of concern. The evidence is consistent with coyote competition during the period of kit fox decline, but the evidence for the following period is not (1c = 0).

25.3.1.6 Prey Abundance: Manipulation of Exposure

To determine the influence of food availability for kit foxes, a supplemental feeding study was conducted in 1988 and 1989. Supplemental feeding at individual occupied dens in 1988 increased survival of pups relative to controls from 10% to 50% and increased survival of adults from 30% to 70% (U.S. DOE, 1993). Results were positive in 1989 as well, but the differences were smaller due to increased survival of unfed foxes. That may be due to heavier coyote control activities in 1989 (U.S. DOE, 1993). This evidence supports prey abundance, but is not strong because the studies occurred after the decline and the manipulation was not of the prey (1 = +).

25.3.1.7 Prey Abundance: Symptoms

25.3.1.7.1 Starvation

Starvation was not reported to be a cause of death in kit fox necropsies. That is negative evidence for a shortage of prey as a cause of mortality (1 = –).

25.3.1.7.2 Reproductive

Male-biased sex ratios of pups, as observed on developed NPR-1, are characteristic of female canids that are in poor condition due to poor nutrition (Zoellick et al., 1987). This symptom supports prey abundance but may occur with other causes (1 = +). Because this symptom occurred on developed NPR-1, it supports prey abundance through the disturbance pathway (1a = +). It does not support the climatic pathway because the climate did not differ between areas of NPR-1 (1b = –).

25.3.1.8 Prey Abundance: Stressor–Response Relationships from Ecological Simulation Models

The most likely demographic mechanism for low prey abundance is poor nutrition and reduced fecundity, but the observed reduction in fecundity was only a minor contributor to the decline (Floit and Barnthouse, 1991). This evidence weakens the case for prey abundance (1 = –).

25.3.1.9 Prey Abundance: Stressor–Response from Other Field Studies

Numerous studies have demonstrated a positive correlation between the abundances of mammalian predators and their prey. In particular, Egoscue (1975) showed that the abundance of kit foxes (*V. m. nevadensis*) in Utah followed the abundances of black-tailed jack rabbits. That population also showed an elevated male:female ratio of pups. This relationship agrees qualitatively with the relationship at the site (1 = +).

25.3.2 Habitat Alteration

On NPR-1, habitat alteration has been thought to result from disturbance associated with oil development (2a—Disturbance) or climatic effects (2b—Climate). The climatic effects are assumed to be reduced plant biomass and production, resulting in reduced habitat quality. In contrast, oil development may act through loss of vegetation, noise, human presence, or other disturbances. Although habitat preferences in terms of vegetation types are known, no habitat model is available for kit foxes that would allow for quantification of the effects of disturbance on habitat quality. The area developed, number of wells drilled, and volume of oil produced are used as surrogates for habitat disturbance. Growing-season precipitation and plant production were used as surrogates for habitat alteration due to climate.

25.3.2.1 Habitat Alteration: Spatial/Temporal Co-Occurrence

25.3.2.1.1 Disturbed versus Undisturbed

The NPR-1 kit fox decline was most severe in the disturbed areas. By 1990, very few foxes in the NPR-1 study area occurred in the developed upland areas; the remaining foxes were found primarily in the flatter undeveloped areas (U.S. DOE, 1993). Hence, the decline spatially co-occurred with cumulative habitat disturbance. This evidence supports disturbance of habitat (2a = +).

25.3.2.1.2 Temporal Co-Occurrence—Disturbance

During the period of decline (1981–1986), oil development continued with a peak in 1982–1983 followed by a relatively low level of drilling. Given the possibility of time lags and cumulative effects, temporal co-occurrence is ambiguous (2a = 0).

25.3.2.1.3 NPR-1 versus NPR-2—Climate

Precipitation was believed to be similar on both developed and undeveloped areas of NPR-1 and on NPR-2, so the differences in the rates and timing of kit fox declines is not accounted for by climatic effects on habitat (2b = –).

25.3.2.2 Habitat Alteration: Temporal Sequence

The period of increased development began in 1974, and drilling appeared to peak in 1976–1978. The beginning of the kit fox decline is uncertain but was no later than the first monitored interval (1981–1982). Hence, the temporal sequence is ambiguous (2 = 0).

25.3.2.3 Habitat Alteration: Stressor–Response Relationships in the Field

25.3.2.3.1 Active Disturbance

The 1981–1986 period of kit fox decline and the full 1981–1990 study period were, in general, also periods of decline in well drilling and oil production. Hence, correlations for active disturbance (i.e., number of wells completed) and kit fox abundance have the wrong sign for the candidate cause. Another approach is to relate the proportional change in kit fox abundance to the number of wells completed in the same year or the previous year, but that yielded no apparent relationships. Hence, the stressor–response relationships weaken the candidate cause (2a = – –).

25.3.2.3.2 Climate

Few data quantify changes in habitat quality that might result from climate and that could be related to fox abundances. However, plant production may be a surrogate for climate-mediated habitat quality. At an undisturbed 32-acre site on NPR-1, annual production declined from 1596 pounds/acre in 1988 to 644 pounds/acre in 1989, and to 85 pounds/acre in 1990 (U.S. DOE, 1993). These data do not correlate well with kit fox abundance in the same year, but they do correlate perfectly ($r^2 = 0.999$) with kit fox abundance in the following year. Although suggestive, correlations based on three data points inspire little confidence; the time series is outside the period of concern, and the period of concern was less arid, so the evidence is ambiguous with respect to the decline (2b = 0).

25.3.2.4 Habitat Alteration: Causal Pathway

25.3.2.4.1 Disturbance

Oil development involves the destruction of vegetation, which diminishes habitat. Noise and human activity also diminish habitat during the period of construction and drilling activity. All steps in this causal pathway were present (2a = ++).

25.3.2.4.2 Climate

The climate was not consistently poor in the period of decline. In particular, while kit foxes steadily declined in the period of concern, precipitation was above average, then below, then above again, and below again (see Figure 25.6). This lack of a relation between precipitation and kit fox abundance weakens the case for climate-induced habitat alteration as a cause (2 = – –).

Vegetation data were available for a later period. Based on visual inspection, vegetation production on NPR-1 declined in undisturbed areas between 1988 and 1991 (U.S. DOE, 1993), but no data are available for the period of decline, so the evidence is ambiguous (2 = 0).

The combined score for the climate-to-habitat pathway is weakly negative (2 = –).

25.3.2.5 Habitat Alteration: Stressor–Response Relationships from Ecological Simulation Models

Because habitat could affect mortality, fecundity, and emigration, the demographic models cannot be used to determine the sufficiency of habitat modification as a cause (2 = 0).

25.3.3 Predators

25.3.3.1 Predators: Spatial/Temporal Co-Occurrence

25.3.3.1.1 Developed versus Undeveloped

Coyotes were more abundant on developed than undeveloped NPR-1 in the period of decline, and the decline was greater on developed NPR-1 (U.S. DOE, 1993). That spatial co-occurrence supports predation (3 = +).

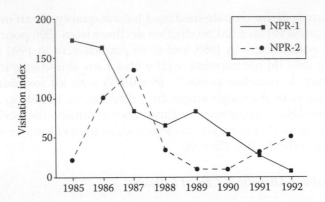

FIGURE 25.7
Winter visitation indices for coyotes on the Elk Hills (NPR-1) and adjacent Buena Vista Hills
(NPR-2), California, 1985–1992.

25.3.3.1.2 Temporal Co-Occurrence on NPR-1

Coyote numbers were lowest when the first survey was conducted on NPR-1
in 1979 (eight observed on 522 miles of transect), but 5 years later, 108 were
observed over those transects (U.S. DOE, 1993). Hence, an increase in coyote
numbers occurred within the same time interval as the observed decline
in kit fox abundance, but the pattern of abundance between those dates in
unknown. Hence, the decline co-occurred with the candidate cause (3 = +).

25.3.3.1.3 NPR-1 versus NPR-2

Coyote abundance on NPR-2 was not known for the period of concern (i.e.,
prior to 1985) (see Figure 25.7). After that period, coyote abundance was
irregular and did not correspond to kit fox abundance patterns except that
both dropped in the late 1980s, after the major kit fox decline (3 = 0).

25.3.3.2 Predators: Temporal Sequence

The low abundance of coyotes in 1979 suggests that an increase in coyote
abundance did not precede the decline in kit foxes, but the timing of the
coyote increase and the beginning of the kit fox decline are unclear. This
evidence is ambiguous (3 = 0).

25.3.3.3 Predators: Covariation of the Stressor and Effects

Between 1979 and 1985, the coyote population on NPR-1 greatly increased and
the kit fox population greatly declined. Then the coyote population declined
from 1985 (when coyote control and regular monitoring, using scent stations,
began) until 1991 (see Figure 25.7) (U.S. DOE, 1993). For 3 years following the
onset of coyote control (1986–1989), the kit fox population stopped declining
(see Figure 25.2). Then from 1989 to 1991, both declined. Because of the switch
from transect surveys to scent stations, correlations with kit fox abundance

cannot be calculated for the period of decline or the entire period of interest. However, the stressor–response relationship is qualitatively correct until 1989. When coyotes increased, kit foxes declined, and when coyotes declined, kit foxes stopped declining. Hence, the stressor–response relationship could not be quantified (3 = NE) and the qualitative association is scored as spatial/temporal co-occurrence, as discussed above.

25.3.3.4 Predators: Causal Pathway

Multiple causal pathways that may associate coyote abundance with oil development were not documented. It is speculated that the absence of shooting and trapping prior to the control program may have allowed the increase in coyote abundance, but this does not explain the initially low numbers. Coyotes may have also benefited from increased road kills to scavenge or from food discarded by workers (Cypher and Spencer, 1998). Those resources inevitably increased with increased oil production activities in the late 1970s and would have been associated with developed areas.

There is some evidence for the causes of the coyote decline. Coyote abundance declined during the period of the control program beginning in 1985. The decline also corresponded to the decline in lagomorph prey and, after 1988, to below average precipitation.

Evidence exists for some steps in the causal pathways to coyote abundance and predation on kit foxes (3 = +).

25.3.3.5 Predators: Evidence of Exposure, Mechanism or Mode of Action

Because coyotes do not consume the foxes that they kill, predation by coyotes was well documented by necropsy of foxes from NPR-1. Coyote-killed foxes were identified by characteristic puncture wounds and associated muscle and bone injuries (Cypher and Spencer, 1998). This evidence for the predation mechanism is clear and consistent (3 = ++).

25.3.3.6 Predators: Manipulation of Exposure

A coyote control program was conducted for 6 years on and around NPR-1 beginning in 1985. The decline of the kit fox population ended in the second year of this period. Overall, 591 coyotes were killed. This evidence supports predation as the cause, but is ambiguous because there is no reference and other factors may confound the effects of coyote control (3 = +).

25.3.3.7 Predators: Stressor–Response Relationships from Ecological Simulation Models

A demographic model of kit foxes on NPR-1 for the period 1980–1986 found that the decline was caused by high mortality, particularly of young-of-the-year foxes (Floit and Barnthouse, 1991). The mortality due to predation

TABLE 25.5

San Joaquin Kit Fox Mortality by Age from Various Causes on
NPR-1 (1980–1986)

Age (Years)	Initial Population	Cause of Death			
		Predation	Vehicle	Other	Unknown
0	152	53	9	4	41
1	79	24	4	0	11
2	54	18	1	0	5
3	42	12	2	2	2
4	28	7	4	0	5

Source: Floit, S. B. and L. W. Barnthouse. 1991. *Demographic Analysis of
a San Joaquin Kit Fox Population.* Oak Ridge, TN: Oak Ridge
National Laboratory. ORNL/TM-11679.

alone was more than sufficient to cause a decline. Although fecundity was
depressed in developed areas relative to undeveloped areas, the population
abundance was insensitive to variance in fecundity. Net emigration from
the developed areas did not significantly contribute to the decline. A less
detailed analysis of an equivalent model for the U.S. Department of Energy
(U.S. DOE, 1993) that extended to 1989 gave qualitatively similar results but
different rates because a period after the decline was included (1986–1989).
In sum, mortality was the mechanism of the decline and predation was the
overwhelming cause of mortality (80%; see Table 25.5). This line of evidence
strongly supports predation as the proximate cause (3 = +++).

25.3.3.8 Predators: Stressor–Response Relationships from Other Field Studies

Coyotes were the cause of 65% of kit fox mortalities on the nearby Carrizo
Plain (Ralls and White, 1995). Coyotes may also be a significant cause of mortality in populations of swift foxes (Scott-Brown et al., 1987) and gray foxes
(Cypher, 1993). Field studies have documented decreases in red fox abundance in apparent response to increased coyote abundance (Harrison et al.,
1989; Major and Sherburne, 1987; Sargeant et al., 1987). This evidence qualitatively supports predation (3 = +).

25.3.4 Toxic Chemicals

The data concerning kit fox exposures and data analyses used for this candidate cause are presented in Suter et al. (1992). That report presents more
results in more detail.

25.3.4.1 Chemicals: Spatial/Temporal Co-Occurrence

Chemicals related to oil production occurred on the developed areas at
much greater concentrations than on undeveloped areas during the period

of population decline. Sources included produced water sumps, oil spills, drilling fluids in sumps or deposited on land, and spills of chemicals used in oil production (Suter, 1988; U.S. DOE, 1993). The arsenical anticorrosion compound W-41 and the hexavalent chromium added to drilling fluids were particular concerns. Arsenic-contaminated water was deposited in six unlined sumps, and arsenic-contaminated wastes were deposited in unlined trenches. Hexavalent chromium was spilled on at least 65 sites. The less toxic trivalent chromium in drilling fluids is widely distributed on the site. This evidence supports toxicants (4 = +).

25.3.4.2 Chemicals: Temporal Sequence

It is hypothesized that increased development after 1976 increased chemical exposures. Until 1986, all wastes were deposited on site, and wastewaters and drilling fluids continued to be deposited in sumps and on land, respectively (U.S. DOE, 1993).

25.3.4.2.1 Arsenic

The arsenical water treatment chemical W-41 was used on NPR-1 from 1922 to 1970. Although arsenic residues persisted at the site, use of arsenical chemicals did not increase immediately before the decline (4 = −).

25.3.4.2.2 Barium

Barite ($BaSO_4$) was used in drilling fluids throughout the period of concern and the years before. Increased drilling before the decline inevitably meant increased use of barite and presumably an accumulation of barite on developed NPR-1 (4 = +).

25.3.4.2.3 Chromium

Lignochromates and hexavalent chromium salts were used in drilling fluids from 1954 to 1983. Hence, it is plausible that chromium exposures increased as drilling increased in the mid-1970s to early 1980s and chromium contamination increased on developed NPR-1 (4 = +).

Overall, this evidence is ambiguous, because there are no data that would provide a temporal sequence from the late 1970s through the early 1980s (4 = 0).

25.3.4.3 Chemicals: Covariation of Stressor and Effect

Chemical exposures were investigated by analyzing the elemental composition of kit fox fur samples. Elements that were not detected by neutron activation analysis in at least half of the samples were excluded, leaving 35 elements. Fox pups were excluded because of relatively low concentrations and the sexes were combined because they did not differ. There were no large or statistically significant correlations of longevity with fur concentrations of any element among the 21 foxes for which both time of death and fur concentration data were available. This evidence weakens toxicants (4 = −).

25.3.4.4 Chemicals: Evidence of Exposure

Analyses of fur samples were also used to determine whether foxes were differentially exposed across sites. Samples came from NPR-1 (49), NPR-2 (12), Camp Roberts (20), and Carrizo Plain (6). Data analysis focused on typical (median concentration) foxes at each site, level of land development, and on foxes with exceptionally high (top decile) concentrations for each element.

Analysis of data for all 35 elements served to indicate the degree of systematic variance among sites in exposure to metals and metalloids. Statistically significant differences among sites were found for all but three elements (chlorine, cobalt, and vanadium) (see Table 25.6). However, most elemental concentrations were not highest on oil fields. Of the 35 elements, Camp Roberts fur had the highest concentrations for 21 and second highest for 6, Elkhorn Plain fur was highest for 7 and second highest for 17, developed NPR-1 fur was highest for 1 and second highest for 6, NPR-2 fur was highest for 4 and second highest for 4 (all NPR-2 foxes were from developed areas), and undeveloped NPR-1 was not highest or second highest for any element, but was lowest for 23 and second lowest for 8. In sum, fur from the undeveloped remote reference sites had the highest concentrations of most detected elements, fur from undeveloped areas on NPR-1 had low concentrations, and sites with extensive oil development had intermediate levels. Hence, although there was a statistically significant positive correlation of fur concentration and percent disturbance of the fox's home range on NPR-1 and NPR-2 combined for 23 elements, consideration of other sites showed that it was attributable to the exceptionally low concentrations for undeveloped NPR-1, not high concentrations in developed locations.

Some elements in fur were associated with oil development and identified as particular hazards.

25.3.4.4.1 Arsenic

Median arsenic levels were higher in fur from developed NPR-1 and NPR-2 than from other sites. Arsenic in fur was strongly positively correlated with percent disturbance, total wells, and new wells in the fox's home ranges. Arsenic concentrations were highly variable among individuals (>2600×) but only moderately variable among site medians (4.3×).

25.3.4.2.2 Barium

Barite is a major constituent of drilling fluids. The median barium concentration in fur from developed NPR-1 was higher than from any other site. The highest individual concentration, and seven of the top 10 concentrations were from foxes from developed NPR-1, but the second and fourth highest were from Camp Roberts. Barium concentrations in fur from both oil fields were significantly positively correlated with percent disturbance and the number of wells.

TABLE 25.6

Ranges of Metal Concentrations (ppm) in Hair of Individual San Joaquin Kit Foxes Sampled on the Elk Hills (NPR-1), Adjacent Buena Vista Hills (NPR-2), Carrizo Plain, and Camp Roberts (Undeveloped Sites), California, Compared With Concentrations in Hair from Other Mammals

Metal	NPR-1 Kit Foxes	NPR-2 Kit Foxes	Undeveloped Site Kit Foxes	Wildlife	Teton Coyotes (Huckabee et al., 1972)	High Exposure Areas	Human Normal	Human Toxic
Aluminum	66.8–1710	110.0–881	68.6–2830				5[a]	
Antimony	0.008–1.4	0.017–0.44	<0.005–0.60				0.03–24[c]	
Arsenic	0.03–4.7	0.15–5.4	<0.01–2.6	<0.2–12[t]	0.09–1.8	0.3–8.9[d,e,f]	0.0–2.0[c]	3[c]
Bromine	3.6–66	8.4–23	1.9–26				30[a]	
Calcium	<67–1000	<67–400	<67–2800				497[g]	
Cerium	<0.3–2.3	0.4–1.5	<0.3–3.0	<1–20[t]	1.9–2.6			
Chromium	<0.1–3.9	0.7–7.7	<0.1–5.8	<0.3–640[p]	0.7–5.8	3.9–4.8[h,i]	0.0–40[c]	
Cobalt	0.15–2.40	0.21–1.15	0.14–1.10				0.1[a]	
Copper	0.015–54	11–23	12–48		23–160	6.9–8.3[c,j]	7.8–120[c]	
Iron	151–1430	282–4150	270–5500	<21–640[b]			26.7[g]	
Magnesium	<40–640	<40–360	<40–660				56.7[g]	
Manganese	0.95–31.70	2.13–27.70	1.74–50.60				0.3[a]	
Mercury	0.21–1.2	0.28–3.9	0.25–10	<0.008–10.7[b]	<0.008–2.8	9.8–117.5[k,l]	0.01–30[c]	50–200[c]
Nickel	<1–7	<1–10	<1–8	0.18–1.7[c]			0.0–11[c]	
Gold	0.0007–0.065	0.0015–0.011	0.0013–0.135	<0.04–0.6[d]	0.002–0.04			
Potassium	<28–360	41–250	<28–1300	5.8–8.3[t,m]				
Rubidium	<0.3–2.9	<0.3–1.5	<0.3–3.7				67.6[g]	
Scandium	0.04–0.46	0.06–0.21	0.05–0.54	<0.05–2[b]	0.005–0.009			
Selenium	0.60–1.8	0.90–3.0	0.50–4.2 0.71–27[n]	0.08–17[n]	0.8–7.83	3.8–12[o,p] 0.89–13[q]	0.3–13[c]	8–30[c]
Silver	<0.1–0.2	<0.1	<0.1–0.3	<0.4–110[p]	0.06–12	0.97–18[r]		
Sodium	4.1–212.0	6.6–98.0	14.0–208.0				309[g]	

continued

TABLE 25.6 (continued)

Ranges of Metal Concentrations (ppm) in Hair of Individual San Joaquin Kit Foxes Sampled on the Elk Hills (NPR-1), Adjacent Buena Vista Hills (NPR-2), Carrizo Plain, and Camp Roberts (Undeveloped Sites), California, Compared With Concentrations in Hair from Other Mammals

Metal	NPR-1 Kit Foxes	NPR-2 Kit Foxes	Undeveloped Site Kit Foxes	Wildlife	Teton Coyotes (Huckabee et al., 1972)	High Exposure Areas	Human Normal	Human Toxic
Titanium	<14–120.0	<14–58.0	<14–114.0				4[a]	
Vanadium	0.3–11.5	0.6–3.2	<0.1–4.4				0.006–2.7[c]	
Zinc	93–220	118–178	87–180	13–6300[b]	91–620		65–200[s]	

Source: Suter, G., II et al. 1992. Results of analysis of fur samples from the San Joaquin kit fox and associated water and soil samples from the Naval Petroleum Reserve No. 1, Tupman, California. Oak Ridge, TN: Oak Ridge National Laboratory, U.S. Department of Energy. ORNL/TM-12244.

[a] Lenihan (1978).
[b] Huckabee et al. (1972).
[c] Jenkins (1979).
[d] Lewis (1972).
[e] Orheim et al. (1974).
[f] Livestock grazing near smelters; reference animals had 0–0.46 ppm.
[g] Barker et al. (1976).
[h] Taylor et al. (1975).
[i] Cotton rats from near cooling towers using chromate corrosion inhibitors; reference rats had 0.39 ppm.
[j] Livestock grazing near smelters; reference animals had 6.8–7.8 ppm.
[k] Doi (1973).
[l] Cats from the vicinity of Minamata, Japan.
[m] Rodents from areas of heavily mineralized soils in Idaho.
[n] Kit foxes from Bakersfield.
[o] Schroeder et al. (1970).
[p] Rats fed nominally toxic levels of Se; control rats had 0.6 ppm.
[q] Kit foxes from the Kesterson Reservoir.
[r] Coyotes from the Kesterson Reservoir.
[s] Petering et al. (1971).

25.3.4.4.3 Chromium

The median chromium concentration in fur from developed NPR-1 was lower than for any other site. Although the median fur concentration and soil concentrations were low, the highest fur concentration was from developed NPR-2, and four of the top 10 concentrations were from developed areas. This suggests that some individual foxes had been exposed to chromium-containing wastes. Chromium concentrations in fur were significantly positively correlated with the number of wells in the home range but not the percent disturbance.

25.3.4.4.4 Sodium

Sodium was used as a marker for produced water, which is primarily a sodium chloride solution. However, neither median nor extreme fur concentrations of sodium were high for developed NPR-1 relative to other sites.

25.3.4.4.5 Vanadium

Vanadium occurs in relatively high concentrations in petroleum and is used as a marker for petroleum in the environment. The median fur concentration was highest for the Carrizo Plain, but six of the top 10 individuals were from developed NPR-1, and the other four of the top 10 were from undeveloped NPR-1, even though undeveloped NPR-1 had the lowest median concentration. Vanadium concentrations were significantly positively correlated with percent development on both NPR-1 and NPR-2. This suggests that some foxes were exposed to petroleum.

To summarize, the median concentrations of arsenic and barium were higher on developed NPR-1 than on other sites, and some foxes appeared to be relatively highly exposed. However, there was considerable overlap of the distribution of concentrations with the other sites. Median chromium and vanadium concentrations from NPR-1 were not higher than other sites, but some foxes were relatively highly exposed. This evidence is taken as positive in that it showed that some foxes were exposed to petroleum or metals in the area where the decline occurred (4 = ++).

25.3.4.5 Chemicals: Causal Pathway

Individual pathways of exposure and lines of evidence are scored separately.

25.3.4.5.1 Soil Concentrations

Soil may be a pathway of exposure through direct ingestion or through the food web. Direct ingestion includes grooming and soil ingested incidentally with prey. However, there were no large or statistically significant correlations between elemental concentrations in random soil samples and percent disturbance in the quarter section from which the sample was taken. Similarly, the differences in fur concentrations among sites were not attributable to those soil concentrations. There were no large or statistically significant positive correlations of soil and fur concentrations at NPR-1 or Camp Roberts. Differences in soil concentrations among sites were small relative to differences in fur concentrations. Hence, neither soil contamination nor natural soil concentrations can account for differences in exposure among foxes. However, this conclusion addresses only soil contamination that is sufficiently widespread to be detected by random soil sampling (4 = –).

25.3.4.5.2 Soil Intake

Differences in exposure to metals in soils may be due to differences in rates of intake rather than differences in concentration. Differences in disturbance between developed and undeveloped NPR-1 may result in increased exposure to soil due to dust, but cannot account for differences among other sites. Differences in prey composition may explain the differences among sites (Suter et al., 1992), but that explanation does not account for the decline of foxes on NPR-1 (4 = –).

25.3.4.5.3 Local Soil Contamination (Wastes)

Local spills and deposits of contaminants were abundant on developed NPR-1. Hence, the evidence for soil as a pathway is positive on the basis of local soil contamination (4 = +).

25.3.4.5.4 Wastewater

Produced waters in open sumps may have been a route of exposure to toxicants due to drinking. Kit foxes are desert animals that do not require drinking water and do not normally drink, but they could consume water from produced water sumps. There is no evidence for wastewaters as a route of exposure (4 = 0).

25.3.4.5.5 Petroleum

Foxes were potentially exposed to petroleum in spills and oil recovery sumps. One kit fox died in spilled oil during the period of study. The evidence for contact with oil as an exposure route is weakly positive (4 = +).

25.3.4.6 Chemicals: Stressor–Response Relationships from Ecological Simulation Models

The demographic model indicated that the decline was caused by mortality, primarily due to predation. There was no evidence that toxicity caused mortality of foxes so it was not the proximate cause (4 = –).

25.3.4.7 Chemicals: Stressor–Response Relationships from Other Field Studies

Livestock have died from drinking produced waters at other oil fields, primarily due to osmotic burden (McCoy and Edwards, 1980). Sump waters on NPR-1 were highly saline; samples contained 1720–14,400 mg/L of sodium and four other metals were found at >1000 mg/L, which is consistent with the McCoy and Edwards (1980) study. However, kit foxes do not require drinking water, and as desert animals, they may not be as sensitive to osmotic stress as livestock. This evidence is ambiguous (4 = 0).

25.3.4.8 Chemicals: Stressor–Response Relationships from Laboratory Studies

Four metals (cadmium, copper, molybdenum, and strontium) were found in produced waters from open sumps at concentrations above drinking water criteria, so these metals are potentially toxic in chronic exposures. However, there is no evidence of exposure. This evidence is ambiguous (4 = 0). Although soils concentrations were available for the site, it was not possible to estimate exposures to these materials for comparison to toxic doses. Soil consumption is inevitable, but unquantifiable. This evidence is ambiguous (4 = 0).

25.3.4.9 Chemicals: Stressor–Response Relationships from Other Studies—Fur

Elemental concentrations in fur that are related to toxic effects are rare. Concentrations in the fur of wildlife from undeveloped areas were taken to be no-effect levels, and concentrations in coyotes from the Grand Teton National Park, Wyoming, were considered particularly relevant. These no-effect concentrations were available for 12 elements, and none of them were exceeded by NPR-1 kit foxes. Concentrations in fur from various contaminated sites were considered to represent potentially toxic levels. Finally, concentrations associated with toxic effects were available for a few elements. Comparisons are presented here for the three elements of concern for which effects or no-effects data were found (see Table 25.6).

25.3.4.9.1 Arsenic

One fox associated with NPR-1 had 26 ppm arsenic in its fur. This is much higher than concentrations in the hair of humans who died of arsenic poisoning (3 ppm; see Table 25.6). However, that fox lived north of NPR-1 along the California aqueduct and may have been exposed to residues of arsenical agrochemicals. That fox was alive and apparently healthy at the time that fur was collected and lived for more than a year after capture. One fox from developed NPR-1 and one from developed NPR-2 also exceeded the 3 ppm level. This suggests that human hair concentrations are not good indicators of toxic exposures to arsenic in kit foxes. No data were found for other wildlife.

25.3.4.9.2 Chromium

Chromium concentrations in NPR-1 kit foxes were within the range of concentrations in Teton coyotes and other wildlife from uncontaminated areas, so they are assumed to be nontoxic.

25.3.4.9.3 Selenium

Selenium concentrations in NPR-1 kit foxes were low relative to concentrations in rats fed toxic doses of selenium, relative to humans experiencing selenium toxicity, and relative to kit foxes and coyotes at Kesterson reservoir where birds experienced severe selenium toxicity. They were also lower than concentrations in Teton coyotes and other wildlife.

This evidence weakens toxic chemicals as a cause, because there was no indication that the observed fur concentrations were related to toxicity (4 = –).

25.3.5 Vehicular Activities

25.3.5.1 Vehicles: Spatial/Temporal Co-Occurrence

The increase in development inevitably increased vehicle traffic, and it seems likely that traffic was greatest in developed areas. Most vehicle deaths involved young-of-the-year foxes and occurred in developed areas (see Table 25.5). This evidence supports vehicular activity as a cause (5 = +).

Other types of accidents were minor. Among all known kit fox mortalities on NPR-1 (1980–1990), one was buried during construction, one was trapped in a pipe, and four died in live traps during the demographic studies (U.S. DOE, 1993). Hence, other accidents are not considered.

25.3.5.2 Vehicles: Evidence of Exposure, Mechanism or Mode of Action

Fifteen percent of identified mortalities of radio-collared kit foxes on NPR-1 during 1980–1988 were due to vehicle collisions based on location and necropsy results (U.S. DOE, 1993). This evidence supports vehicular activity as a cause (5 = ++).

25.3.5.3 Vehicles: Causal Pathway

Vehicular activity was not quantified, but it inevitably increased on the site due to increased oil development activities. This supports the causal pathway (5 = +).

25.3.5.4 Vehicles: Stressor–Response Relationships from Ecological Simulation Models

A demographic model of kit foxes on NPR-1, between 1981 and 1986, found that the decline was caused by high mortality, particularly of young-of-the-year foxes (Floit and Barnthouse, 1991). A less detailed analysis of an equivalent model, but for the period 1981–1989, gave qualitatively similar results (U.S. DOE, 1993). Hence, early mortality was the cause of the decline, and vehicular strikes were responsible for approximately 15% of identified mortality (see Table 25.5). This line of evidence supports accidents as a contributing proximate cause (5 = +).

25.3.5.5 Vehicles: Stressor–Response from Other Field Studies

The 15% of total mortality on NPR-1 due to vehicle strikes was higher than in most other studies where vehicular strikes rarely exceed 10% of mortalities (Bjurlin and Cypher, 2003). This evidence strengthens vehicles as a cause of the decline (5 = +).

25.3.6 Disease

25.3.6.1 Disease: Spatial Co-Occurrence

Necropsies provided little evidence of possible disease-induced mortality at either NPR-1 or NPR-2 between 1980 and 1995 (Cypher et al., 2000). However, it is possible that an increased frequency of nonlethal disease may have weakened foxes, thereby causing increased predation on developed areas. The absence of evidence of co-occurrence weakens disease as a cause (6 = –).

25.3.6.2 Disease: Causal Pathway

The elements of the hypothesized causal pathway (humans with pets and coyotes) were present, but transport of pathogens onto NPR-1 was not documented, so the pathways remain hypothetical (6 = 0).

25.3.6.3 Disease: Evidence of Exposure, Mechanism, or Mode of Action

A serological survey for pathogens was conducted in 1981–1982 and 1984 (McCue and O'Farrell, 1986, 1988), and serum chemistry was analyzed (McCue and O'Farrell, 1992). Canine parvovirus antibodies were found in nearly all foxes, regardless of development. Antibodies for other pathogens were rare and data were insufficient to make comparisons between levels of development. The investigators presumed that if foxes were highly exposed to pathogens it would be reflected in changes in hematological parameters. Sufficient data on hematology were gathered in 1981–1982 to make comparisons between levels of development, but no differences in either mean or extreme values were found (McCue and O'Farrell, 1987). This evidence greatly weakens disease as a cause (6 = – –).

25.3.6.4 Disease: Stressor–Response from Other Field Studies

Mortality due to disease is hard to detect, and Cypher et al. (2000) found no documentation of epizootics in kit foxes. However, field studies have documented decreases in the abundance of other fox species associated with diseases (Nicholson and Hill, 1984). This evidence qualitatively supports disease (6 = +).

25.4 Conclusions

25.4.1 Proximate Causes

Having analyzed the evidence for each candidate cause in the prior sections, the next step is to synthesize the results. First, the consistency of the evidence was evaluated across types of evidence for each candidate cause. That is, was the evidence all positive, all negative or mixed? All candidate causes except predation had inconsistent evidence. The second criterion was the existence of an explanation for the inconsistencies. Explanations were developed for the inconsistencies in three candidate causes (habitat modification, prey abundance, and vehicular activity) that involved converting them from candidate causes to components of the web of antecedent causation.

After the evidence for each candidate cause was summarized, the evidence was compared across candidate causes to determine the one best supported (see Table 25.7). First, the candidate causes that could be eliminated were identified, then the most likely cause from among those that remain was identified, and finally the other candidate causes were reconsidered.

Although the evidence was inconsistent, disease (Candidate Cause 6) was clearly eliminated, because the evidence from the site was negative. Very few of the trapped foxes were observed to be diseased, little evidence of disease was found during necropsies, and neither serological nor hematological analyses showed evidence of an epizootic that would account for the decline. Disease has caused population declines in other places, but this supporting evidence would be relevant only if there is some positive evidence from the site.

In contrast, evidence for predation (Candidate Cause 3) as the principal proximate cause was consistent and strong. Predation by coyotes is the major cause of death in kit foxes, and a demographic analysis showed that the decline was due to high mortality, with little influence from low fecundity or high emigration (Floit and Barnthouse, 1991).

Evidence for vehicular accidents (Candidate Cause 5) was also positive, but the mortality rate due to accidents was much lower than for predation and not sufficient to account for the decline. Hence, it was concluded to be a contributing cause.

The evidence for environmental contaminants (Candidate Cause 4) was inconsistent and complex. Contaminants from oil development were present and potential routes of exposure were identified, but only two chemicals, arsenic and barium, were elevated in the fur of most foxes from developed NPR-1 relative to reference sites. Arsenic levels in three foxes reached levels that indicate acute toxicity in humans, but those foxes appeared healthy when captured and their longevity was not apparently reduced. Barium is

TABLE 25.7

Comparison of the Strength of Evidence for the Candidate Causes. Types of Evidence With No Evidence for Any Candidate Cause Were Excluded

	1. Prey		2. Habitat		3.	4.	5.	6.
Types of Evidence	a. Disturbance	b. Climate	a. Disturbance	b. Climate	Predation	Toxics	Accidents	Disease
Evidence that uses data from the case								
Spatial/temporal co-occurrence (pathway independent)	++	+	+	–	+	+	+	–
Spatial/temporal co-occurrence (by pathway)	+	–						
Temporal sequence	0		0	NE	0	0	NE	NE
Evidence of exposure or biological mechanism (pathway independent)	++		NE	NE	++	++	++	– –
Evidence of exposure or biological mechanism (by pathway)	–	+						
Causal pathway[a]	+	–	++	–	+	+	+	0
Covariation of stressor and effect (pathway independent)	+++		NE		NE	NE	NE	NE
Covariation of stressor and effect (by pathway)	–	+	– –	0				
Manipulation of exposure	+		NE	NE	+	NE	NE	NE
Symptoms, starvation[b]	–		NE	NE	NE	NE	NE	NE
Symptoms, reproductive[b] (pathway independent)	+		NE		NE	NE	NE	NE
Symptoms, reproductive[b] (by pathway)	+	–	NE	NE				
Stressor-response relationships from ecological simulation models	–		0		+++	–	+	– –

continued

TABLE 25.7 (continued)

Comparison of the Strength of Evidence for the Candidate Causes. Types of Evidence With No Evidence for Any Candidate Cause Were Excluded

Types of Evidence	1. Prey		2. Habitat		3. Predation	4. Toxics	5. Accidents	6. Disease
	a. Disturbance	b. Climate	a. Disturbance	b. Climate				
Evidence that uses data from elsewhere								
Evidence of mechanism or mode of action	+	+	+	+	+	+	+	+
Stressor-response relationships from other field studies	+	NE	NE	NE	0	0	+	+
Stressor-response relationships from laboratory studies	NE	NE	NE	NE	–	0	NE	NE
Evaluating multiple lines of evidence								
Consistency of evidence	–	–	–	– –	+++	–	–	–
Explanation of the evidence	C^c	C^c	C^c	– –	NA	–	C^c	–

^a An additional causal pathway for prey abundance, competition for prey by coyotes, was ambiguous.
^b The categories of symptoms apply only to prey abundance.
^c The explanation of the evidence makes the candidate cause a contributor to another cause.

much less toxic, and although fur levels on NPR-1 were high, these levels overlapped with fur from reference sites. One fox died after becoming coated in oil. In sum, there was no evidence that toxic exposures could account for the high mortality rates that caused the decline.

The availability and utilization of lagomorph prey (Candidate Cause 1) were strongly related to kit fox abundance, but clinical symptoms of poor condition or starvation were not observed in trapped animals or during necropsies. Prey availability can affect fecundity, and females on developed areas produced fewer pups, but the demographic analysis indicated that variance in kit fox fecundity did not significantly contribute to variance in kit fox abundance. Hence, prey availability does not appear to be a significant proximate cause. However, it may be a contributing factor in other sources of mortality. That is, fewer large prey and greater use of small prey implies more time spent hunting and greater exposure to coyotes and vehicles.

The evidence for habitat alteration (Candidate Cause 2) was ambiguous. The area devegetated is known, but the quality of habitat provided by the vegetated and devegetated areas and the effects of human activities on habitat utility for kit foxes are unknown. The fact that emigration from the developed areas exceeded emigration from the undeveloped areas suggests that habitat quality was lower in developed areas.

The available evidence indicates that the cause of the kit fox decline in the early-to-mid-1980s was increased predation by coyotes. Predation by coyotes was the major cause of death in kit foxes, and a demographic analysis showed that the decline was due to high mortality, with little influence from low fecundity or high emigration. The mechanism is mortality which is shared with vehicular accidents. However, the mortality rate due to accidents is much lower than for predation, so accidents contributed but were not sufficient to account for the decline.

25.4.2 Sources

Although causal analysis must begin by identifying the proximate cause, identifying its source is useful for planning management actions. Hence, we asked why coyote abundance and associated mortality increased in the early 1980s.

Climate is a potential source of habitat alteration and reduced abundance of lagomorph prey. This region is semiarid and a few drier-than-average years can reduce the fecundity and survival of lagomorph prey. However, the period of concern was not especially or consistently dry. The year with the second highest precipitation in the 30-year record occurred during the decline (see Figure 25.6). In addition, climate would be the same for developed and undeveloped areas and for both NPR-1 and NPR-2. Hence, just as climate can be eliminated as the cause of kit fox decline via the habitat or prey causal pathways (1b and 2b), it cannot be the cause of increased coyote numbers in the early 1980s. The later dry period of 1988–1990 shows that low precipitation can produce a clear signal: plant production and abundance of lagomorphs, coyotes, and kit foxes all declined on both NPRs. Therefore, climate can be eliminated as the source of the decline.

Disturbance due to oil development and production is a source of habitat alteration and reduced prey abundance. Evidence for the effects of disturbance comes primarily from comparisons of developed and undeveloped areas of NPR-1. The decline in both kit foxes and lagomorphs was greater on developed than undeveloped NPR-1. Although the mechanism is unclear, it seems likely that some aspect of active oil development contributed to the coyote-caused decline. However, it is possible that the differences in the demographics of kit foxes and lagomorphs between developed and undeveloped areas were due to natural differences.

Counterintuitively, disturbance may also be a source of increased coyote abundance. Coyotes were more abundant on developed than undeveloped NPR-1 during the decline. Prior to the coyote control program, site development may have improved coyote habitat by keeping hunters off the site and by providing sources of fresh water, discarded food, and road kills to be scavenged. Cypher and Spencer (1998) suggested that the availability of anthropogenic food resources may have increased coyote abundance and predation of kit foxes on the Elk Hills.

Diseases in coyotes may also be sources of changes in coyote abundance. The low observed abundance of coyotes in 1979 may have been due to disease. Between 1972 and 1983, the prevalence of antibodies against canine parvovirus in wild coyotes captured in three western states coincided with the epizootic of the disease in domestic dogs (Thomas et al., 1984). Canine parvovirus is a significant potential pathogen for wild canids, and it was believed to be linked with declines in coyote numbers (Cypher et al., 2000). There is no known evidence of a parvovirus epizootic in coyotes in the San Joaquin Valley, but kit foxes tested on NPR-1 carried parvovirus antibodies. It is possible that the increase in coyotes was a rebound from the parvovirus epizootic and that kit foxes are resistant. However, that hypothesis suggests that the high abundance of coyotes in 1985 reflected the peak of a population that oscillates over long time periods. That would suggest in turn that kit foxes may be rare in the developed areas of NPR-1 except during periods of coyote epizootics.

The final conceptual model for the cause of the kit fox decline is presented in Figure 25.8. The proximate cause is predation. The mechanism is mortality, primarily of young foxes, which is shared with vehicular accidents, so accidents are a contributor but are not sufficient. The source of the increased predation is much less clear. However, the availability of prey and other food are likely contributors. The coyote control program is a likely source of the decline in coyote abundance that ended the kit fox decline, but reduced prey abundance may have also contributed to the coyote decline.

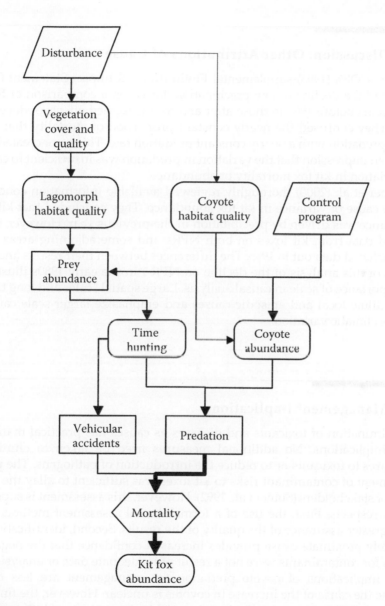

FIGURE 25.8
The final conceptual model for the cause of the kit fox decline. The thickness of the arrow lines indicates the degree of confidence in the causal connection.

25.5 Discussion: Other Attributions of Cause

The U.S. DOE's (1993) supplemental Environmental Impact Statement (EIS) attributed the decline to low precipitation, based on a comparison of the 3 and 5 years before 1981 to those after and to unspecified effects of development. They confused the nearly constant proportion of mortality that was due to predation with a nearly constant predation rate. This error created the mistaken impression that the variation in predation was insufficient to cause the variation in kit fox mortality or abundance.

Cypher et al. (2000) thoroughly reviewed available information concerning the cause of variance in kit fox abundance. They concluded that kit fox abundance was driven by precipitation in the previous year. However, they lumped data from kit foxes on both NPRs and some adjoining areas and they included data out to 1995. The differences between their results and the results of this analysis of the decline on NPR-1 in the early 1980s illustrate the importance of scale in causal analysis. Large spatial scales and long time-scales dilute local and episodic causes and emphasize larger scale causes, such as climatic variation.

25.6 Management Implications

The elimination of toxicants and diseases as causes has practical management implications. No additional measures need be taken to eliminate exposures to toxicants or to reduce the introduction of pathogens. The prior assessment of contaminant risks to kit foxes was sufficient to allay the concerns of stakeholders (Suter et al., 1992). However, this assessment is superior in two respects. First, the use of a formal causal assessment method provides greater assurance of the quality of the results. Second, identification of the likely proximate cause provides increased confidence that the negative results for contaminants were not a result of inadequate data or analysis.

The implications of coyote predation for management are less clear, because the cause of the increase in coyotes is unclear. However, the finding that precipitation is not absolutely or invariably determinate of kit fox abundance should encourage management actions. These might include revegetation to increase prey abundance, preservation of kit fox dens that provide cover from predators, coyote control, and in extreme situations, supplemental feeding. All of these were practiced on NPR-1 for some time and to some degree, but it is not clear how successful they were. The endangered status of kit foxes could justify adaptive management studies to determine the most efficacious practices. However, privatization of the site ended the U.S. DOE's monitoring and assessment program.

Glossary

The terms in this glossary are defined as we use them in this book related to causal assessment.

Agent: A physical, chemical, or biological entity that may affect a biotic system. This term is similar to but more general than stressor in that it does not imply harm. For example, dissolved oxygen and woody debris are agents; low dissolved oxygen and reduced woody debris may be stressors.

Agent, causal: An agent that directly induces the effect in a causal relationship.

Alteration: (1) The characteristic of a causal relationship that the entity is changed by the interaction with the cause. (2) A change in an entity that has interacted with a cause.

Analogy: An inference from similarity of known attributes to similarity in other attributes. In causal assessments, similar causes are expected to have similar effects and similar effects are expected to have similar causes.

Analysis, causal: A process by which data and other information are organized and evaluated, using quantitative and logical techniques to generate evidence concerning the likely cause of an observed condition. The analytical component of causal assessments is synonymous with the step in the framework: derive evidence.

Antecedence: The characteristic of a causal relationship that connects it to processes that precede it.

Antecedent: An agent, event, or process that precedes another.

Assemblage: A group of organisms in a habitat that are sampled and enumerated together and in the same way.

Assessment, causal: (1) The process of determining causes based on scientific evidence. (2) The product of such an assessment process.

Assessment, environmental: (1) A process of generating and presenting scientific information to inform an environmental management decision. Causal assessments are one type of environmental assessment. (2) The product of an environmental assessment process.

Association: The degree to which one variable (e.g., representing a cause) co-occurs or covaries with another (e.g., representing an effect).

Bioassessment (biological assessment): Evaluation of ecosystem condition using biological surveys and other direct measurements of resident biota.

CADDIS: The Causal Analysis/Diagnosis Decision Information System, a web-based technical support system for implementing the Stressor

435

Identification process for determining environmental causes (at www.epa.gov/caddis).

Case: (1) The situation that is the subject of a causal assessment; for example, the case may be an affected stream reach or an area of forest with many dying trees. (2) The set of evidence relevant to a candidate cause, for example, "the lack of co-occurrence weakens the case for fine sediments."

Causality: The concept that effects have causes.

Causation: (1) The act of something causing an effect. (2) A relationship between events involving a process connection in which a causal agent (i.e., a stressor) affects an entity (e.g., an organism).

Causation, direct: The induction of an effect through a single cause–effect relationship; for example, the direct effect of an herbicide may be reduced algal production. Compare this to indirect causation.

Causation, general: The capability of an agent, event, or process to produce the prescribed effects.

Causation, indirect: The induction of effects through a series of cause-effect relationships, such that the impaired biological resource may not even be exposed to the initial cause. For example, the direct effect of an herbicide may be reduced algal production, which may indirectly lead to reduced herbivore and predator populations. Compare with direct causation.

Causation, specific: (1) An instance of causation. (2) The relationship between an agent, process, or event and an effect in a particular situation.

Cause: An event, an agent, or a set of events or agents that interact with a susceptible entity resulting in an identified biological effect.

Cause, candidate: A proposed cause of an environmental effect that is sufficiently credible to be analyzed.

Cause, complex: A cause that has multiple components. The individual components of a complex cause are necessary but not sufficient by themselves.

Cause, indirect: A cause that acts by inducing an effect that, through one or more further cause–effect relationships, ultimately results in the biological effect of concern. Indirect causes eventually lead to the direct (i.e., proximate) cause and then the effect.

Cause, likely: The candidate cause that is best supported by and best explains the evidence.

Cause, plural: A set of causes that co-occur and each is capable of inducing the effect without the others. The individual components of a plural cause are sufficient but not necessary.

Cause, proximate: The cause that induces the effect through direct exposure. Compare to an indirect cause.

Cause, ultimate: The action or policy that is responsible for creating or sustaining a source.

Characteristic, causal: An attribute that serves to identify a causal relationship.

Co-occurrence: (1) The characteristic of a causal relationship that the cause and effect are collocated in space and time. (2) An instance of collocation in space and time.

Coherence: The quality of a body of evidence that its constituent pieces are logically linked together, thus forming a reasonable explanation.

Confounder: Stressors or influencing factors that interfere with the ability to quantify the contribution of a specific cause to an observed biological effect.

Confounding: Bias in the statistical representation of a causal relationship due to the presence of a confounder.

Consilience: Consistency of a hypothesis with prior independently derived knowledge.

Correlation: A statistical relationship between two or more variables such that systematic changes in the value of one variable are accompanied by systematic changes in the other.

Corroboration: Supporting evidence for a candidate cause from one or more independent studies providing similar results.

Diagnosis: (1) The identification of a cause by recognizing characteristic signs and symptoms. (2) Differential diagnosis is identification of a disease by comparing all diseases that might plausibly account for the known symptoms.

Ecoepidemiology: The study of the nature and causes of past or ongoing effects in ecological systems.

Effect: (1) In general, an effect is some change in an entity that inevitably follows a cause. A biological effect is the biological result of exposure to a causal agent or event. This term is similar to response, but emphasizes the agent that acts (e.g., the effect of cadmium) rather than the receptor that responds to it (e.g., the response of trout). (2) In practice, an effect is an observed discrepancy of an entity from its expected or nominal condition (e.g., the number of species in a biotic community relative to reference communities) that prompts a causal assessment.

Elimination: Definitive rejection of a candidate cause based on strong evidence that an expected association between that cause and the effect does not occur.

Entity, affected: The thing that has been changed by the causal agent. The entity and an attribute that has been changed constitute the effect.

Entity, susceptible: Something that could display the effect of interest in response to a particular candidate cause.

Epidemiology, environmental: The study of the nature and causes of past or ongoing effects on humans in the environment (see ecoepidemiology).

Evidence: Information linking causes to effects that informs beliefs regarding causation.

Evidence, body of: All the available evidence used to determine causation.

Evidence, piece of: The basic unit of evidence; examples include the results of a toxicity test or a stream survey.

Evidence, type of: A category of evidence that provides a logically distinct way to support, weaken, or refute the case for a candidate cause.

Exposure: The co-occurrence or contact of an agent with an organism, population, or community such that interaction has the potential to occur.

Exposure-response: The relationship between the intensity, frequency, or duration of exposure to an agent or stressor and the intensity, frequency, or duration of the biological response. Equivalent terms include concentration–response, dose–response, and stressor–response.

Impairment: A detrimental effect on a population, community, or ecosystem that is sufficient to prompt a management or regulatory action.

Inference: (1) The act of reasoning from evidence. (2) A result of such reasoning.

Information: Data or other facts used to derive evidence.

Interaction: (1) The characteristic of causal relationships that a causal agent contacts, impinges upon, or enters a susceptible entity in a way that initiates the effect. (2) An instance of contact or impingement of a causal agent on a susceptible entity that initiates an effect.

Judgment, expert: A method of inference based on the knowledge and skill of qualified assessors rather than a formal analysis.

Manipulation: A modification of environmental factors by human actions that change the exposure of an entity to an agent (e.g., shutting down an effluent source, fencing cattle from a stream, or caging fish in a contaminated lake).

Mechanism: A process by which a cause induces an effect.

Mechanism of action: A description of the specific process by which a cause induces an effect. In contrast to the similar term mode of action, mechanism of action usually describes events at a lower level of organization than the effect of concern (e.g., blocking of acetylcholine receptors) (see mode of action).

Mesocosms: Outdoor or indoor facilities with controlled physico-chemical conditions and multiple species used to simulate natural ecosystems.

Mode of action: A phenomenological description of how a cause induced an effect (e.g., paralysis) (see mechanism of action).

Model, conceptual: A graphic depiction of the causal network linking sources and effects, that is used to identify candidate causes, organize the assessment, and ultimately to communicate why some pathways are unlikely and others are very likely.

Model, simulation: A mathematical representation of a system based on knowledge of its components and their mutual influences.

Model, statistical: A mathematical representation of a system derived by fitting a function to data from the system (also called an empirical model).

Pragmatism: The philosophy that thinking is for doing. The pragmatic philosophy of science posits that scientific truth is derived from encounters with nature, and it is ultimately what the scientific community agrees upon in the long term based on weighing multiple pieces of evidence.

Reasonable explanation: (1) A statement or account that coherently explains a body of evidence. (2) Informed reasons for apparent inconsistencies in a body of evidence that provides coherence.

Refutation: The logical process of demonstrating the impossibility of a candidate cause, thus allowing it to be eliminated from further consideration.

Relationship, causal: The connection between a cause and an effect (i.e., not just an association).

Response: The biological result of an exposure to an agent or stressor. This term is synonymous with effect, but emphasizes the receptor that responds (e.g., the response of trout) rather than the agent that acts upon it (e.g., the effect of cadmium).

Site: A location in an ecosystem, such as a stream reach, a watershed, or an area of grassland, where measurements or observations are taken.

Site, affected: A site where an effect has been shown to occur.

Site, comparison: A site which has some property that makes it relevant for generating evidence by comparing it to the affected site. In general, they are sites where the effect or a candidate cause is known to occur or to not occur.

Source: An origination point, area, or entity that releases or emits an agent that may be an indirect cause or a proximate cause.

Stakeholders: People or organizations with an interest in the outcome of an assessment.

Strength: The degree to which evidence demonstrates a large difference or a high degree of association between a cause and effect relative to background levels. It is a component of the weight of evidence.

Stressor: A potentially adverse causal agent.

Stressor–response: The relationship between the intensity, frequency, or duration of exposure to an agent or stressor and the intensity, frequency, or duration of the biological response. Equivalent terms include concentration–response, dose–response, and exposure–response.

Sufficiency: (1) The characteristic of a causal relationship that the agent or event must be adequate to induce the effect in susceptible entities. (2) An occurrence of enough of an agent or process to affect a susceptible entity.

Supports: Suggests that evidence is consistent with expectations concerning a candidate cause (see weakens).

Symptom: A property of affected organisms, populations, communities, or ecosystems that is indicative of a specific cause or a few causes.

Symptomology: A set of symptoms that is indicative of a specific cause or a few causes.

Time order: (1) The characteristic of a causal relationship that the cause precedes the effect. (2) The sequence, in time, of the occurrence of a candidate cause and the effect of concern. It is sometimes called temporality or temporal sequence.

Weakens: Suggests that evidence is contrary to expectations concerning a candidate cause (see supports).

Weigh: Consider the logical implications, reliability, and quality of the body of evidence to assess the likelihood of a cause–effect relationship.

Weight: (1) (noun) The importance of a piece or category of evidence. (2) (verb) Assign importance to a piece or category of evidence.

Weight of evidence: The relative degree of support for a candidate cause or other conclusion provided by evidence. The result of weighing the body of evidence.

List of Abbreviations and Acronyms

ΣTU	sum of toxic units
2,3,7,8 TCDD	2,3,7,8-tetrachlorodibenzo-p-dioxin
AMD	acid mine drainage
ANCOVA	analysis of covariance
BACI	before-after-control-impact
BACIP	before-after-control-impact-pairs
BKME	bleached kraft mill effluent
BLM	biotic ligand model
BMP	best management practices
CA	correspondence analysis
CADDIS	Causal Analysis/Diagnosis Decision Information System
CART	classification and regression tree
CCC	criterion continuous concentration
CCME	Canadian Council of Ministers of the Environment
CMC	criterion maximum concentration
CREM	Council for Regulatory Environmental Modeling
CWA	Clean Water Act
DA	Department of Agriculture
DAG	directed acyclic graph
DDT	dichlorodiphenyltrichloroethane
DGT	diffusive gradient in thin film
DOE	Department of Energy
DO	dissolved oxygen
EC_{50}	median effective concentration
EEM	environmental effects monitoring
EIS	environmental impact statement
EPA	Environmental Protection Agency
EPT	Ephemeroptera, Plecoptera, and Trichoptera
EROD	ethoxyresorufin-O-deethylase
FFG	functional feeding group
GIS	geographic information system
GLEMEDS	Great Lakes embryo mortality, edema, and deformity syndrome
GS	Geological Survey
HBI	Hilsenhoff Biotic Index
INUS	insufficient but necessary parts of unnecessary but sufficient set
ISA	impervious surface area
LC_{50}	median lethal concentration
LOEC	lowest-observable-effect concentration

LOWESS   locally weighted scatterplot smoothing
LTER   long-term ecological research
LWD   large woody debris
Max   maximum
M-D   mean difference
MEDEP   Maine Department of Environmental Protection
Min   minimum
MPCA   Minnesota Pollution Control Agency
NE   no evidence
NESS   necessary element of a sufficient set
NHD   National Hydrography Dataset
NMS   nonmetric multidimensional scaling
NOEC   no-observable-effect concentration
NPDES   National Pollutant Discharge Elimination System
NPR   Naval Petroleum Reserve
NPS   National Park Service
NRBS   Northern River Basins Study
NREI   Northern Rivers Ecosystem Initiative
P   probability
PAH   polycyclic aromatic hydrocarbon
PCA   principal components analysis
PCB   polychlorinated biphenyl
PME   pulp mill effluent
POCIS   polar organic chemical integrative samplers
POTW   publicly owned treatment works
QA   quality assurance
Q-Q   quantile-quantile
RBP   rapid bioassessment protocol
REF   reference location
RR   relative risk
SI   stressor identification
SPEAR   species at risk
SPMD   semipermeable membrane devices
S-R   stressor-response
SRP   soluble reactive phosphorus
SSD   species sensitivity distribution
TDS   total dissolved solids
TIE   toxicity identification evaluation
TMDL   total maximum daily load
TN   total nitrogen
TSS   total suspended solids
TU   toxic unit
UTM   Universal Transverse Mercator
VTG   vitellogenin
WET   whole effluent toxicity

WQC	water quality criterion
WVDEP	West Virginia Department of Environmental Protection
WVSCI	West Virginia Stream Condition Index
WYSIATI	what you see is all there is

References

Acevedo, M. F., M. Alban, K. L. Dickson et al. 1997. Estimating pesticide exposure in tidal streams of Leadenwah Creek, South Carolina. *J Toxicol Environ Health* 52:295–316.

Admiraal, W., C. Barranguet, S. A. M. van Beusekom et al. 2000. Linking ecological and ecotoxicological techniques to support river rehabilitation. *Chemosphere* 41:289–295.

Agresti, A. 2002. *Categorical Data Analysis*, 2nd ed. Hoboken, New Jersey: Wiley-Interscience.

Ahlers, J., C. Riedhammer, M. Vogliano et al. 2006. Acute to chronic ratios in aquatic toxicity—Variation across trophic levels and relationship with chemical structure. *Environ Toxicol Chem* 25:2937–2945.

Alexander, G. G. and J. D. Allan. 2007. Ecological success in stream restoration: Case studies from the midwestern United States. *Environ Manage* 40 (2):245–255.

Alexander, A. C., J. M. Culp, K. Liber, and A. J. Cessna. 2007. Effects of insecticide exposure on feeding inhibition in mayflies and oligochaetes. *Environ Toxicol Chem* 26:1726–1732.

Alexander, A. C., K. S. Heard, and J. M. Culp. 2008. Emergent body size of mayfly survivors. *Freshwat Biol* 53:171–180.

Allan, D. J. 1995. *Stream Ecology: Structure and Function of Running Waters*. London: Chapman & Hall.

Allan, J. D. and M. M. Castillo. 2007. *Stream Ecology: Structure and Function of Running Waters*, 2nd ed. Dordrecht, the Netherlands: Springer.

Allan, I. J., B. Vrana, R. Greenwood et al. 2006. A "toolbox" for biological and chemical monitoring requirements for the European Union's Water Framework Directive. *Talanta* 69:302–322.

Allan, J. D., L. L. Yuan, P. Black et al. 2012. Investigating the relationships between environmental stressors and stream condition using Bayesian belief networks. *Freshw Biol* 57:58–73.

Allert, A. L., J. F. Fairchild, R. J. DiStefano et al. 2008. Effects of lead–zinc mining on crayfish (*Orconectes hylas*) in the Black River Watershed, Missouri, USA. *Freshw Crayfish* 16:99–113.

Allert, A. L., J. F. Fairchild, R. J. DiStefano et al. 2009. Ecological effects of lead mining on Ozark streams: In situ toxicity to woodland crayfish (*Orconectes hylas*). *Ecotoxicol Environ Saf* 72:1207–1219.

Alvarez, D. A. 2013. Development of semipermeable membrane devices (SPMDs) and polar organic chemical integrative samplers (POCIS) for environmental monitoring. *Environ Toxicol Chem* 32 (10):2179–2181.

Amiard-Triquet, C., J.-C. Amiard, and Rainbow, P. S. 2012. *Ecological Biomarkers: Indicators of Ecotoxicological Effects*. Boca Raton: Taylor & Francis/CRC Press.

Anderson, D. R. 2008. *Model Based Inference in the Life Sciences: A Primer on Evidence*. New York: Springer.

Anderson, C. 2013. How to give a killer presentation: Lessons from TED. *Harvard Business Review*. Boston, MA: Harvard Business School Publishing Corporation.

Anderson, D. R., K. P. Burnham, and W. L. Thompson. 2000. Null hypothesis testing: Problems, prevalence, and an alternative. *J Wildlife Manage* 64:912–923.

Ankley, G. T., D. M. DiToro, D. J. Hansen, and W. J. Berry. 1996. Technical basis and proposal for deriving sediment quality criteria for metals. *Environ Toxicol Chem* 15:2056–2066.

Argent, D. G. and P. A. Flebbe. 1999. Fine sediment effects on brook trout eggs in laboratory streams. *Fish Res* 39 (3):253–262.

Arnett, E. B., M. M. P. Huso, M. R. Schirmacher, and J. P. Hayes. 2011. Altering turbine speed reduces bat mortality at wind-energy facilities. *Frontiers Ecol Environ* 9 (4):209–214.

Ashford, J. R. 1981. General models for the joint action of drugs. *Biometrics* 37:457–474.

ASTM. 2013. *Standard Guide for Conducting in-situ Field Bioassays with Caged Bivalves*. West Conshohocken, PA: ASTM International.

Austin, P. C. 2011. An introduction to propensity score methods for reducing the effects of confounding in observational studies. *Multivariate Behav Res* 46 (3):399–424.

Baerwald, E. F., G. H. D'Amours, B. J. Klug, and R. M. R. Barclay. 2008. Barotrauma is a significant cause of bat fatalities at wind turbines. *Curr Biol* 18 (16):R695–R696.

Bailar, J. 2005. Redefining the confidence interval. *Hum Ecol Risk Assess* 11:169–177.

Bailey, R., R. Norris, and T. Reynoldson. 2004. *Bioassessment of Freshwater Ecosystems using the Reference Condition Approach*. Dordrecht, the Netherlands: Kluwer Academic Publishers.

Baird, D. J., S. S. Brown, L. Lagadic et al. 2007a. In situ-based effects measures: Determining the ecological relevance of measured responses. *Integr Environ Assess Manage* 3:259–267.

Baird, D. J., G. A. Burton, J. M. Culp, and L. Maltby. 2007b. Summary and recommendations from a SETAC Pellston workshop on in situ measures of ecological risk. *Integr Environ Assess Manage* 3:275–278.

Baker, R. and R. King. 2010. A new method for detecting and interpreting biodiversity and ecological community thresholds. *Meth Ecol Evol* 1:25–37.

Barata, C., J. Damasio, M. A. Lopez et al. 2007. Combined use of biomarkers and in situ bioassays in *Daphnia magna* to monitor environmental hazards of pesticides in the field. *Environ Toxicol Chem* 26:370–379.

Barbour, M. and J. Gerritsen. 2006. Key features of bioassessment development in the United States of America. In *Biological Monitoring of Rivers. Applications and Perspectives*, edited by G. Ziglio, M. Siligardi, and G. Flaim. New York: Wiley.

Barbour, M. T., J. M. Diamond, and C. O. Yoder. 1996. Biological assessment strategies: Applications and limitations. In *Whole Effluent Toxicity Testings: An Evaluation of Methods and Prediction of Receiving System Impacts*, edited by D. R. Grothe, K. L. Dickson, and D. K. Reed-Judkins. Pensacola, FL: CRC Press.

Barbour, M., J. Gerritsen, B. Snyder, and J. Stribling. 1999. *Rapid Bioassessment Protocols for Use in Streams and Wadeable Rivers: Periphyton, Benthic Macroinvertebrates, and Fish.*, 2nd ed. Washington, DC: U.S. Environmental Protection Agency, Office of Water. EPA/841/B-99/002.

Barker, D. H., A. C. Rencher, B. M. Mittal et al. 1976. Metal concentrations in human hair from India (Pilani, Rajasthan). In *Trace Substances in Environmental Health*, Vol 10, edited by D. Hemphill. Columbia, MO: University of Missouri Press.

Barnthouse, L. W. 2007. Population Modeling. In *Ecological Risk Assessment*, edited by G. W. Suter II. Boca Raton, FL: CRC Press.

Barnthouse, L. W., D. Marmorek, and C. N. Peters. 2000. Assessment of multiple stresses at regional scales. In *Multiple Stressors in Ecological Risk and Impact Assessment: Approaches to Risk Estimation*, edited by S. A. Ferenc and J. A. Foran. Pensacola, FL: SETAC Press.

Bartell, S. M. 2006. Biomarkers, bioindicators, and ecological risk assessment—A brief review and evaluation. *Environ Bioindicators* 1:60–73.

Bartell, S. M. 2007. Ecosystem effects modeling. In *Ecological Risk Assessment*, edited by G. W. Suter II. Boca Raton, FL: CRC Press.

Bartell, S. M., R. A. Pastorok, H. R. Akcakaya et al. 2003. Realism and relevance of ecological models used in chemical risk assessment. Human and ecological risk assessment. *Human Ecol Risk Assess* 9:907–938.

Barwick, D. H., J. W. Foltz, and D. M. Rankin. 2004. Summer habitat use by rainbow trout and brown trout in Jocassee Reservoir. *North Am J Fish Manage* 24 (2):735–740.

Bednarek, A. T. and D. D. Hart. 2005. Modifying dam operations to restore rivers: Ecological responses to Tennessee river dam mitigation. *Ecol Appl* 15 (3):997–1008.

Beissinger, S. R. and D. R. McCollough. 2002. *Population Viability Analysis*. Chicago, IL: University of Chicago Press.

Beketov, M. A., K. Foit, R. B. Schafer et al. 2009. SPEAR indicates pesticide effects in streams—Comparative use of species- and family-level biomonitoring data. *Environ Pollut* 157 (6):1841–1848.

Belanger, S. E., J. B. Barnum, D. M. Woltering et al. 1994. Algal periphyton structure and function in response to consumer chemicals in stream mesocosms. In *Aquatic Mesocosm Studies in Ecological Risk Assessment*, edited by R. L. Graney, J. H. Kennedy, and J. H. Rodgers. Boca Raton, FL: CRC Press.

Belden, J. and M. J. Lydy. 2000. Impact of atrazine on organophosphate insecticide toxicity. *Environ Toxicol Chem* 19:2266–2274.

Belden, J., R. J. Gilliom, and M. J. Lydy. 2007a. How well can we predict the toxicity of pesticide mixtures to aquatic life? *Integr Environ Assess Manage* 3:364–372.

Belden, J., R. J. Gilliom, and M. J. Lydy. 2007b. Relative toxicity and occurrence patterns of pesticide mixturess in streams draining agricultural watersheds dominated by corn and soybean production. *Integr Environ Assess Manage* 3:90–100.

Bellucci, C., G. Hoffman, and S. Cormier. 2010. *An Iterative Approach for Identifying the Causes of Reduced Benthic Macroinvertebrate Diversity in the Willimantic River, Connecticut*. U.S. Cincinnati, OH: U.S. Environmental Protection Agency, National Center for Environmental Assessment, Office of Research and Development. EPA/600/R-08/144.

Benke, A. C. and J. B. Wallace. 2003. Influence of wood on invertebrate communities in streams and rivers. *Am Fish Soc Symp* 37:149–177.

Benke, A. C., T. C. Vanarsdall, D. M. Gillespie, and F. K. Parrish. 1984. Invertebrate productivity in a sub-tropical blackwater river—The importance of habitat and life-history. *Ecol Monogr* 54 (1):25–63.

Bernhardt, E. S., M. A. Palmer, J. D. Allan et al. 2005. Ecology—Synthesizing US river restoration efforts. *Science* 308 (5722):636–637.

Bernhardt, E. S., E. B. Sudduth, M. A. Palmer et al. 2007. Restoring rivers one reach at a time: Results from a survey of US river restoration practitioners. *Restoration Ecol* 15 (3):482–493.

Berstein, R. 2010. *The Pragmatic Turn*. Cambridge: Polity Press.

Bervoets, L., K. Van Campenhout, H. Reynders et al. 2009. Bioaccumulation of micro-pollutants and biomarker responses in caged carp (*Cyprinus carpio*). *Ecotoxicol Environ Saf* 72:720–728.

Besser, J. M., W. G. Brumbaugh, T. W. May, and C. J. Schmitt. 2007. Biomonitoring of lead, zinc, and cadmium in streams draining lead-mining and non-mining areas, Southeast Missouri, USA. *Environ Monit Assess* 129:227–241.

Besser, J. M., W. G. Brumbaugh, A. L. Allert et al. 2009. Ecological impacts of lead min-ing on Ozark streams: Toxicity of sediment and pore water. *Ecotoxicol Environ Saf* 72:516–526.

Beyer, W. N. and J. P. Meador. 2011. *Environmental Contaminants in Biota: Interpreting Tissue Concentrations*, 2nd ed. Boca Raton, FL: CRC Press.

Beyer, W. N., J. C. Franson, L. N. Locke, R. K. Stroud, and L. Sileo. 1998a. Retrospective study of the diagnostic criteria in a lead-poisoning survey of waterfowl. *Arch Environ Contam Toxicol* 35 (3):506–512.

Beyer, W. N., D. J. Audet, A. Morton, J. K. Campbell, and L. LeCaptain. 1998b. Lead exposure of waterfowl ingesting Coeur d'Alene River Basin sediments. *J Environ Qual* 27:1533–1538.

Beyer, W. N., D. J. Audet, D. J. Heinz, D. J. Hoffman, and D. Day. 2000. Relation of waterfowl poisoning to sediment lead concentrations in the Coeur d'Alene River Basin. *Ecotoxicology* 9:207–218.

Biggs, B. J. F. 2000. Eutrophication of streams and rivers: Dissolved nutrient–chlorophyll relationships for benthic algae. *J North Am Benthol Soc* 19 (1):17–31.

Birge, W. J., J. A. Black, T. M. Short, and A. G. Westerman. 1989. A comparative ecolog-ical and toxicological investigation of a secondary wastewater treatment plant effluent and its receiving stream. *Environ Toxicol Chem* 8:437–450.

Bivand, R. S. 2014. CRAN Task View: Analysis of spatial data. R Foundation for Statistical Computing. http://cran.r-project.org/web/views/Spatial.html (accessed February 1, 2014).

Bivand, R. S., E. J. Pebesma, and Gomez-Rubio. 2008. *Applied Spatial Data Analysis with R*. New York: Springer.

Bjurlin, C. D. and B. L. Cypher. 2003. Effects of roads on San Joaquin kit foxes: A review and synthesis of existing data. In *International Conference on Ecology and Transportation*, edited by C. L. Irwin, P. Garrett, and K. P. McDermott. Raleigh, NC: North Carolina State University.

Black, S. 2013. *Mystery Bee Kill: Causes Being Sought*. Portland, OR: Xerces Society for Invertebrate Conservation.

Blakely, T. J., J. S. Harding, A. R. McIntosh, and M. J. Winterbourn. 2006. Barriers to the recovery of aquatic insect communities in urban streams. *Freshw Biol* 51 (9):1634–1645.

Blankenship, K. 1994. Survey finds broad support for bay cleanup. http://www.bayjournal.com/94-06/survey.htm.

Blankenship, J. and D. F. Dansereau. 2000. The effect of animated node-link displays on information recall. *J Exp Educ* 68 (4):293–308.

Blazer, V. S., L. R. Iwanowicz, D. D. Iwanowicz et al. 2007. Intersex (testicular oocytes) in smallmouth bass from the Potomac River and selected nearby drainages. *J Aquat Anim Health* 19 (4):242–253.

Blus, L. J., C. J. Henny, D. J. Hoffman, and R. A. Grove. 1991. Lead toxicosis in tundra swans near a mining and smelting complex in northern Idaho. *Arch Environ Contam Toxicol* 21 (4):549–555.

Bogan, A. E. 1993. Fresh-water bivalve extinctions (mollusca, unionoida)—A search for causes. *Am Zool* 33 (6):599–609.

Bohannon, J. 2013. Who's afraid of peer review? *Science* 342 (6154):60–65.

Booth, D. B. and C. R. Jackson. 1997. Urbanization of aquatic systems: Degradation thresholds, stormwater detection, and the limits of mitigation. *J Am Water Res Assoc* 33 (5):1077–1090.

Bostrom, A., B. Fischhoff, and M. G. Morgan. 1992. Characterizing mental models of hazardous processes: A methodology and an application to radon. *J Social Iss* 48:85–100.

Bothwell, M. L. 1992. Eutrophication of rivers by nutrients in treated kraft pulp mill effluent. *Water Pollut Res J Canada* 27:447–472.

Bothwell, M. L. 1993. Artificial streams in the study of algal/nutrient dynamics. In *Research in Artificial Streams: Applications, Uses, and Abuses*, ed. G. A. Lamberti and A. D. Steinman, pp. 313–384. *J North Am Benthol Soc* 12:327–333.

Bowling, G., J. Laversee, P. F. Landrum, and J. P. Giesy. 1983. Acute mortality of anthracene contaminated fish exposed to sunlight. *Aquatic Toxicol* 3:79–90.

Box, G. E. P. 1966. Use and abuse of regression. *Am Statist J* 8:625–629.

Boyle, T. P. and J. F. Fairchild. 1997. The role of mesocosm studies in ecological risk analysis. *Ecol Appl* 7:1099–1102.

Braun, C. E. 2005. *Techniques for Wildlife Investigations and Management*. Bethesda, MD: The Wildlife Society.

Bray, J. and J. Curtis. 1957. An ordination of the upland forest communities of southern Wisconsin. *Ecol Monogr* 27 (4):324–349.

British Columbia. 1993. *Ambient Water Quality Criteria for Polycyclic Aromatic Hydrocarbons (PAHs)*. Victoria, BC: British Columbia Ministry of Environment.

Brock, T. C., R. M. Roijackers, R. Rollon, F. Bransen, and L. Van der Heyden. 1995. Effects of nutrient loading and insecticide application on the ecology of Elodea-dominated freshwater microcosms: II. Responses of macrophytes, periphyton and macroinvertebrate grazers. *Arch Hydrobiol* 134:53–74.

Brock, T., R. Van Wijngaarden, and G. Van Geest. 2000. *Ecological risks of pesticides in freshwater ecosystems. Part 2: Insecticides*. Wageningen, the Netherlands: Alterra, Green World Research.

Brock, T. C., I. Roessink, J. D. Belgers, F. Bransen, and S. J. Maund. 2009. Impact of a benzoyl urea insecticide on aquatic macroinvertebrates in ditch meso-cosms with and without non-sprayed sections. *Environ Toxicol Chem* 28:2191–2205.

Brown, L. R., T. F. Cuffney, J. F. Coles et al. 2009. Urban streams across the USA: Lessons learned from studies in 9 metropolitan areas. *J North Am Benthol Soc* 28 (4):1051–1069.

Bruner, J. S. and M. C. Potter. 1964. Interference in visual recognition. *Science* 144 (3617):424–425.

Bunge, M. 1979. *Causality and Modern Science*, 3rd ed. New York: Dover Publications.

Burdick, S. M. and J. E. Hightower. 2006. Distribution of spawning activity by anadromous fishes in an Atlantic slope drainage after removal of a low-head dam. *Trans Am Fish Soc* 135 (5):1290–1300.

Burnham, K. P. and D. R. Anderson. 2002. *Model Selection and Multimodel Inference*, 2nd ed. New York: Springer.

Burton Jr, G. A. 1991. Assessing the toxicity of fresh-water sediments. *Environ Toxicol Chem* 10:1585–1627.

Burton Jr, G. A., M. S. Greenberg, C. D. Rowland et al. 2005. In situ exposures using caged organisms: A multi-compartment approach to detect aquatic toxicity and bioaccumulation. *Environ Pollut* 134 (1):133–144.

Cade, B. S. and B. R. Noon. 2003. A gentle introduction to quantile regression for ecologists. *Frontiers Ecol Environ* 1 (8):412–420.

Cairns, J. J. and J. Pratt. 1993. A history of biological monitoring using benthic macroinvertebrates, pp. 10–27. In *Freshwater Biomonitoring and Benthic Macroinvertebrates*, edited by D. M. Rosenberg and V. H. Resh. New York: Chapman & Hall.

Campaner, P. and M. Galavotti. 2007. Plurality in causality. In *Thinking About Causes: From Greek Philosophy to Modern Physics* edited by P. Machamer and G. Wolters. Pittsburg, PA: University of Pittsburgh Press.

Campbell, D. 1982. Evolutionary epistemology. In *Learning Development and Culture: Essays in Evolutionary Epistemology*, edited by H.C. Plotkin. London: Wiley.

Canadian Council of Ministers of the Environment (CCME). 2013. *Canadian Environmental Quality Guidelines (CEQG online)*. Water Quality Guidelines for the Protection of Aquatic Life.

Cao, Y. and C. M. Hawkins. 2005. Simulating biological impairment to evaluate the accuracy of ecological indicators. *J Appl Ecol* 42:954–965.

Carey, J., O. Cordeiro, and B. Brownlee. 2001. Distribution of contaminants in the water, sediment and biota in the Peace, Athabasca and Slave River basins: Present levels and predicted future trends. Northern River Basins Study Report No. 3. Edmonton, Alberta: Environment Canada.

Carlisle, D., M. Meador, S. Moulton II, and P. Ruhl. 2007. Estimation and application of indicator values for common macroinvertebrate genera and families of the United States. *Ecol Indicat* 7 (1):22–33.

Carpenter, S. R. 1996. Microcosm experiments have limited relevance for community and ecosystem ecology. *Ecology* 77 (3):677–680.

Carpenter, S. R. 1999. Microcosm experiments have limited relevance for community and ecosystem ecology: Reply. *Ecology* 80 (3):1085–1088.

Carpenter, S. R., J. J. Cole, T. E. Essington et al. 1998a. Evaluating alternative explanations in ecosystem experiments. *Ecosystems* 1 (4):335–344.

Carpenter, S., N. Caraco, D. Correll et al. 1998b. Nonpoint pollution of surface waters with phosphorus and nitrogen. *Ecol Appl* 8 (3):559–568.

Carstensen, J., D. Krause-Jensen, S. Markager, K. Timmermann, and J. Windolf. 2013. Water clarity and eelgrass responses to nitrogen reductions in the eutrophic Skive Fjord, Denmark. *Hydrobiologia* 704 (1):293–309.

Cartwright, N. 2003. *Causation: One Word, Many Things*, Tech. Rpt. 07/03. London: London School of Economics.

Cartwright, N. 2007. *Hunting Causes and Using Them: Approaches in Philosophy and Economics*. Cambridge: Cambridge Univeristy Press.

Case, E. 2013. *Bee Deaths a Result Of Pesticide Safari; Count Upped to 50,000 Dead Insects.* The Oregonian. Oregon Live LLC. The Oregonian. Oregon Live LLC.

Cash, K. J., W. Gibbons, K. Munkittrick, S. S. Brown, and C. J. 2000. Fish health in the Peace, Athabasca and Slave river systems. *J Aquat Ecosyst Stress Recovery* 8:77–86.

Cash, K. J., J. M. Culp, M. G. Dubé et al. 2003. Integrating mesocosm experiments with field and laboratory studies to generate weight-of-evidence risk assessments for ecosystem health. *Aquatic Ecosyst Health Manage* 6 (2):177–183.

Cassee, F. R., J. P. Groten, P. J. van Bladeren, and V. J. Feron. 1998. Toxicological evaluation and risk assessment of chemical mixtures. *CRC Crit Rev Toxicol* 28 (1):73–101.

Center for Watershed Protection (CWP). 2003. *Impacts of Impervious Cover on Aquatic Systems*. Watershed Protection Research Monograph 1. Ellicott City, MD: Center for Watershed Protection.

Chamberlin, T. C. 1995. Historical essay—The method of multiple working hypotheses (reprinted from Science, 1890). *J Geol* 103 (3):349–354.

Chambers, P. A., G. J. Scrimgeour, and A. Pietroniro. 1997. Winter oxygen conditions in ice-covered rivers: The impact of pulp mill and municipal effluents. *Can J Fish Aquat Sci* 54:2796–2806.

Chambers, P. A., A. R. Dale, G. J. Scrimgeour, and M. L. Bothwell. 2000. Nutrient enrichment of northern rivers in response to pulp mill and municipal discharges. *J Aquat Ecosyst Stress Recov* 8 (1):53–66.

Chatfield, C. 2004. *The Analysis of Time Series: An Introduction*, 6th ed. New York: Chapman & Hall/CRC.

Cheng, P. W. 1997. From covariation to causation: A causal power theory. *Psychol Revew* 104 (2):367–405.

Chessman, B. C. and P. K. McEvoy. 1998. Towards diagnostic biotic indices for river macroinvertebrates. *Hydrobiologia* 364:169–182.

Christman, V. D., J. R. Voshell Jr, D. G. Jenkins et al. 1994. Ecological development and biometry of untreated pond mesocosms. In *Aquatic Mesocosm Studies in Ecological Risk Assessment*, edited by R. L. Graney, J. H. Kennedy, and J. H. Rodgers. Boca Raton, FL: CRC Press.

Churchland, A. K., R. Kiani, and M. N. Shadlen. 2008. Decision-making with multiple alternatives (vol 11, p. 693, 2008). *Nat Neurosci* 11 (7):851–851.

City of Austin. 2005. *PAHs in Austin, Texas Sediments and Coal-Tar Based Pavement Sealants Polycyclic Aromatic Hydrocarbons*. Austin, TX: Watershed Protection and Development Review Department, Environmental Resources Management Division.

City of Austin. 2006. Ordinance No. 2005117-070. http://www.austintexas.gov/faq/coal-tar-ban-details (accessed February 4, 2011).

Clark, J. L. and W. H. Clements. 2006. The use of in situ and stream microcosm experiments to assess population- and community- level responses to metals. *Environ Toxicol Chem* 25 (9):2306–2312.

Clement, R., S. Suter, E. Reiner, D. McCurvin, and D. Hollinger. 1989. Concentrations of chlorinated dibenzo-*p*-dioxins and dibenzifurans in effluents and centrifuged particulates from Ontario pulp and paper mills. *Chemosphere* 19:649–654.

Clements, W. H. 2004. Small-scale experiments support causal relationships between metal contamination and macroinvertebrate community responses. *Ecol Appl* 14 (3):954–967.

Clements, W. H., D. M. Carlisle, J. Lazorchak, and P. Johnson. 2000. Heavy metals structure benthic communities in Colorado mountain streams. *Ecol Appl* 10:626–638.

Clements, W. H., D. M. Carlisle, L. A. Courtney, and E. A. Harrahy. 2002. Integrating observational and experimental approaches to demonstrate causation in stream biomonitoring studies. *Environ Toxicol Chem* 21 (6):1138–1146.

Cleveland, W. 1993. *Visualizing Data*. Summit, NJ: Hobart Press.

Clews, E. and S. J. Omerod. 2009. Improving biodiagnostic monitoring using simple combinations of standard biotic indices. *River Res Applic* 25:348–361.

Cochran, W. G. 1957a. Analysis of covariance: Its nature and uses. *Biometrics* 13:261–281.

Cochran, W. G. 1957b. *Experimental Designs*. New York: Wiley.

Cochran, W. G. 1965. The planning of observational studies with human populations with discussion. *J Roy Statist Soc A* 128:234–265.

Cochran, W.G. and G. M. Cox. 1957. *Experimental Designs*, 2nd edition. New York: Wiley.

Coffey, D. B., S. M. Cormier, and J. Harwood. 2014. Using field-based species sensitivity distributions to infer multiple causes. *Hum Ecol Risk Assess* 20:402–432.

Cohen, S. E. 1997. *MacKenzie Basin Impact Study Final Report*. Ottawa: Atmospheric Environment Service, Environment Canada.

Collins, J., N. Hall, and L. A. Paul. 2004. *Causation and Counterfactuals*. Cambridge, MA: MIT Press.

Connolly, N. M., M. R. Crossland, and R. G. Pearson. 2004. Effect of low dissolved oxygen on survival, emergence, and drift of tropical stream macroinvertebrates. *J North Am Benthol Soc* 23 (2):251–270.

Cooper, J. E. and M. E. Cooper. 2013. *Wildlife Forensic Investigation: Principles and Practice*. Boca Raton, FL: CRC Press.

Cormier, S. M. and G. W. Suter II. 2008. A framework for fully integrating environmental assessment. *Environ Manage* 42 (4):543–556.

Cormier, S. M. and G. W. Suter II. 2011. Sources of data for water quality criteria. *Environ Toxicol Chem* 32:254.

Cormier, S. M. and G. W. Suter II. 2013a. A method for deriving water-quality benchmarks using field data. *Environ Toxicol Chem* 32 (2):255–262.

Cormier, S. M. and G. W. Suter II. 2013b. A method for assessing causation of field exposure–response relationships. *Environ Toxicol Chem* 32 (2):272–276.

Cormier, S. M., E. Lin, M. Millward et al. 2000. Using regional exposure criteria and upstream reference data to characterize spatial and temporal exposures to chemical contaminants. *Environ Toxicol Chem* 19 (4):1127–1135.

Cormier, S. M., S. B. Norton, G. W. Suter II, D. Altfater, and B. Counts. 2002. Determining the causes of impairments in the Little Scioto River, Ohio, USA: Part 2. Characterization of causes. *Environ Toxicol Chem* 21 (6):1125–1137.

Cormier, S. M., J. F. Paul, R. L. Spehar et al. 2008. Using field data and weight of evidence to develop water quality criteria. *Integr Environ Assess Manage* 4 (4):490–504.

Cormier, S. M., G. W. Suter II, and S. B. Norton. 2010. Causal characteristics for eco-epidemiology. *Hum Ecol Risk Assess* 16 (1):53–73.

Cormier, S. M., G. W. Suter II, and L. Zheng. 2013a. Derivation of a benchmark for freshwater ionic strength. *Environ Toxicol Chem* 32 (2):263–271.

Cormier, S. M., G. W. Suter II, L. Zheng, and G. Pond. 2013b. Assessing causation of the extirpation of stream macroinvertebrates by a mixture of ions. *Environ Toxicol Chem* 32 (2):177–287.

Courtney, L. A. and W. H. Clements. 2002. Assessing the influence of water and substratum quality on benthic macroinvertebrate communities in a metal-polluted stream: An experimental approach. *Freshw Biol* 47 (9):1766–1778.

Cox, T. and J. Rutherford. 2000. Thermal tolerances of two stream invertebrates exposed to diurnally varying temerature. *NZ J Mar Freshw Res* 34 (2):203–208.

Crane, M. and L. Maltby. 1991. The lethal and sublethal responses of Gammarus pulex to stress: Sensitivity and sources of variation in an in situ bioassay. *Environ Toxicol Chem* 10 (10):1331–1339.

Crane, M., C. Attwood, D. Sheahan, and S. Morris. 1999. Toxicity and bioavailability of the organophosphorus insecticide pirimiphos methyl to the freshwater amphipod *Gammarus pulex* L. in laboratory and mesocosm systems. *Environ Toxicol Chem* 18 (7):1456–1461.

Crane, M., G. A. Burton, J. M. Culp et al. 2007. Review of aquatic in situ approaches for stressor and effect diagnosis. *Integr Environ Assess Manage* 3 (2):234–245.

Croce, B. and R. M. Stagg. 1997. Exposure of Atlantic salmon parr (*Salmo salar*) to a combination of resin acids and a water soluble fraction of diesel fuel oil: A model to investigate the chemical causes of Pigmented Salmon Syndrome. *Environ Toxicol Chem* 16 (9):1921–1929.

Crossland, N. O., G. C. Mitchell, and P. B. Dorn. 1992. Use of outdoor artificial streams to determine threshold toxicity concentrations for a petrochemical effluent. *Environ Toxicol Chem* 11 (1):49–59.

Cuffney, T., M. Bilger, and A. Haigler. 2007. Ambiguous taxa: Effects on the characterization and interpretation of invertebrate assemblages. *J N Am Benthol Soc* 26:286–307.

Culp, J. M. and D. J. Baird. 2006. Establishing cause–effect relationships in multi-stressor environments. In *Methods in Stream Ecology* edited by G. A. Lamberti and F. R. Hauer. New York: Elsevier.

Culp, J. M., R. B. Lowell, and K. J. Cash. 2000a. Integrating in situ community experiments with field studies to generate weight-of-evidence risk assessments for large rivers. *Environ Toxicol Chem* 19:1167–1173.

Culp, J. M., C. L. Podemski, K. J. Cash, and R. B. Lowell. 2000b. A research strategy for using stream microcosms in ecotoxicology: Integrating experiments at different levels of biological organization with field data. *J Aquat Ecosyst Stress Recov* 7 (2):167–176.

Culp, J. M., K. J. Cash, and F. J. Wrona. 2000c. Integrated assessment of ecosystem integrity of large northern rivers: The Northern River Basins Study example. *J Aquat Ecosyst Stress Recov* 8 (1):1–5.

Culp, J. M., C. L. Podemski, and K. J. Cash. 2000d. Interactive effects of nutrients and contaminants from pulp mill effluents on riverine benthos. *J Aquat Ecosyst Stress Recov* 8 (1):67–75.

Culp, J. M., K. J. Cash, N. E. Glozier, and R. B. Brua. 2003. Effects of pulp mill effluent on benthic assemblages in mesocosms along the Saint John River, Canada. *Environ Toxicol Chem* 22 (12):2916–2925.

Culp, J., D. Armanini, M. Dunbar et al. 2010. Incorporating traits in aquatic biomonitoring to enhance causal diagnosis and prediction. *Integr Environ Assess Manage* 7 (2):187–197.

Cummins, K. 1974. Structure and function of stream ecosystems. *BioScience* 24 (11):631–641.

Cypher, B. L. 1993. Food item use by three sympatric canids in Southern Illinois. *Trans Illinois State Acad Sci* 86:139–144.

Cypher, B. L. and K. A. Spencer. 1998. Competitive interactions between coyotes and San Joaquin kit foxes. *J Mammal* 79:204–214.

Cypher, B. L., G. D. Warrick, M. R. M. Otten et al. 2000. Population dynamics of San Joaquin kit foxes at the Naval Petroleum Reserves in California. *Wildl Monogr* 145:1–43.

Davis, P. and T. Simon. 1995. *Biological Assessment and Criteria: Tools for Water Resource Planning and Decision Making*. Boca Raton, FL: Lewis Publishers.

Davies, S. P. and D. L. Tsomides. 2002. *Methods for Biological Sampling and Analysis of Maine's Rivers and Streams*. Augusta, ME: Maine Department of Environmental Protection.

Davies, J. J. L., A. Jenkins, D. T. Monteith, C. D. Evans, and D. M. Cooper. 2005. Trends in surface water chemistry of acidified UK Freshwaters, 1988–2002. *Environ Pollut* 137 (1):27–39.

Dawes, R. M. 2001. *Everyday Irrationality*. Boulder, CO: Westview Press.

Day, D. D., W. N. Beyer, D. J. Hoffman et al. 2003. Toxicity of lead-contaminated sediment to mute swans. *Arch Environ Contam Toxicol* 60:1332–1344.

De'ath, G. and K. E. Fabricius. 2000. Classification and regression trees: A powerful yet simple technique for ecological data analysis. *Ecology* 81 (11):3178–3192.

de Bruyn, A. M. H., D. J. Marcogliese, and J. B. Rasmussen. 2003. The role of sewage in a large river food web. *Can J Fish Aquat Sci* 60 (11):1332–1344.

DeGasperi, C. L., H. B. Berge, K. R. Whiting et al. 2009. Linking hydrologic alteration to biological impairment in urbanizing streams of the Puget Lowland, Washington, USA. *J Am Water Res Assoc* 45 (2):512–533.

Degerman, E., M. Appelberg, and P. Nyberg. 1992. Effects of liming on the occurrence and abundance of fish populations in acidified swedish lakes. *Hydrobiologia* 230 (3):201–212.

Dehijia, R. H. and S. Whahba. 2002. Propensity score-matching methods for nonexperimental causal studies. *Rev Econ Stat* 84:151–161.

DeMott, R. P., T. D. Gauthier, J. M. Wiersema, and G. Crenson. 2010. Polycyclic aromatic hydrocarbons (PAHs) in Austin sediments after a ban on pavement sealers. *Environ Forensics* 11 (4):372–382.

DeNicola, D. M. and M. G. Stapleton. 2002. Impact of acid mine drainage on benthic communities in streams: The relative roles of substratum vs. aqueous effects. *Environ Pollut* 119 (3):303–315.

Dennison, W. C., T. R. Lookingbill, T. J. B. Carruthers, J. M. Hawkey, and S. L. Carter. 2007. An eye-opening approach to developing and communicating integrated environmental assessments. *Frontiers Ecol Environ* 5 (6):307–314.

deNoyelles, F., S. L. Dewey, D. G. Huggins, and W. D. Kettle. 1994. Aquatic mesocosms in ecological effects testing: Detecting direct and indirect effects of pesticides. In *Aquatic Mesocosm Studies in Ecological Risk Assessment*, edited by R. L. Graney, J. H. Kennedy, and J. H. Rodgers. Boca Raton, FL: CRC Press.

Detenbeck, N. E., P. W. DeVore, G. J. Niemi, and A. Lima. 1992. Recovery of temperate-stream fish communities from disturbance: A review of case studies and synthesis of theory. *Environ Manage* 16:33–53.

de Zwart, D. and L. Posthuma. 2005. Complex mixture toxicity for single and multiple species: Proposed methodologies. *Environ Toxicol Chem* 24 (10):2665–2676.

de Zwart, D., S. D. Dyer, L. Posthuma, and C. P. Hawkins. 2006. Predictive models attribute effects on fish assemblages to toxicity and habitat alteration. *Ecol Appl* 16 (4):1295–1310.

de Zwart, D., L. Posthuma, M. Gevrey, P. van der Ohe, and E. de Dekere. 2009. Diagnosis of ecosystem impairment in a multiple-stress context—How to formulate effective river basin management plans. *Integr Environ Assess Manage* 5:38–49.

Diamond, J. and C. Daley. 2000. What is the relationship between whole effluent toxicity and instream biological condition? *Environ Toxicol Chem* 19 (1):158–168.

Di Toro, D. M. and J. A. McGrath. 2000. Technical basis for narcotic chemicals and polycyclic aromatic hydrocarbon criteria. I. Mixtures and sediments. *Environ Toxicol Chem* 19:1971–1982.

Dixon, K. R. 2012. *Modeling and Simulation in Ecotoxicology with Applications in MATLAB and Simulink*. Boca Raton, FL: CRC Press.

Diz, H. 1997. *Chemical and biological treatment of acid mine drainage for the removal of heavy metals and acidity*. PhD Thesis. Blacksburg, VA: Virginia Polytechnic Institute.

Doi, R. 1973. Environmental mercury pollution and its influence in the cities of Japan. *Annu Rep Tokyo Metropolitan Res Inst Environ Protection* 3:257–261.

Dowe, P. 1991. Process causality and asymmetry. *Erkenntnis* 37:179–196.

Dowe, P. 2000. *Physical Causation*. Cambridge, UK: Cambridge Unversity Press.

Downs, P. W. and G. M. Kondolf. 2002. Post-project appraisals in adaptive management of river channel restoration. *Environ Manage* 29 (4):477–496.

Draper, N. and H. Smith. 1998. *Applied Regression Analysis*, 3rd ed. New York: Wiley.

Drenner, R. W. and A. Mazumder. 1999. Microcosm experiments have limited relevance for community and ecosystem ecology: Comment. *Ecology* 80 (3):1081–1085.

Driscoll, K. S. B. and R. M. Burgess. 2007. An overview of the development, status, and application of equilibrium partitioning sediment benchmarks for PAH mixtures. *Hum Ecol Risk Assess* 13 (2):286–301.

Dubé, M. G., J. Culp, and G. J. Scrimgeour. 1997. Nutrient limitation and herbivory: Processes influences by bleached kraft pulp mill effluent. *Can J Fish Aquat Sci* 54:2584–2595.

Dubé, M. G., J. M. Culp, K. J. Cash et al. 2002. Artificial streams for environmental effects monitoring (EEM): Development and application in Canada over the past decade. *Water Qual Res J Can* 37 (1):155–180.

Eddington, A. S. 1928. *The Nature of the Physical World*. New York: Macmillan.

Egoscue, H. J. 1975. Population dynamics of the kit fox in western Utah. *S Calif Acad Sci Bull* 74:122–127.

Eisler, R. 1987. *Polycyclic aromatic hydrocarbon hazards to fish, wildlife, and invertebrates: A synoptic review*. Biological Report 86(1.11). Laurel, MD: U.S. Fish and Wildlife Service, Patuxent Wildlife Research Center.

Entrekin, S. A., J. L. Tank, E. J. Rosi-Marshall, T. J. Hoellein, and G. A. Lamberti. 2008. Responses in organic matter accumulation and processing to an experimental wood addition in three headwater streams. *Freshw Biol* 53 (8):1642–1657.

Environment Canada. 1992a. *Biological test method: Chronic toxicity test using the cladoceran Ceriodaphnia dubia*. Ottawa: Environment Canada.

Environment Canada. 1992b. *Biological test method: Growth inhibition test using the freshwater algal Selenastrum capricornutum*. Ottawa: Environment Canada.

Environment Canada. 1992c. *Biological test method: Test of larval growth and survival using fathead minnow*. Ottawa: Environment Canada.

Environment Canada. 2002. *Metal mining guidance document for aquatic environmental effects monitoring*. Gatineau: Environment Canada.

Environmental Law Institute (ELI). 2007. *Mitigation of Impacts to Fish and Wildlife Habitat: Estimating Costs and Identifying Opportunities*. Washington, DC: Environmental Law Institute.

Escher, B. I. and J. Hermens. 2002. Modes of action in ecotoxicology: Their role in body burdens, species sensitivity, QSARs, and mixture effects. *Environ Sci Technol* 36:4201–4217.

Federal Geographic Data Commission (FGDC). 2014. Geoplatform.gov. Federal Geographic Data Commission. http://www.geoplatform.gov (accessed February 1, 2014).

Feld, C. K., S. Birk, D. C. Bradley et al. 2011. From natural to degraded rivers and back again: A test of restoration ecology theory and practice. *Adv Ecol Res.* 44: 119–209.

Fenton, N. and M. Neil. 2013. *Risk Assessment and Decision Analysis with Bayesian Networks.* Baca Raton, FL: CRC Press.

Ferrington Jr, L. C., M. A. Blackwood, C. A. Wright, T. M. Anderson, and D. S. Goldhammer. 1994. Sediment transfers and representativeness of mesocosm test fauna. In *Aquatic Mesocosm Studies in Ecological Risk Assessment*, edited by R. L. Graney, J. H. Kennedy, and J. H. Rodgers. Boca Raton, FL: CRC Press.

Feynman, R. 2001. Cargo cult science: Some remarks on science, pseudoscience, and learning how not to fool yourself. In *The Pleasure of Finding Things Out: The Best Short Works of Richard Feynman*, edited by J. Robbins. Cambridge, MA: Perseus.

Fielding, A. and J. Bell. 1997. A review of methods for the assessment of prediction errors in conservation presence/absence models. *Environ Conserv* 24 (1):38–49.

Fisher, R. 1937. *The Design of Experiments.* London: MacMillan.

Fisher, K. 1999. *Revegetation of fluvial tailings deposits on the Arkansas River near Leadville, Colorado.* MS Thesis. Fort Collins: C. S. University.

Fisher, K., J. Brummer, W. Leininger, and D. Heil. 2000. Interactive effects of soil amendments and depth of incorporation on Geyer willow. *J Environ Qual* 29:1786–1793.

Flagler, R. B. 1998. *Recognition of Air Pollution Injury to Vegetation: A Pictorial Atlas*, 2nd ed. Edited by J. S. Jacobson and A. C. Hill. Pittsburg, PA: Air & Waste Management Association.

Fleeger, J. W., K. R. Carman, and R. M. Nisbet. 2003. Indirect effects of contaminants in aquatic ecosystems. *Sci Total Environ* 317 (1):207–233.

Floit, S. B. and L. W. Barnthouse. 1991. *Demographic analysis of a San Joaquin kit fox population.* Oak Ridge, TN: U. S. Department of Energy, Oak Ridge National Laboratory. ORNL/TM-11679.

Florida Department of Environmental Protection (FDEP). 2005. *Statistical Analysis of Surface Water Quality Specific Conductance Data.* Bureau of Laboratories, Division of Resource Assessment and Management, Biology Section.

Folke, C., S. Carpenter, B. Walker et al. 2004. Regime shifts, resilience, and biodiversity in ecosystem management. *Annu Rev Ecol Evol Syst* 35:557–581.

Forbes, V. E. and T. L. Forbes. 1994. *Ecotoxicology in Theory and Practice*, Vol. 2. London: Chapman & Hall.

Forbes, V. E. and P. Calow. 2002. Species sensitivity distributions revisited: A critical appraisal. *Hum Ecol Risk Assess* 8 (3):473–492.

Forbes, V. E., A. Palmqvist, and L. Bach. 2006. The use and misuse of biomarkers in ecotoxicology. *Environ Toxicol Chem* 25:272–280.

Forbes, V. E., P. Calow, V. Grimm et al. 2011. Adding value to ecological risk assessment with population modeling. *Hum Ecol Risk Assess* 17:287–299.

Forrest, J. and S. E. Arnott. 2006. Immigration and zooplankton community responses to nutrient enrichment: A mesocosm experiment. *Oecologia* 150 (1):119–131.

Fox, G. A. 1991. Practical causal inference for ecoepidemiologists. *J Toxicol Environ Health, Part A Curr Iss* 33 (4):359–373.

Frankfort, H. 1946. *Before Philosophy: The Intellectual Adventure of Ancient Man*. London: Harmondsworth.

Freeman, W. 2000. Perception of time and causation through the kinesthesia of intentional action. *Gen Psychol Spec Iss* 1:18–34.

Friend, M. and J. C. Franson. 1999. *Field Guide to Wildlife Diseases*. Washington: U.S. Geological Survey.

Fulton, E. A., J. S. Link, I. C. Kaplan et al. 2011. Lessons in modelling and management of marine ecosystems: The Atlantis experience. *Fish Fisheries* 12 (2):171–188.

Gagnon, C., F. Gagné, P. Turcotte et al. 2006. Exposure of caged mussels to metals in primary-treated municipal wastewater plume. *Chemosphere* 62:998–1010.

Galli, J. and R. Dubose. 1990. *Water temperature and freshwater biota: An overview*. Washington, DC: Department of Environmental Programs, Metropolitan Washington Council of Governments.

Genkai-Kato, M., H. Mitsuhashi, Y. Kohmatsu et al. 2005. A seasonal change in the distribution of a stream-dwelling stonefly nymph reflects oxygen supply and water flow. *Ecol Res* 20:223–226.

Gentile, J. H., K. R. Solomon, J. B. Butcher et al. 1999. Linking stressors and ecological responses. In *Multiple Stressors in Ecological Risk and Impact Assessment*, edited by J. A. Foran and S. A. Ferenc. Pensacola, FL: SETAC Press.

Germano, J. 1999. Ecology, statistics, and the art of misdiagnosis: The need for a paradigm shift. *Environ Rev* 7 (4):167–190.

Gerritsen, J., J. Burton, and M. T. Barbour. 2000a. *A stream condition index for West Virginia wadeable streams*. Owings Mills, MD: Tetra Tech, Inc.

Gerritsen, J., B. Jessup, E. W. Leppo, and J. White. 2000b. *Development of Lake Condition Indexes (LCI) for Florida*. Tallahassee, FL: Florida Department of Environmental Protection.

Gerritsen, J., L. Zheng, J. Burton et al. 2010. *Inferring Causes of Biological Impairment in the Clear Fork Watershed, West Virginia*. Cincinnati, OH: U.S. Environmental Protection Agency, Office of Research and Development, National Center for Environmental Assessment. EPA/600/R-08/146.

Gibbons, W. N. and K. R. Munkittrick. 1994. A sentinal monitoring framework for identifying fish population responses to industrial discharges. *J Aquat Ecosyst Health* 3:227–237.

Gibbons, W. N., K. R. Munkittrick, and W. D. Taylor. 1998. Monitoring aquatic environments receiving industrial effluents using small fish species 1: Response of spoonhead sculpin (*Cottus ricei*) downstream of a bleached-kraft pulp mill. *Environ Toxicol Chem* 17 (11):2227–2237.

Giddings, J. M., G. W. Suter II, L. W. Barnthouse, and A. S. Hammons. 1981. *Methods for Ecological Toxicology: A Critical Review of Laboratory Multispecies Tests*. Ann Arbor, MI: Ann Arbor Science Publishers.

Giddings, J. M., R. C. Biever, R. L. Helm, G. L. Howick, and F. J. DeNoyelles. 1994. The fate and effects of Guthion (azinphos methyl) in mesocosms. In *Aquatic Mesocosm Studies in Ecological Risk Assessment*, edited by R. L. Graney, J. H. Kennedy, and J. H. Rodgers. Boca Raton, FL: CRC Press.

Gigerenzer, G. and U. Hoffrage. 1995. How to improve Bayesian reasoning without instruction: Frequency formats. *Psychol Rev* 102 (4):684–704.

Gilbertson, M., T. Kubiak, J. Ludwig, and G. Fox. 1991. Great-Lakes embryo mortality, edema, and deformities syndrome (GLEMEDS) in colonial fish-eating birds—similarity to chick-edema disease. *J Toxicol Environ Health* 33 (4):455–520.

Gladwell, M. 2007. *Blink*. New York: Little, Brown and Company.

Glass, G. 1976. Primary, secondary and meta-analysis of research. *Edu Res* 5 (10):3–8.

Gleick, J. 2003. *Isaac Newton*. New York: Pantheon Books.

Glymour, C. 2001. *The Mind's Arrows: Bayes Net and Graphical Causal Models in Psychology*. Cambridge, MA: MIT Press.

Glymour, C. and S. Greenland. 2008. Causal diagrams. In *Modern Epidemiology*, edited by K. J. Rothman, S. Greenland, and T. L. Lash. Philadelphia: Wolters Kluwer.

Goodchild, M. F., B. O. Parks, and L. T. Steyaert. 1993. *Environmental Modeling with GIS*. New York: Oxford University Press.

Gordon, G. E. 1988. Receptor models. *Environ Sci Technol* 22:1132–1142.

Gordon, C. 2005. *Temperature and Toxicology: An Integrative, Comparative and Environmental Approach*. Boca Raton, FL: Taylor & Francis.

Graham, M. H. 2003. Confronting multicollinearity in ecological multiple regression. *Ecology* 84:2809–2815.

Graney, R. L., J. H. Kennedy, and J. H. Rodgers. 1994. Introduction. In *Aquatic Mecosocosm Studies in Ecological Risk Assessment*, edited by R. L. Graney, J. H. Kennedy, and J. H. Rodgers. Boca Raton, FL: CRC Press.

Graney, R. L., J. P. Giesy, and J. R. Clark. 1995. Field studies. In *Fundamentals of Aquatic Toxicology: Effects, environmental fate, and risk assessment*, edited by G. M. Rand. Washington, DC: Taylor & Francis.

Green, R. H. 1979. *Sampling Design and Statistical Methods for Environmental Biologists*. New York: Wiley Interscience.

Greenland, S. 2008. Introduction to regression modeling. In *Modern Epidemiology*, edited by K. J. Rothman, S. Greenland, and T. L. Lash. Philadelphia: Wolters Kluwer.

Greenland, S., J. Pearl, and E. L. Robinson. 1999. Causal diagrams for epidemiological research. *Epidemiology* 10:37–48.

Gregory, S. V., F. J. Swanson, W. A. McKee, and K. W. Cummins. 1991. An ecosystem perspective of riparian zones. *Bioscience* 41:540–551.

Grosell, M. and C. M. Wood. 2002. Copper uptake across rainbow trout gills: Mechanisms of apical entry. *J Exp Biol* 205 (Pt 8):1179–1188.

Grosell, M., C. Nielsen, and A. Bianchini. 2002. Sodium turnover rate determines sensitivity to acute copper and silver exposure in freshwater animals. *Comp Biochem Physiol C Toxicol Pharmacol* 133 (1–2):287–303.

Guckert, J. B. 1993. Artificial streams in ecotoxicology. In *Research in Artificial Streams: Applications, Uses, and Abuses*, edited by G.A. Lamberti and A.D. Steinman, pp. 313–384. *J N Am Benthol Soc* 12:350–356.

Gunderson, L. and L. Pritchard. 2002. *Resilience and the Behavior of Large-Scale Systems*. Washington, DC: Island Press.

Guzelian, P. S., M. S. Victoroff, N. C. Halmes, R. C. James, and C. P. Guzelian. 2005. Evidence-based toxicology: A comprehensive framework for causation. *Hum Exp Toxicol* 24 (4):161–201.

Haake, D., T. Wilton, K. Krier, A. Stewart, and S. Cormier. 2010a. Causal assessment of biological impairment in the Little Floyd River, Iowa, USA. *Hum Ecol Risk Assess* 16:116–148.

Haake, D., T. Wilton, K. Krier et al. 2010b. *Stressor Identification in an Agricultural Watershed: Little Floyd River, Iowa*. Cincinnati, OH: U.S. Environmental Protection Agency, Office of Research and Development, National Center for Environmental Assessment. EPA/600/R-08/131.

Haase, P., D. Hering, S. C. Jaehnig, A. W. Lorenz, and A. Sundermann. 2013. The impact of hydromorphological restoration on river ecological status: A comparison of fish, benthic invertebrates, and macrophytes. *Hydrobiologia* 704 (1):475–488.

Haddon, M. 2001. *Modeling and Quantitative Methods in Fisheries*. New York: Chapman & Hall.

Hagerthey, S. E., S. B. Norton, K. Schiff et al. 2013. The Salinas case study. In *Causal Assessment Evaluation and Guidance for California*, edited by K. Schiff, D. Gillett, A. Rehn, and M. Paul. Long Beach, CA: Southern California Coastal Water Research Project.

Hahn, G. and W. Meeker. 1991. *Statistical Intervals: A Guide for Practitioners*. Hoboken, NJ: Wiley.

Hall Jr., L. W. and J. M. Giddings. 2000. The need for multiple lines of evidence for predicting site-specific ecological effects. *Hum Ecol Risk Assess* 6 (4):679–710.

Hamers, T., J. Legler, L. Blaha et al. 2013. Expert opinion on toxicity profiling—Report from a NORMAN Expert Group Meeting. *IEAM* 9:185–191.

Hames, R. S., J. D. Lowe, S. B. Swarthout, and K. V. Rosenberg. 2006. Understanding the risk to neotropical migrant bird species of multiple human-caused stressors: Elucidating processes behind the patterns. *Ecol Soc* 11 (1):24. http://www.ecologyandsociety.org/vol11/iss1/art24/.

Hansen, J. A., J. Lipton, P. G. Welsh et al. 2002. Relationship between exposure duration, tissue residues, growth, and mortality in rainbow trout (*Oncorhynchus mykiss*) juveniles sub-chronically exposed to copper. *Aquat Toxicol* 58:175–188.

Hanson, M. L., H. Sanderson, and K. R. Solomon. 2003. Variation, replication, and power analysis of *Myriophyllum* spp. microcosm toxicity data. *Environ Toxicol Chem* 22 (6):1318–1329.

Harding, J. S., E. F. Benfield, P. V. Bolstad, G. S. Helfman, and E. B. D. Jones. 1998. Stream biodiversity: The ghost of land use past. *Proc Natl Acad Sci USA* 95 (25):14843–14847.

Harrell, F. E. 2001. *Regression Modeling Strategies: With Applications to Linear Models, Logistic Regression and Survival Analysis*. New York: Springer.

Harris, C. E., T. P. O'Farrell, P. M. McCue, and T. T. Kato. 1987. *Capture–recapture estimation of San Joaquin kit fox population size on Naval Petroleum Reserve #1, Kern County, California*. Santa Barbara, CA: U. S. Department of Energy Topical Report No.10282-2149.

Harrison, I. 1997. Case Based Reasoning. Artificial Intelligence Applications Institute, University of Edinburgh. http://www.aiai.ed.ac.uk/links/cbr.html (accessed February 4, 2014).

Harrison, D. J., J. A. Bissonette, and J. S. Sherburne. 1989. Spatial relationships between coyotes and red foxes in Eastern Maine. *J Wildl Manage* 53:181–185.

Harwood, J. J. and R. A. Stroud. 2012. A survey on the utility of the USEPA CADDIS stressor identification procedure. *Environ Monit Assess* 184 (6):3805–3812.

Harwood, A. D., J. You, and M. J. Lydy. 2009. Temperature as a toxicity identification evaluation tool for pyrethroid insecticides: Toxicokinetic confirmation. *Environ Toxicol Chem* 28 (5):1051–1058.

Hawkins, C., R. Norris, J. Gerritsen et al. 2000. Evaluation of the use of landscape classifications for the prediction of freshwater biota: Synthesis and recommendation. *J N Am Benthol Soc* 19:541–556.

Heath, C. and D. Heath. 2008. *Made to Stick*. New York: Random House.

Hedges, L. and I. Olkin. 1985. *Statistical Methods for Meta-Analysis*. San Diego, CA: Academic Press.

Hegarty, M. and M. A. Just. 1993. Constructing mental models of machines from text and diagrams. *J Memory Language* 32:717–742.

Heimbach, F., J. Berndt, and W. Pflueger. 1994. Fate and biological effects of an herbicide on two artificial pond ecosystems of different size. In *Aquatic Mesocosm Studies in Ecological Risk Assessment*, edited by R. L. Graney, J. H. Kennedy, and J. H. Rodgers. Boca Raton, FL: CRC Press.

Heinis, L. J. and M. L. Knuth. 1992. The mixing, distribution and persistence of esfenvalerate within littoral enclosures. *Environ Toxicol Chem* 11 (1):11–25.

Helfrich, L. A. and S. A. Smith. 2009. *Fish Kills: Their Causes and Prevention*. Blacksburg, VA: Virginia Cooperative Extension. Publication 420-252.

Hellyer, G., P. Leinenbach, J. Hollister et al. 2011. Geospatial Tools. EPA Watershed Central Wiki. U.S. Environmental Protection Agency. http//water.epa.gov/type/watersheds/datait/watershedcentral/wiki.cfm (accessed January 1, 2014).

Helsel, D. and R. Hirsch. 1992. Statistical methods in water resources. *Studies in Environmental Science 49*. New York: Elsevier.

Hem, J. D. 1985. *Study and interpretation of the chemical characteristics of natural water*. Water-Supply Paper 2254. Richmond, VA: U.S. Geological Survey.

Hempel, C. 1965. *Aspects of Scientific Explanation*. New York: Free Press.

Henley, W., M. Patterson, R. Neves, and A. Lemly. 2000. Effects of sedimentation and turbidity on lotic food webs: A concise review for natural resource managers. *Rev Fish Sci* 8 (2):125–139.

Hering, D., A. Borja, L. Carvalho, and C. K. Feld. 2013. Assessment and recovery of European water bodies: Key messages from the WISER project. *Hydrobiologia* 704 (1):1–9.

Heugens, E. H., A. J. Hendriks, T. Dekker, N. M. van Straalen, and W. Admiraal. 2001. A review of the effects of multiple stressors on aquatic organisms and analysis of uncertainty factors for use in risk assessment. *Crit Rev Toxicol* 31 (3):247–284.

Hewitt, L. M., M. G. Dube, S. C. Ribey et al. 2005. Investigation of cause in pulp and paper environmental effects monitoring. *Water Qual Res J Can* 40 (3):261–274.

Hicks, M., K. Whittington, J. Thomas et al. 2010. *Causal Assessment of Biological Impairment in the Bogue Homo River, Mississippi Using the U.S. EPA's Stressor Identification Methodology*. Cincinnati, OH: U.S. Environmental Protection Agency, National Center for Environmental Assessment.

Higgins, J. P. T. and S. Green. 2008. *Cochrane Handbook for Systematic Reviews of Interventions*. Chichester, UK: John Wiley and Sons.

Hill, A. B. 1965. The environment and disease: Association or causation? *Proc R Soc Med* 58:295–300.

Hillborn, R. and M. Mangel. 1997. *The Ecological Detective: Confronting Models with Data*. Princeton, NJ: Princeton University Press.

Hilsenhoff, W. L. 1987. An improved biotic index of organic stream pollution. *Great Lakes Entomol* 20 (1):31–39.

Hinck, J. E., V. S. Blazer, C. J. Schmitt, D. M. Papoulias, and D. E. Tillitt. 2009. Widespread occurrence of intersex in black basses (*Micropterus* spp.) from US rivers, 1995–2004. *Aquat Toxicol* 95 (1):60–70.

Hitchcock, C. 2007. How to be a causal pluralist. In *Thinking About Causes: From Greek Philosophy to Modern Physics*, edited by P. Machamer and G. Wolters. Pittsburgh, PA: University of Pittsburgh Press.

Hobbie, J., S. Carpenter, N. Grimm, J. Gosz, and T. Seastedt. 2003. The US long term ecological research program. *Bioscience* 53 (1):21–32.

Hoff, D., W. Lehmann, A. Pease et al. 2010. *Predicting the toxicities of chemicals to aquatic animal species*. US Environmental Protection Agency Report.

Holling, C. S. 1978. Adaptive Environmental Management and Assessment. London: John Wiley and Sons.

Holmstrup, M., A. M. Bindesbøl, G. J. Oostingh et al. 2010. Interactions between effects of environmental chemicals and natural stressors: A review. *Sci Total Environ* 408 (18):3746–3762.

Homer, C., J. Dewitz, J. Fry et al. 2007. Completion of the 2001 National Land Cover Database for the conterminous United States. *Photogrammetric Eng Remote Sens* 73 (4):337–341.

Hoornbeek, J., E. Hansen, E. Ringquist, and R. Carlson. 2008. *Implementing Total Maximum Daily Loads: Understanding and Fostering Successful Results*. Kent, OH: Center for Public Administration and Public Policy, Kent State University.

Horizons Systems Corporation. 2012. NHDPlus Version 2. Horizon System Corporation. http://horizon-systems.com/NHDPlus/NHDPlusV2_home.php (accessed February 1, 2014).

Howick, G. L., F. deNoyelles Jr., J. M. Giddings, and R. L. Graney. 1994. Earthen ponds vs. fiberglass tanks as venues for assessing the impact of pesticides on aquatic environments: A parallel study with sulprofos. In *Aquatic Mesocosm Studies in Ecological Risk Assessment*, edited by R. L. Graney, J. H. Kennedy, and J. H. Rodgers. Boca Raton, FL: CRC Press.

Hoy, C. W., G. P. Head, and F. R. Hall. 1998. Spatial heterogeneity and insect adaptation to toxins. *Annu Rev Entomol* 43 (1):571–594.

Huckabee, J. W., F. O. Cartan, G. P. Head, and F. R. Hall. 1972. *Environmental influence on trace elements in hair of 15 species of mammals*. Oak Ridge, TN: U. S. D. o. E. Oak Ridge National Laboratory. ORNL/TM-3747.

Huffman, J. E. and J. R. Wallace. 2011. *Wildlife Forensics: Methods and Applications*. Hoboken, NJ: Wiley.

Huitima, B. E. 2011. *The Analysis of Covariance and Alternatives: Statistical Methods for Experiments, Quasi-experiments, and Single-Case Studies*, 2nd ed. New York: Wiley.

Hume, D. 1748. *An Inquiry Concerning Human Understanding*. Amherst, NY: Prometheus Books.

Hurlbert, S. H. 1984. Pseudoreplication and the design of ecological field experiments. *Ecol Monogr* 54:187–211.

Huston, M. 1997. Hidden treatments in ecological experiments: Re-evaluating the ecosystem function and biodiversity. *Oecologia* 110:449–460.

HydroQual, Inc. 2007. *Biotic Ligand Model Windows Interface, Version 2.2.3. User's Guide and Reference Manual*. Mahwah, NJ: HydroQual, Inc.

Hyerle, D. 2000. *A Field Guide to Using Visual Tools*. Lyme, New Hampshire: Designs for Thinking.

Hynes, H. 1960. *The Biology of Polluted Waters*. Liverpool, UK: Liverpool University Press.

Hynes, H. 1970. *The Ecology of Running Waters*. Toronto, Canada: Toronto Press.

Ivorra, N., J. Hettelaar, G. M. J. Tubbing et al. 1999. Translocation of microbenthic algal assemblages used for in situ analysis of metal pollution in rivers. *Arch Environ Contam Toxicol* 37:19–28.

Jaag, O. and H. Ambühl. 1964. The effect of the current on the composition of biocoenoses in flowing water Streams. *Adv Water Pollut Res* 1:39–49.

James, F. C. and C. E. McCulloch. 1990. Multivariate analysis in ecology and systematics: panacea or Pandora's box? *Ann Rev Syst Ecol* 21:129–166.

Jansson, R., H. Backx, A. J. Boulton et al. 2005. Stating mechanisms and refining criteria for ecologically successful river restoration: A comment on Palmer et al. (2005). *J Appl Ecol* 42 (2):218–222.

Jarvinen, A. W. and G. T. Ankley. 1999. *Linkages of Effects to Tissue Residues: Development of a Comprehensive Data Base for Aquatic Organisms Exposed to Inorganic and Organic Chemicals*. Pensacola, FL: SETAC Press.

Jarvinen, A. W. and G. T. Ankley. 2009. Toxicity/Residue Database. http://www.epa. gov/med/Prods_Pubs/tox_residue.htm (accessed February 9, 2014).

Jenkins, D. W. 1979. *Toxic trace metals in mammalian hair and nails*. Las Vegas, NV: U.S. Environmental Protection Agency. EPA-600/4-79-049.

Jenkins, K. M. and A. J. Boulton. 2007. Detecting impacts and setting restoration targets in arid-zone rivers: Aquatic micro-invertebrate responses to reduced floodplain inundation. *J Appl Ecol* 44 (4):823–832.

Jenkins, J., K. Roy, C. Driscoll, and C. Buerkett. 2007. *Acid Rain in the Adirondacks: An Environmental History*. Ithaca, NY: Cornell University Press.

Jho, E. H., J. An, and K. Nam. 2011. Extended biotic ligand model for prediction of mixture toxicity of Cd and Pb using single metal toxicity data. *Environ Toxicol Chem* 30 (7):1697–1703.

Joffe, M. M. and P. R. Rosenbaum. 1999. Invited commentary: Propensity scores. *Am J Epidemiol* 150 (4):327–333.

Johnson, D. 1999. The insignificance of statistical significance testing. *J Wildl Manage* 63 (3):763–772.

Johnson, P. C., J. H. Kennedy, R. G. Morris, F. E. Hambleton, and R. L. Graney. 1994. Fate and effects of cyfluthrin (pyrethroid insecticide) in pond mesocosms and concrete microcosms. In *Aquatic Mesocosm Studies in Ecological Risk Assessment*, edited by R. L. Graney, J. H. Kennedy, and J. H. Rodgers. Boca Raton, FL: CRC Press.

Jongman, R. H. G., C. J. F. ter Braak, and O. F. R. van Tongeren. 1987. *Data Analysis in Community and Landscape Ecology*. Wageningen, the Netherlands: Pudoc.

Josephson, J. R. and S. G. Josephson. 1996. *Abductive Inference*. Cambridge: Cambridge University Press.

Joy, J. and B. Patterson. 1997. *Total maximum daily load evaluation report for the Yakima River*. Olympia, WA: Washington State Department of Ecology.

Kahneman, D. 2011. *Thinking, Fast and Slow*. New York: Farrar, Straus and Giroux.

Kapo, K. E., G. A. Burton Jr., D. de Zwart, L. Posthuma, and S. D. Dyer. 2008. Quantitative lines of evidence for screening-level diagnostic assessment of regional fish community impacts: A comparison of spatial database evaluation methods. *Environ Sci Technol* 42 (24):9412–9418.

Kefford, B. J., P. J. Papas, and D. Nugegoda. 2003. Relative salinity tolerance of macroinvertebrates from the Barwon River, Victoria, Australia. *Mar Freshw Res* 54 (6):755–765.

Kennen, J. G. 1999. Relation of macroinvertebrate community impairment to catchment characteristics in New Jersey streams. *J Am Water Res Assoc* 35 (4):939–955.

Kersting, K. 1994. Functional endpoints in field testing. In *Freshwater Field Tests for Hazard Assessment of Chemicals*, edited by I. R. Hill, F. Heimbach, P. Leeuwangh, and P. Matthiessen. Boca Raton, FL: CRC Press.

Kidd, K. A., P. J. Blanchfield, K. H. Mills et al. 2007. Collapse of a fish population after exposure to a synthetic estrogen. *Proc Natl Acad Sci* 104 (21):8897–8901.

Kilgour, B., K. Somers, and D. Matthews. 1998. Using the normal range as a criterion for ecological significance in environmental monitoring and assessment. *EcoScience* 5:542–550.

Kimball, K. D. and S. A. Levin. 1985. Limitations of laboratory bioassays: The need for ecosystem-level testing. *Bioscience* 35:165–171.

Knowlton, M. F. and J. R. Jones. 2000. Non-algal seston, light, nutrients and chlorophyll in Missouri reservoirs. *Lake Reserv Manage* 16:322–332.

Knowlton, M. and J. R. Jones. 2006. Natural variability in lakes and reservoirs should be recognized in setting nutrient criteria. *Lake Reserv Manage* 22 (2):161–166.

Koel, T. and J. Peterka. 1995. Survival to hatching of fishes in sulfate–saline waters, Devils Lake, North Dakota. *Can J Fish Aquat Sci* 52 (3):464–469.

Kravitz, M. 2011. *Stressor Identification (SI) at Contaminated Sites: Upper Arkansas River, Colorado (Final)*. Washington, DC: U.S. Environmental Protection Agency, Office of Research and Development, National Center for Environmental Assessment. EPA/600/R-08/029.

Kristensen, P. 2004. The DPSIR Framework. Paper read at Comprehensive/Detailed Assessment of the Vulnerability of Water Resources to Environmental Change in Africa Using River Basin Approach, September 27–29, 2004, Nairobi, Kenya.

Kuehl, D. W., B. C. Butterworth, W. M. Devita, and C. P. Sauer. 1987. Environmental contamination by polychlorinated dibenzo-*p*-dioxins and dibenzofurans associated with pulp and paper mill discharge. *Biomed Environ Mass Spectrom* 14 (8):443–447.

Kutner, M., C. Nachtscheim, J. Neter, and W. Li. 2004. *Applied Linear Regression Models*. New York: McGraw-Hill.

Lachin, J. M. 2000. *Biostatistical Methods: The Assessment of Relative Risk*. New York: Wiley.

Lamberti, G. A. 1993. Grazing experiments in artificial streams. In *Research in Artificial Streams: Applications, Uses, and Abuses*, edited by G.A. Lamberti and A.D. Steinman, pp. 313–384. *J N Am Benthol Soc* 12:337–342.

Lamberti, G. A. and A. D. Steinman. 1993. Conclusions. In *Research in Artificial Streams: Applications, Uses, and Abuses*, edited by G.A. Lamberti and A.D. Steinman, pp. 313–384. *J N Am Benthol Soc* 12:370.

La Point, T. W., M. T. Barbour, D. L. Borton et al. 1996. Discussion synopsis: Field assessments. In *Whole-effluent Toxicity Testing: An Evaluation of Methods and Predictability of Receiving System Responses*, edited by D. R. Grothe, K. L. Dickson, and D. K. Reed. Pensacola, FL: CRC Press.

Larkin, J. H. and H. A. Simon. 1987. Why a diagram is (sometimes) worth ten thousand words. *Cognit Sci* 11:65–99.

Laskowski, R. 1995. Some good reasons to ban the use of NOEC, LOEC, and related concepts in ecotoxicology. *Oikos* 73: 140–144.

Laskowski, R., A. J. Bednarska, P. E. Kramarz et al. 2010. Interactions between toxic chemicals and natural environmental factors—A meta-analysis and case studies. *Sci Total Environ* 408 (18):3763–3774.

Lawrence, G. B., W. C. Shortle, M. B. David et al. 2012. Early indications of soil recovery from acidic deposition in US Red Spruce Forests. *Soil Sci Soc Am J* 76 (4):1407–1417.

Lawton, J. H. 1996. The Ecotron facility at Silwood Park: The value of "big bottle" experiments. *Ecology* 77 (3):665–669.

Legendre, P. and L. Legendre. 2012. *Numerical Ecology*, 3rd English ed. Amsterdam, the Netherlands: Elsevier.

Lehman, J. T. 1986. Control of eutrophication in Lake Washington: Case study. In *Ecological Knowledge and Environmental Problem Solving: Concepts and Case Studies*. Washington, DC: National Academy Press.

Lehrer, J. 2009. *How We Decide*. New York: Houghton Mifflin Harcourt Publishing Company.

Lemly, A. D. 2000. Using bacterial growth on insects to assess nutrient impacts in streams. *Environ Monit Assess* 63 (3):431–446.

Lenat, D. R. and V. H. Resh. 2001. Taxonomy and stream ecology—the benefits of genus- and species-level identification. *J N Am Benthol Soc* 20:287–298.

Lenihan, J. 1978. Hair as a mirror of the environment. In *Measuring and Monitoring the Environment*, edited by J. Lenihan and W. W. Fletcher. Glasgow: Blackie.

Leslie, A. M. and S. Keeble. 1987. Do six-month-old infants perceive causality? *Cognition* 25:265–288.

Lessard, J. L. 2000. *The influence of temperature changes on fish and macroinvertebrates below dams in Michigan*. M.S. Thesis. East Lancing, MI: Michigan State University.

Lessard, J. L. and D. B. Hayes. 2003. Effects of elevated water temperature on fish and macroinvertebrate communities below small dams. *River Res Appl* 19 (7):721–732.

Lewis, T. R. 1972. Effects of air pollution on livestock and animal products. In *Helena Valley, Montana Area Pollution Study*. Research Triangle Park, NC: U.S. Environmental Protection Agency.

Lewis, D. 1973. Causation. *J Phil* 71 (17):556–567.

Liber, K., K. R. Solomon, N. K. Kaushik, and J. H. Carey. 1994. Impact of 2,3,4,6-tetrachlorophenol (DIATOX (R)) on plankton communities in limnocorrals. In *Aquatic Mesocosm Studies in Ecological Risk Assessment*, edited by R. L. Graney, J. H. Kennedy, and J. H. Rodgers. Boca Raton, FL: CRC Press.

Liber, K., W. Goodfellow, P. Den Besten et al. 2007. In situ-based effects measures: Considerations for improving methods and approaches. *Integr Environ Assess Manage* 3 (2):246–258.

Liess, M. 2014. SPEAR. Department of System Ecotoxicology at the Helmholtz Center for Environmental Research. http://www.systemecology.eu/spear/ (accessed February 9, 2014).

Liess, M. and P. C. von der Ohe. 2005. Analyzing effects of pesticides on invertebrate communities in streams. *Environ Toxicol Chem* 24 (4):954–965.

Lindberg, T. T., E. S. Bernhardt, R. Bier et al. 2011. Cumulative impacts of mountaintop mining on an Appalachian watershed. *Proc Natl Acad Sci USA* 108 (52):20929–20934.

Linkov, I., D. Loney, S. Cormier, F. K. Satterstrom, and T. Bridges. 2009. Weight-of-evidence evaluation in environmental assessment: Review of qualitative and quantitative approaches. *Sci Total Environ* 407 (19):5199–5205.

Linkov, I., S. Cormier, J. Gold, F. K. Satterstrom, and T. Bridges. 2012. Using our brains to develop better policy. *Risk Anal* 32 (3):374–380.

Lipton, P. 2004. *Inference to the Best Explanation*. London: Routledge.

Long, E. R. and P. M. Chapman. 1985. A sediment quality triad - measures of sediment contamination, toxicity and infaunal community composition in Puget-Sound. *Mar Pollut Bull* 16:405–415.

Long Creek Watershed Management District (LCWMD). 2014. Long Creek Watershed Management District. http://restorelongcreek.org/ (accessed January 9, 2014).

Lorenz, K. 1965. *Evolution and modification of behavior.* Chicago, IL: University of Chicago Press.

Lorenz, K. 2009. Kant's doctrine of the a priori in the light of contemporary biology. In *Philosophy after Darwin: Classic and Contemporary Readings,* edited by M. Ruse. Princeton, NJ: Princeton University Press.

Lorenz, A. W. and C. K. Feld. 2013. Upstream river morphology and riparian land use overrule local restoration effects on ecological status assessment. *Hydrobiologia* 704 (1):489–501.

Lowell, R. B., J. M. Culp, and M. G. Dube. 2000. A weight-of-evidence approach for Northern river risk assessment: Integrating the effects of multiple stressors. *Environ Toxicol Chem* 19 (4):1182–1190.

Lozano, S. J., S. L. O'Halloran, K. W. Sargent, and J. C. Brazner. 1992. Effects of esfenvalerate on aquatic organisms in littoral enclosures. *Environ Toxicol Chem* 11 (1):35 47.

Ludwig, J. A. and J. F. Reynolds. 1988. *Statistical Ecology.* New York: Wiley.

Lukacs, P. M., W. L. Thompson, W. L. Kendall et al. 2007. Concerns regarding a call for pluralism of information theory and hypothesis testing. *J Appl Ecol* 44 (2):456–460.

Mack, A. and I. Rock. 1998. *Inattentional Blindness.* Cambridge, MA: MIT Press.

Mackay, D. and N. Mackay. 2007. Mathematical models of chemical transport and fate. In *Ecological Risk Assessment,* edited by G. W. Suter II. Boca Raton, FL: CRC Press.

Mackie, J. 1965. Causes and conditions. *Am Phil Q* 2 (4):245–255.

Mackie, J. 1974. *The Cement of the Universe: A Study of Causation.* Oxford, UK: Clarendon Press.

MacLock, R., B. Lyons, and E. Ellehoj. 1996. *Environmental Overview of the Northern Rivers Basins.* Northern River Basins Study Report No. 8. Edmonton, Alberta: Environment Canada.

Mahler, B. J., P. C. Van Metre, T. J. Bashara, J. T. Wilson, and D. A. Johns. 2005. Parking lot sealcoat: An unrecognized source of urban polycyclic aromatic hydrocarbons. *Environ Sci Technol* 39 (15):5560–5566.

Maine Department of Environmental Protection (MEDEP). 2002. *A biological, physical, and chemical assessment of two urban streams in southern Maine: Long Creek & Red Brook.* Portland, ME: Maine Department of Environmental Protection.

Maine Department of Environmental Protection (MEDEP). 2009. General Permit: Post Construction Discharge of Stormwater in the Long Creek Watershed. http://www.maine.gov/dep/water/wd/long_creek/ (accessed January 9, 2009).

Major, J. T. and J. A. Sherburne. 1987. Interspecific relationships of coyotes, bobcats, and red foxes in western Maine. *J Wildl Manage* 51:606–616.

Maltby, L., S. A. Clayton, H. Yu et al. 2000. Using single-species toxicity tests, community-level responses, and toxicity identification evaluations to investigate effluent impacts. *Environ Toxicol Chem* 19 (1):151–157.

Maltby, L., N. Blake, T. Brock, and P. J. Van den Brink. 2005. Insecticide species sensitivity distributions: Importance of test species selection and relevance to aquatic ecosystems. *Environ Toxicol Chem* 24 (2):379–388.

Martel, A., D. Pathy, J. Madill et al. 2001. Decline and regional extirpation of freshwater mussels (Unionidae) in a small river system invaded by *Dreissena polymorpha*: The Rideau River, 1993–2000. *Can J Zool* 79 (12):2181–2191.

Massachusetts Department of Environmental Protection (MDEP). 2011. *Large volume ethanol spills: Environmental impacts and response options.* Salem, NH: Shaw's Environmental and Infrastructure Group.

Materna, E. 2001. Issue Paper 4. *Temperature Interaction: Prepared as part of EPA Region 10 temperature water quality guidance development project.* Seattle, WA: U.S. Environmental Protection Agency.

Maxted, J. 1996. *The use of percent impervious cover to predict the ecological condition of wadable nontidal streams in Delaware.* Paper read at Assessing the Cumulative Impacts of Watershed Development on Aquatic Ecosystems and Water Quality Conference, March 19–21, 1996. Chicago, IL.

Mayer, R. and R. Moreno. 2003. Nine ways to reduce cognitive load in multimedia learning. *Edu Psychol* 38 (1):43–52.

McCarty, L. S. and C. J. Borgert. 2006. Review of the toxicity of chemical mixtures: Theory, policy and regulatory practice. *Reg Toxicol Pharmacol* 45:119–143.

McClelland, W. T. and M. A. Brusven. 1980. Effects of sedimentation on the behavior and distribution of riffle insects in a laboratory stream. *Aquatic Insects* 2 (3):161–169.

McCoy, C. P. and W. C. Edwards. 1980. Sodium ion poisoning in livestock from oil field wastes. *Bovine Practitioner* 15:152–154.

McCubbin, N. and J. Folke. 1993. *A review of literature on pulp and paper mill effluent characteristics in the Peace and Athabasca River basins.* Northern River Basins Study Report No. 15. Edmonton, Alberta: Environment Canada.

McCue, P. M. and T. P. O'Farrell. 1986. Serologic survey for disease in endangered San Joaquin kit fox, *Vulpes macrotis mutica*, inhabiting the Elk Hills Naval Petroleum Reserve, Kern County, California Santa Barbara, CA. EG&G Energy Measurements, Inc. U. S. Department of Energy Topical Report No. EGG 10282-2110.

McCue, P. M. and T. P. O'Farrell. 1987. Hematologic values of the endangered San Joaquin kit fox, *Vulpes macrotis mutica*. *J Wildl Dis* 23 (1):144–151.

McCue, P. M. and T. P. O'Farrell. 1988. Serological survey for selected diseases in the endangered San Joaquin kit fox (*Vulpes macrotis mutica*). *J Wildl Dis* 24 (2):274–281.

McCue, P. M. and T. P. O'Farrell. 1992. Serum chemistry values of the endangered San Joaquin kit fox (*Vulpes macrotis mutica*). *J Wildl Dis* 28 (3):414–418.

McDonald, J. 2008. *Handbook of Biological Statistics.* Baltimore, MD: Sparky House Publishing.

McIntire, C. D. 1993. Historical and other perspectives of laboratory stream research. In *Research in Artificial Streams: Applications, Uses, and Abuses*, ed. G. A. Lamberti and A. D. Steinman, pp. 313–384. *J N Am Benthol Soc* 12:318–323.

McLain, R. J. and R. G. Lee. 1996. Adaptive management: Promises and pitfalls. *Environ Manage* 20 (4):437–448.

McLoughlin, P., T. Coulson, and T. Clutton-Brock. 2008. Cross-generational effects of habitat and density on life history in red deer. *Ecology* 89:3317–3326.

McPherson, C., P. M. Chapman, A. M. H. DeBruyn, and L. Cooper. 2008. The importance of benthos in weight of evidence sediment assessments—A case study. *Sci Total Environ* 394:252–264.

McWilliam, R. A. and D. J. Baird. 2002. Postexposure feeding depression: A new toxicity endpoint for use in laboratory studies with *Daphnia magna*. *Environ Toxicol Chem* 21 (6):1198–1205.

Meals, D. W., S. A. Dressing, and T. E. Davenport. 2010. Lag time in water quality response to best management practices: A review. *J Environ Qual* 39 (1):85–96.

Melcher, A., H. Kremser, A. Zitek et al. 2012. Meta-analysis of restoration projects to improve riverine fish populations, pp. 185–227 in *WISER Deliverable D5.1–2: Driver-Pressure Impact and Response-Recovery Chains in European Rivers: Observed and Predicted Effect on BQEs*. http://www.wiser.eu/download/D5.1-2. pdf (accessed February 1, 2014).

Menand, L. 2001. *The Metaphysical Club, A Story of Ideas in America*. New York: Farrar, Straus and Giroux.

Menzie, C., M. Henning, J. Cura et al. 1996. A weight-of-evidence approach for evaluating ecological risks: Report of the Massachusetts Weight-of-Evidence Work Group. *Hum Ecol Risk Assess* 2 (2):277–304.

Merritt, R.W. and K.W. Cummins. 1996. *An Introduction to the Aquatic Insects of North America*. Dubuque, IA: Kendall Hunt.

Meyer, F. P. and L. A. Barclay. 1990. *Field Manual for the Investigation of Fish Kills*. Washington, DC: U.S. Fish and Wildlife Service. Resource Pub. 177.

Michaels, D. 2008. *Doubt is Their Product*. New York: Oxford University Press.

Michotte, A. 1946. *La Perception de la Causalite* [English translation by T. Miles and E. Miles, *The Perception of Causality*, Oxford, UK: Basic Books]. Louvain: Institut Superieur de Philosphie.

Mika, S., J. Hoyle, G. Kyle et al. 2010. Inside the "black box" of river restoration: Using catchment history to identify disturbance and response mechanisms to set targets for process-based restoration. *Ecol Soc* 15 (4):8. http://www.ecologyand society.org/vol15/iss/art8/.

Mill, J. 1843. *A System of Logic, Ratiocinative and Inductive: Being a Connected View of the Principles of Evidence and the Methods of Scientific Investigation*. Indianapolis, IN: Liberty Fund.

Miller, G. 1956. The magical number seven, plus or minus two. *Psychol Rev* 63:81–97.

Miller, S. W., P. Budy, and J. C. Schmidt. 2010. Quantifying macroinvertebrate responses to in-stream habitat restoration: Applications of meta-analysis to river restoration. *Restoration Ecol* 18 (1):8–19.

Miller, K. A., J. A. Webb, S. C. de Little, and M.J. Stewardson. 2013. Environmental flows can reduce the encroachment of terrestrial vegetation into river channels: A systematic literature review. *Environ Manage* 52 (5):1202–1212.

Miltner, R. 2010. A method and rationale for deriving nutrient criteria for small rivers and streams in Ohio. *Environ Manage* 45:842–855.

Miltner, R. J. and E. T. Rankin. 1998. Primary nutrients and the biotic integrity of rivers and streams. *Freshw Biol* 40 (1):145–158.

Minnesota Pollution Control Agency (MPCA). 2009. *Groundhouse river total maximum daily loads for fecal coliform and biota (sediment) impairments*. Final Report.

Mithen, S. 1998. *The Prehistory of the Mind: A Search for the Origins of Art, Religion and Science*. London: Phoenix.

Mittelstrass, J. 2007. The concept of causality in Greek thought. In *Thinking About Causes: From Greek Philosophy to Modern Physics*. Pittsburg, PA: University of Pittsburgh Press.

Moerke, A. H. and G. A. Lamberti. 2003. Responses in fish community structure to restoration of two Indiana streams. *N Am J Fish Manage* 23 (3):748–759.

Mohr, S., M. Fieibicke, T. Ottenstroer et al. 2005. Enhanced experimental flexibility and control in ecotoxicological mesocosm experiments—A new outdoor and indoor pond and stream system. *Environ Sci Pollut Res Int* 12:5–7.

Monmonier, M. 1993. *Mapping It Out: Expository Cartography for the Humanities and Social Sciences*. Chicago, IL: University of Chicago Press.

Monteith, D. T., A. G. Hildrew, R. J. Flower et al. 2005. Biological responses to the chemical recovery of acidified fresh waters in the UK. *Environ Pollut* 137 (1):83–101.

Morgan, S. L. and C. Winship. 2007. *Counterfactuals and Causal Inference: Methods and Principles for Social Research*. Cambridge, UK: Cambridge University Press.

Morley, S. A. and J. R. Karr. 2002. Assessing and restoring the health of urban streams in the Puget Sound basin. *Conserv Biol* 16 (6):1498–1509.

Morris, W. F. and D. F. Doak. 2002. *Quantitative Conservation Biology: Theory and Practice of Population Viability Analysis*. Sunderland, MA: Sinauer Associates.

Morse, C. C., A. D. Huryn, and C. Cronan. 2003. Impervious surface area as a predictor of the effects of urbanization on stream insect communities in Maine, USA. *Environ Monit Assess* 89 (1):95–127.

Morton, M. G., K. L. Dickson, W. T. Waller et al. 2000. Methodology for the evaluation of cumulative episodic exposure to chemical stressors in aquatic risk assessment. *Environ Toxicol Chem* 4:1213–1221.

Mosisch, T. D., S. E. Bunn, and P. M. Davies. 2001. The relative importance of shading and nutrients on algal production in subtropical streams. *Freshw Biol* 46 (9):1269–1278.

Mount, D. R., D. D. Gulley, R. Hockett, T. D. Garrison, and R. H. Everest. 1997. Statistical models to predict the toxicity of major ions to *Ceriodaphnia dubia*, *Daphnia magna*, and *Pimephales promelas* (fathead minnows). *Environ Toxicol Chem* 16 (10):2009–2019.

Munkittrick, K. R. and D. G. Dixon. 1989a. A holistic approach to ecosystem health assessment using fish population characteristics. *Hydrobiologia* 188 (1):123–135.

Munkittrick, K. R. and D. G. Dixon. 1989b. Use of white sucker (*Catostomus commersoni*) populations to assess the health of aquatic ecosystems exposed to low-level contaminant stress. *Can J Fish Aquat Sci* 46 (8):1455–1462.

Munkittrick, K. R., M. E. McMaster, G. Van Der Kraak et al. 2000. *Development of Methods for Effects-Driven Cumulative Effects Assessment Using Fish Populations: Moose River Project*. Pensacola, FL: SETAC Press.

Munkittrick, K. R., C. J. Arens, R. B. Lowell, and G. P. Kaminski. 2009. A review of potential methods of determining critical effect size for designing environmental monitoring programs. *Environ Toxicol Chem* 28 (7):1361–1371.

Murphy, B. L. and R. D. Morrison. 2002. *Introduction to Environmental Forensics*, edited by B. L. Murphy and R. D. Morrison. San Diego, CA: Academic Press.

Mutz, M. 2000. Influences of woody debris on flow patterns and channel morphology in a low energy, sand-bed stream reach. *Int Rev Hydrobiol* 85 (1):107–121.

Myers, R. 1990. *Classical and Modern Regression with Applications*, 2nd ed. Belmont, CA: Duxbury Press.

National Academy of Sciences (NAS). 2011. *Sustainability and the U.S EPA*. Washington, DC: National Academies Press.

National Park Service (NPS). 1997. Benzo (b and k) fluoranthene. In *Environmental Contaminants Encyclopedia*, edited by R. J. Irwin. Fort Collins, CO: Water Resources Division.

Nebeker, A. V. 1972. Effect of low oxygen concentration on survival and emergence of aquatic insects. *Trans Am Fish Soc* 4:675–679.

Nesbit, J. C. and O. O. Adesope. 2006. Learning with concept and knowledge maps: A meta-analysis. *Rev Edu Res* 76 (3):413–448.

Newcombe, C. P. and D. D. MacDonald. 1991. Effects of suspended sediments on aquatic ecosystems. *N Am J Fish Manage* 11:72–82.

Newman, M. and W. Clements. 2008. *Ecotoxicology: A Comprehensive Treatment*. Boca Ration, FL: CRC Press.

Newman, M. C., D. R. Ownby, L. C. Mezin et al. 2000. Applying species-sensitivity distributions in ecological risk assessment: Assumptions of distribution type and sufficient numbers of species. *Environ Toxicol Chem* 19:508–515.

Newman, M., Y. Zhao, and J. Carriger. 2007. Coastal and estuarine ecological risk assessment: The need for a more formal approach to stressor identification. *Hydrobiologia* 577:31–40.

Neyman, J. and E. Pearson. 1933. On the problem of the most efficient tests of statistical hypotheses. *Philos Trans R Soc Lond A* 231:289–337.

Nicholson, W. and E. P. Hill. 1984. Mortality in gray foxes from east-central Alabama. *J Wildl Manage* 48:1429–1432.

Niemi, G. J., P. DeVore, N. Detenbeck et al. 1990. Overview of case studies on recovery of aquatic systems from disturbance. *Environ Manage* 14:571–587.

Nimick, D. A., C. H. Gammons, T. E. Cleasby et al. 2003. Diel cycles in dissolved metal concentrations in streams: Occurrence and possible causes. *Water Res Res* 39 (9):1247, doi:1029/2002WR001571.

Niroula, D. R. and C. G. Saha. 2010. The incidence of color blindness among some school children of Pokhara, Western Nepal. *Nepal Med Coll J* 12 (1):48–50.

Nisbett, R. and L. Ross. 1980. *Human Inference: Strategies and Shortcomings of Social Judgement*. Englewood Cliffs, NJ: Prentice-Hall.

Niyogi, S. and C. M. Wood. 2004. Biotic ligand model, a flexible tool for developing site-specific water quality guidelines for metals. *Environ Sci Technol* 38 (23):6177–6192.

Norberg-King, T. J., L. W. Ausley, D. T. Burton et al. 2005. *Toxicity Reduction and Toxicity Identification Evaluations for Effluents, Ambient Waters, and Other Aqueous Media*. Pensacola, FL: Society of Environmental Toxicology and Chemistry.

Nordin, R. N. 1985. *Water Quality Criteria for Nutrients and Algae (Technical Appendix)*. Victoria, BC: British Columbia Ministry of the Environment.

Norris, R. H., J. A. Webb, S. J. Nichols, M. J., Stewardson, and E. T. Harrison. 2012. Analyzing cause and effect in environmental assessments: using weighted evidence from the literature. *Freshw Sci* 31 (1): 5–21.

Northern River Basins Study Board (NRBS). 1996. *Northern River Basins Study Report to the Ministers*. Edmonton, Alberta: Alberta Environment Protection.

Norton, S. B., S. M. Cormier, M. Smith, and R. C. Jones. 2000. Can biological assessments discriminate among types of stress? A case study from the Eastern Corn Belt Plains ecoregion. *Environ Toxicol Chem* 19 (4):1113–1119.

Norton, S. B., S. M. Cormier, G. W. Suter II et al. 2002a. Determining probable causes of ecological impairment in the Little Scioto River, Ohio, USA: Part 1. Listing candidate causes and analyzing evidence. *Environ Toxicol Chem* 21 (6):1112–1124.

Norton, S. B., S. M. Cormier, M. Smith, R. C. Jones, and M. Schubauer-Berigan. 2002b. Predicting levels of stress from biological assessment data: Empirical models from the Eastern Corn Belt Plains, Ohio, USA. *Environ Toxicol Chem* 21 (6):1168–1175.

Norton, S. B., L. Rao, G. W. Suter II, and S. M. Cormier. 2003. Minimizing cognitive errors in site-specific causal assessments. *Hum Ecol Risk Assess* 9 (1):213–229.

Norton, S. B., P. J. Boon, S. Gerould et al. 2004. Formulating assessment questions. In *Ecological Assessment of Aquatic Resources: Linking Science to Decision-Making*, edited by M. T. Barbour, S. B. Norton, H. R. Preston, and K. W. Thornton. Pensacola, FL: SETAC Press.

Norton, D. J., J. D. Wickham, T. G. Wade et al. 2009. A method for comparative analysis of recovery potential in impaired waters restoration planning. *Environ Manage* 44 (2):356–368.

O'Donnell, A. M. 1994. Learning from knowledge maps: The effects of map orientation. *Contemporary Edu Psychol* 19:33–44.

O'Donnell, A. M., D. F. Dansereau, and R. H. Hall. 2002. Knowledge Maps as Scaffolds for Cognitive Processing. *Edu Pyschol Rev* 14 (1):71–86.

Okebukola, P. A. 1990. Attaining meaningful learning of concepts in genetics and ecology: An examination of the potency of the concept-mapping technique. *J Res Sci Teaching* 27:493–504.

Orheim, R. M., L. Lippman, C. J. Johnson, and H. H. Bovee. 1974. Lead and arsenic levels of dairy bovine in proximity to a copper smelter. *Environ Lett* 7 (3):229–236.

Oris, J. T. and J. P. Giesy Jr. 1985. The photoenhanced toxicity of anthracene to juvenile sunfish (*Lepomis* spp.). *Aquat Toxicol* 6 (2):133–146.

Oris, J. T. and J. P. Giesy Jr. 1987. The photo-induced toxicity of polycyclic aromatic hydrocarbons to larvae of the fathead minnow (*Pimephales promelas*). *Chemosphere* 16 (7):1395–1404.

Osenberg, C. W., R. J. Schmitt, S. J. Holbrook, K. E. Abu-Saba, and A. R. Flegal. 1994. Detection of environmental impacts: Natural variability, effect size, and power analysis. *Ecol Appl* 4:16–30.

Palace, V. P., C. Doebel, C. L. Baron et al. 2005. Caging small-bodied fish as an alternative method for environmental effects monitoring (EEM). *Water Qual Res J Can* 40:328–333.

Palmer, M. A. 2009. Reforming watershed restoration: Science in need of application and applications in need of science. *Estuaries Coasts* 32 (1):1–17.

Palmer, M. A., R. F. Ambrose, and N. L. Poff. 1997. Ecological theory and community restoration ecology. *Restoration Ecol* 5 (4):291–300.

Palmer, M. A., E. S. Bernhardt, J. D. Allan et al. 2005. Standards for ecologically successful river restoration. *J Appl Ecol* 42 (2):208–217.

Palmer, M. A., H. L. Menninger, and E. Bernhardt. 2010. River restoration, habitat heterogeneity and biodiversity: A failure of theory or practice? *Freshw Biol* 55:205–222.

Panov, V. and D. McQueen. 1998. Effects of temperature on individual growth rate and body size of a freshwater amphipod. *Can J Zool* 76 (6):1107–1116.

Paquin, P. R., J. W. Gorsuch, S. Apte et al. 2002. The biotic ligand model: A historical overview. *Comp Biochem Physiol Part C* 133:3–35.

Park, R. A., J. S. Clough, and M. C. Wellman. 2008. AQUATOX: Modeling environmental fate and ecological effects in aquatic ecosystems. *Ecol Modell* 213 (1):1–15.

Pastershank, G. and D. Muir. 1995. *Contaminants in environmental samples: PCDDs and PCDFs downstream of bleached kraft mills, Peace and Athabasca Rivers, 1992.* Edmonton, Alberta: Environment Canada. Northern River Basins Study Report No. 44.

Pastershank, G. and D. Muir. 1996. *Environmental contaminants in fish: Polychlorinated biphenyls, organochlorine pesticides and chlorinated phenols, Peace and Athabasca Rivers 1992 to 1994.* Edmonton, Alberta: Environment Canada. Northern River Basins Study Report No. 101.

Patterson, M. E., D. F. Dansereau, and D. Newbern. 1992. Effects of communication aids and strategies on cooperative teaching. *J Edu Psychol* 84:453–461.

Paul, M. J. and J. L. Meyer. 2001. Streams in the urban landscape. *Annu Rev Ecol Syst* 32:333–365.

Pearl, J. 2009. *Causality: Models, Reasoning, and Inference,* 2nd ed. Cambridge, UK: Cambridge University Press.

Peckarsky, B. L., P. R. Fraissinet, M. A. Penton, and D. J. Conklin Jr. 1990. *Freshwater macroinvertebrates of northeastern North America.* Ithaca, NY: Cornell University Press.

Petering, H. G., D. W. Yeager, and S. O. Witherup. 1971. Trace metal content of hair. *Arch Environ Health Int J* 23 (3):202–207.

Piattelli-Palmarini, M. 1994. *Inevitable Illusions: How Mistakes of Reason Rule our Minds.* Translated by M. Piattelli-Palmarini. Edited by K. Botsford. New York: Wiley.

Pinker, S. 1997. *How the Mind Works.* New York: W.W. Norton.

Pinker, S. 2008. *The Stuff of Thought: Language as a Window into Human Nature.* New York: Penguin Group.

Plafkin, J., M. Barbour, K. Porter, S. Gross, and R. Hughes. 1989. *Rapid Bioassessment Protocols for Use in Streams and Rivers: Benthic Macroinvertebrates and Fish.* Washington, DC: U.S. Environmental Protection Agency, Office of Water. EPA/444/4–89/001.

Pollard, A. I. and L. Yuan. 2006. Community response patterns: Evaluating benthic invertebrate composition in metal-polluted streams. *Ecol Appl* 16 (2):645–655.

Pollard, A. I. and L. L. Yuan. 2010. Assessing the consistency of response metrics of the invertebrate benthos: A comparison of trait- and identity-based measures. *Freshw Biol* 55 (7):1420–1429.

Ponader, K., D. Charles, and T. Belton. 2007. Diatom-based TP and TN inference models and indices for monitoring nutrient enrichment of New Jersey streams. *Ecol Indic* 7:79–93.

Pond, G. J. 2004. *Effects of surface mining and residential land use on headwater stream biotic integrity in the eastern Kentucky coalfield region.* Frankfort, KY: Kentucky Department of Environmental Protection, Division of Water.

Pope, C., N. Mays, and J. Popay. 2007. *Synthesizing Qualitative and Quantitative Health Evidence: A Guide to Methods.* Maidenhead, UK: Open University Press.

Posthuma, L. and D. De Zwart. 2006. Predicted effects of toxicant mixtures are confirmed by changes in fish species assemblages in Ohio, USA, Rivers. *Environ Toxicol Chem* 25:1094–1105.

Posthuma, L., G. W. Suter II, and T. Traas. 2002. *Species Sensitivity Distributions in Ecotoxicology.* Boca Raton, FL: Lewis Publishers/CRC Press.

Prasad, A., L. Iverson, and A. Liaw. 2006a. Random forests for modeling the distribution of tree abundances. *Ecosystems* 9:181–199.

Prasad, A., L. Iverson, and A. Liaw. 2006b. Newer classification and regression tree techniques: Bagging and random forests for ecological prediction. *Ecosystems* 9:181–199.

Preston, B. L. and J. Shackelford. 2002. Multiple stressor effects on benthic biodiversity of Chesapeake Bay: Implications for ecological risk assessment. *Ecotoxicology* 11 (2):85–99.

Quinn, T. J. and R. B. Deriso. 1999. *Quantitative Fish Dynamics*. Oxford: Oxford University Press.

Rainbow, P. S. 2002. Trace metal concentrations in aquatic invertebrates: Why and so what? *Environ Pollut* 120 (3):497–507.

Ralls, K. and P. J. White. 1995. Predation on San Joaquin kit foxes by larger canids. *J Mammal* 76 (3):723–729.

Rand, G. M., J. R. Clark, and C. M. Holmes. 2000. Use of outdoor freshwater pond microcosms: II. Responses of biota to pyridaben. *Environ Toxicol Chem* 19 (2):396–404.

Raymundo, L. J., C. S. Couch, and C. D. Harvell. 2008. *Coral Disease Handbook: Guidance for Assessment, Monitoring and Management*. St Lucia, Australia: The Coal Reef Targeted Research and Capacity Building for Management Program.

Reichenbach, H. 1956. *The Direction of Time*. Berkeley, CA: University of California Press.

Reimchen, T. E. 1987. Human color vision deficiencies and atmospheric twilight. *Soc Biol* 34:1–11.

Reiter, R. 1987. A theory of diagnosis from first principles. *Artif Intell* 32:57–95.

Relyea, R. A. 2006. The effects of pesticides, pH, and predatory stress on amphibians under mesocosm conditions. *Ecotoxicology* 15 (6):503–511.

Relyea, C., G. Minshall, and R. Danehy. 2000. Stream insects as bio-indicators of fine sediment. In *Watershed Management 2000 Conference Proceedings*. Sponsored by the Water Environmental Federation and the British Columbia Water and Wastes Association, Vancouver, BC, July 9–12. Alexandria, VA: Water Environment Federation.

Richards, R. A. 1999. A case history of effective fishery management: Chesapeake Bay striped bass. *N Am J Fish Manage* 19 (2):356–375.

Richter, E. D. and R. Laster. 2004. The precautionary principle, epidemiology and the ethics of delay. *Int J Occup Med Environ Health* 17 (1):9–16.

Richter, B. D., D. P. Braun, M. A. Mendelson, and L. L. Master. 1997. Threats to imperiled freshwater fauna. *Conserv Biol* 11 (5):1081–1093.

Riva-Murray, K., R. W. Bode, P. J. Phillips, and G. L. Wall. 2002. Impact source determination with biomonitoring data in New York State: Concordance with environmental data. *Northeastern Naturalist* 9 (2):127–162.

Roberts, R. 2012. *Fish Pathology*. Hoboken, NJ: Wiley-Blackwell.

Roberts, L., G. Boardman, and R. Voshell. 2009. Benthic macroinvertebrate susceptibility to trout farm effluents. *Water Environ Res* 81 (2):150–159.

Robinson, D. H. and K. A. Kiewra. 1995. Visual Argument: Graphic Organizers are Superior to Outlines in Improving Learning from Text. *J Edu Psychol* 87:455–467.

Rohr, J. R. and P. W. Crumrine. 2005. Effects of an herbicide and an insecticide on pond community structure and processes. *Ecol Appl* 15 (4):1135–1147.

Rohr, J. R., J. L. Kerby, and A. Sih. 2006. Community ecology as a framework for predicting contaminant effects. *Trends Ecol Evol* 21 (11):606–613.

Roni, P., K. Hanson, and T. Beechie. 2008. Global review of the physical and biological effectiveness of stream habitat rehabilitation techniques. *N Am J Fish Manage* 28 (3):856–890.

Roosenburg, W. 2000. Hypothesis testing, decision theory, and common sense in resource management. *Conserv Biol* 14 (4):1208–1210.

Rosenbaum, P. R. and D. B. Rubin. 1983. The central role of the propensity score in observational studies for causal effects. *Biometrika* 70 (1):41–55.

Rothman, K. J. and S. Greenland. 1998. *Modern Epidemiology*, 2nd Edition. Philadelphia, PA: Lippincott, Williams & Wilkins.

Rothman, K. J., S. Greenland, and T. L. Lash. 2008. *Modern Epidemiology* 3rd ed. Philadephia, PA: Lippincott Williams and Wilkins.

Rubin, D. B. 2007. The design versus the analysis of observational studies for causal effects: Parallels with the design of randomized trials. *Statist Med* 26 (1):20–36.

Rumps, J. M., S. L. Katz, K. Barnas et al. 2007. Stream restoration in the Pacific Northwest: Analysis of interviews with project managers. *Restoration Ecol* 15 (3):506–515.

Ruse, M. 1989. The view from somewhere. A critical defense of evolutionary epistemology. In *Issues in Evolutionary Epistemology* edited by K. Halweg and C. A. Hooker. Albany NY. SUNY Press.

Russell, B. 1948. *Human Knowledge, Its Scope and Limits, Part V: Probability*. London: George Allen & Unwin Ltd.

Russo, R. and J. Williamson. 2007. Interpreting causality in the health sciences. *Int Studies Phil Sci* 21 (2):157–170.

Russom, C. L., S. P. Bradbury, S. J. Broderius, D. E. Hammermeister, and R. A. Drummond. 1997. Predicting modes of toxic action from chemical structure: Acute toxicity in the fathead minnow (*Pimephales promelas*). *Environ Toxicol Chem* 16:948.

Rysgaard, S., N. Risgaard-Petersen, N. Sloth, K. Jensen, and L. Nielsen. 1994. Oxygen regulation of nitrification and denitrification in sediments. *Limnol Oceanogr* 39 (7):1643–1652.

Sackett, D. L. 1997. Evidence-based medicine. *Sem Perinatol* 21 (1):3–5.

Salmon, W. 1984. *Scientific Explanation and the Causal Structure of the World*. Princeton, NJ: Princeton University Press.

Salmon, W. 1994. Causality without counterfactuals. *Phil Sci* 61 (2):297–312.

Salmon, W. 1998. *Causality and Explanation*. Oxford: Oxford University Press.

Sargeant, A. B., S. H. Allen, and J. O. Hastings. 1987. Spatial relations between sympatric coyotes and red foxes in North Dakota. *J Wildl Manage* 51:285–293.

SAS Institute, Inc. (SAS). 2008. *SAS/STAT 9.2 User's Guide*. I. Cary, NC: SAS Institute.

Scarfe, A. 2012. Kant and Hegel's responses to Hume's skepticism concerning causality and evolutionary epistemological perspective. *Cos Hist J Nat Soc Phil* 8 (1):227–288.

Scheff, P. A. and R. A. Wadden. 1993. Receptor modeling of volatile organic compounds. 1. Emission inventory and validation. *Environ Sci Technol* 27 (4):617–625.

Scheffer, M. and S. R. Carpenter. 2003. Catastrophic regime shifts in ecosystems: Linking theory to observation. *Trends Ecol Evol* 18 (12):648–656.

Schindler, D. W. 1998. Whole-ecosystem experiments: Replication versus realism: The need for ecosystem-scale experiments. *Ecosystems* 1 (4):323–334.

Schlenk, D., R. Handy, S. Steinert, M. H. Depledge, and W. Benson. 2008. Biomarkers. In *The Toxicology of Fishes*, edited by R. T. Di Giulio and D. E. Hinton. Boca Raton, FL: CRC Press.

Schnoor, J. 1996. *Environmental Modeling: Fate and Transport of Pollutants in Water, Air and Soil*. New York: Wiley Interscience.

Scholefield, D., T. Le Goff, J. Braven et al. 2005. Concerted diurnal patterns in riverine nutrient concentrations and physical conditions. *Sci Total Environ* 344 (1–3):201–210.

Schroeder, H. A., D. V. Frost, and J. J. Balassa. 1970. Essential trace metals in man: Selenium. *J Chronic Dis* 23 (4):227–243.

Schulz, R., G. Thiere, and J. M. Dabrowski. 2002. A combined microcosm and field approach to evaluate the aquatic toxicity of azinphosmethyl to stream communities. *Environ Toxicol Chem* 21 (10):2172–2178.

Scott-Brown, J. M., S. Herrero, and J. Reynolds. 1987. Swift fox. In *Wild Furbearer Management and Conservation in North America*, edited by M. Novak, J. A. Baker, M. E. Obband, and B. Malloch. Toronto, Ont.: Ministry of Natural Resources.

Scrimgeour, G. J. and P. A. Chambers. 2000. Cumulative effects of pulp mill and municipal effluents on epilithic biomass and nutrient limitation in a large northern river ecosystem. *Can J Fish Aquat Sci* 57 (7):1342–1354.

Scrimgeour, G. J., J. M. Culp, M. L. Bothwell, F. J. Wrona, and M. H. McKee. 1991. Mechanisms of algal patch depletion: Importance of consumptive and nonconsumptive losses in mayfly-diatom systems. *Oecologia* 85 (3):343–348.

Scrimgeour, G. J., J. M. Culp, and F. J. Wrona. 1994. Feeding while avoiding predators: Evidence for a size-specific trade-off by a lotic mayfly. *J N Am Benthol Soc* 13: 368–378.

Scrivner, J. H., T. P. O'Farrell, and T. T. Kato. 1987. *Diet of the San Joaquin kit fox, Vulpes macrotis mutica, on Naval Petroleum Reserve #1, Kern County, California, 1980–1984*. Santa Barbara, CA, EG&G Energy Measurements, Inc. U.S. Department of Energy Topical Report No. 10282-2168.

Setty, K. E., K. C. Schiff, and S. B. Weisberg. 2012. *Forty Years after the Clean Water Act: A Retrospective Look at the Southern California Coastal Ocean*. Costa Mesa, CA: Southern California Coastal Water Research Project. Technical Report 727.

Shannon, E. E. and P. L. Brezonik. 1972. Limnological characteristics of north and central Florida lakes. *Limnol Oceanogr* 17:97–110.

Shaw, J. L. and J. H. Kennedy. 1996. The use of aquatic field mesocosm studies in risk assessment. *Environ Toxicol Chem* 15 (5):605–607.

Shaw, E. and J. Richardson. 2001. Direct and indirect effects of sediment pulse duration on stream invertebrate assemblages and rainbow trout (*Oncorhynchus mykiss*) growth and survival. *Can J Fish Aquat Sci* 58 (11):2213–2221.

Shermer, M. 2002. *Why People Believe Weird Things*. New York: Henry Holt and Company, L.L.C.

Shipley, B. 2000. *Cause and Correlation in Biology*. Cambridge: Cambridge University Press.

Sime, P. 2005. St. Lucie Estuary and Indian River Lagoon conceptual ecological model. *Wetlands* 25 (4):898–907.

Simpson, J. M., J. W. Santo Domingo, and D. J. Reasner. 2002. Microbial source tracking: State of the science. *Environ Sci Technol* 36:5279–5288.

Skelly, J. M., D. D. Davis, W. Merrill et al. 1990. *Diagnosing Injury to Eastern Forest Trees*. Research Triangle Park, NC: U.S. Forest Service.

Smart, G. 1978. Investigations of the toxic mechanisms of ammonia to fish–gas exchange in rainbow trout (*Salmo gairdneri*) exposed to acutely lethal concentrations. *J Fish Biol* 12 (1):93–104.

Smeets, E. and R. Weterings. 1999. *Environmental Indicators: Typology and Overview.* Copenhagen: European Environment Agency. Technical Report No. 25.

Smith, H. L. 1997. Matching with multiple controls to estimate treatment effects in observational studies. *Sociol Methodol* 27 (1):325–353.

Smithson, J. 2007. *West Virginia Stream/River Survey Design 2007–2111.* Charleston, WV: West Virginia Department of Environmental Protection, Division of Water and Waste Management.

Smock, L. A., G. M. Metzler, and J. E. Gladden. 1989. Role of debris dams in the structure and functioning of low-gradient headwater streams. *Ecology* 70 (3):764–775.

Snedecor, G. and W. Cochran. 1989. *Statistical Methods,* 8th ed. Ames, IA: The Iowa State University Press.

Society of Professional Journalists (SPJ). 1996. *SPJ Code of Ethics.* Indianapolis, IN: Society of Professional Journalists.

Soil and Water Conservation Society of Metro Halifax (SWCSMH). 2010. Taxa Tolerance Values. Soil and Water Conservation Society of Metro Halifax. http:/lakes. chebucto.org/ZOOBENTH/BENTHOS/tolerance.html (accessed 6/28/2012).

Sokal, R. R. and F. J. Rohlff. 1995. *Biometry: The Principles and Practice of Statistics in Biological Research,* 3rd ed. New York: Freeman.

Spears, B. L., J. A. Hansen, and D. J. Audet. 2007. Blood lead concentrations in waterfowl utilizing Lake Coeur d'Alene, Idaho. *Arch Environ Contam Toxicol* 52 (1):121–128.

Spears, B., I. Gunn, S. Meis, and L. May. 2011. Analysis of cause-effect-recovery chains for lakes recovering from eutrophication. WISER Deliverable D6.4-2.

Spirtes, P., C. Glymour, and R. Scheines. 2000. *Causation, Prediction, and Search.* Cambridge, MA: MIT Press.

Stanley, E. H. and M. W. Doyle. 2003. Trading off: The ecological removal effects of dam removal. *Frontiers Ecol Environment* 1 (1):15–22.

Starfield, A. M. 1997. A pragmatic approach to modeling for wildlife management. *J Wildl Manage* 61 (2):261–270.

Starfield, A. M. and A. L. Beloch. 1986. *Building Models for Conservation and Wildlife Management.* New York: Macmillan.

State of Maine. 2014. Maine Office of GIS. http://www.maine.gov/megis/ (accessed December 29, 2005).

Stewart, G. B., H. R. Bayliss, D. A. Showler, W. J. Sutherland, and A. S. Pullin. 2009. Effectiveness of engineered in-stream structure mitigation measures to increase salmonid abundance: A systematic review. *Ecol Appl* 19 (4):931–941.

Stewart-Oaten, A. 1995. Rules and judgments in statistics: Three examples. *Ecology* 76 (6):2001–2009.

Stewart-Oaten, A. 1996. Goals in environmental monitoring. In *Detecting Environmental Impacts,* edited by R. J. Schmitt and C. W. Osenberg. New York: Academic Press.

Stewart-Oaten, A. and J. R. Bence. 2001. Temporal and spatial variation in environmental impact assessment. *Ecol Monogr* 71 (2):305–339.

Stewart-Oaten, A., W. W. Murdoch, and K. R. Parker. 1986. Environmental impact assessment: "Pseudoreplication" in time? *Ecology* 67 (4):929–940.

Stoddard, J., L. D. Peck, A. Olsen et al. 2005. *Western Streams and Rivers Statistical Summary.* Washington, DC: U.S. Environmental Protection Agency, Environmental Monitoring and Assessment Program (EMAP). EPA/620/R-05/006.

Stoddard, J. L., D. P. Larsen, C. P. Hawkins, R. K. Johnson, and R. H. Norris. 2006. Setting expectations for the ecological condition of streams: The concept of reference condition. *Ecol Appl* 16 (4):1267–1276.

Strumm, W. and J. Morgan. 1996. *Aquatic Chemistry, Chemical Equilibria and Rates in Natural Waters*, 3rd ed. New York: Wiley.

Stuart, E. A. 2010. Matching methods for causal inference: A review and a look forward. *Statist Sci* 25 (1):1.

Suding, K. N., K. L. Gross, and G. R. Houseman. 2004. Alternative states and positive feedbacks in restoration ecology. *Trends Ecol Evol* 19 (1):46–53.

Susser, M. 1986. Rules of inference in eidemiology. *Regul Toxicol Pharmacol* 6 (2):116–128.

Suter, G. W., II. 1988. *Investigations of relationships between oil field materials and practices and wildlife: Progress report—May 1988–October 1988*. Oak Ridge, TN: U. S. Department of Energy Oak Ridge National Laboratory. ORNL/M-659.

Suter, G. W., II. 1990. Use of biomarkers in ecological risk assessment. In *Biomarkers of Environmental Contamination*, edited by J. F. McCarthy and L. L. Shugart. Ann Abor: Lewis Publishers.

Suter, G. W., II. 1996. Abuse of hypothesis testing statistics in ecological risk assessment. *Hum Ecol Risk Assess* 2 (2):331–349.

Suter, G. W., II. 1998. Retrospective assessment, ecoepidemiology, and ecological monitoring. In *Handbook of Environmental Risk Assessment and Management*, edited by P. Calow. Oxford: Blackwell Scientific.

Suter, G. W., II. 1999. Developing conceptual models for complex ecological risk assessments. *Hum Ecol Risk Assess* 5 (2):375–396.

Suter, G. W., II. 2007. *Ecological Risk Assessment*, 2nd ed. Boca Raton, FL: CRC Press.

Suter, G. W., II. 2012. A Chronological History of Causation for Environmental Scientists. U. S. Environmental Protection Agency. http://www.epa.gov/caddis/si_history.html (accessed February 1, 2014).

Suter, G. W., II and S. M. Cormier. 2008. A theory of practice for environmental assessment. *Integr Environ Assess Manage* 4 (4):478–85.

Suter, G. W., II and S. M. Cormier. 2011. Why and how to combine evidence in environmental assessments: Weighing evidence and building cases. *Sci Total Environ* 409 (8):1406–1417.

Suter, G. W., II and S. M. Cormier. 2012. Two roles for environmental assessors: Technical consultant and advisor. *Hum Ecol Risk Assess* 18 (6):1153–1155.

Suter, G. W., II and S. M. Cormier. 2013a. A method for assessing the potential for confounding applied to ionic strength in central Appalachian streams. *Environ Toxicol Chem* 32 (2):288–295.

Suter, G. W., II and S. M. Cormier. 2013b. Pragmatism: A practical philosophy for environmental scientists. *Integr Environ Assess Manage* 9 (2):181–184.

Suter, G. W., II and T. P. O'Farrell. 2008. *Analysis of the Causes of a Decline in the San Joaquin Kit Fox Population on the Elk Hills, Naval Petroleum Reserve #1, California*. Cincinnati, OH: U.S. Environmental Protection Agency, National Center for Environmental Assessment. EPA/600/R-08/130.

Suter, G. W., II, A. E. Rosen, J. J. Beauchamp, and T. T. Kato. 1992. Results of analysis of fur samples from the San Joaquin kit fox and associated water and soil samples from the Naval Petroleum Reserve No. 1, Tupman, California. Oak Ridge, TN: Oak Ridge National Laboratory, U.S. Department of Energy. ORNL/TM-12244.

Suter, G. W., II, B. Efroymson, B. Sample, and D. Jones. 2000. *Ecological Risk Assessment for Contaminated Sites*. Boca Raton, FL: Lewis Publishers.

Suter, G. W., II, S. Norton, and S. Cormier. 2010a. The science and philosophy of a method for assessing environmental causes. *Hum Ecol Risk Assess* 16 (1):19–34.

Suter, G. W., II, P. Shaw-Allen, L. Yuan, and S. Cormier. 2010b. Basic Analyses: Regression Analysis. U.S. Environmental Protection Agency, Office of Research and Development. http://www.epa.gov/caddis/da_basic_2.html (accessed February 1, 2014).

Suter, G. W., II, S. B. Norton, and S. M. Cormier. 2002. A methodology for inferring the causes of observed impairments in aquatic ecosystems. *Environ Toxicol Chem* 21 (6):1101–1111.

Sutherland, K. P., J. W. Porter, and C. Torres. 2004. Disease and immunity in Caribbean and Indo-Pacific zooxanthellate corals. *Mar Ecol Progr Ser* 266:273–302.

Sutherland, K. P., J. W. Porter, J. W. Turner et al. 2010. Human sewage identified as likely source of white pox disease of the threatened Caribbean elkhorn coral, *Acropora palmata*. *Environ Microbiol* 12 (5):1122–1131.

Sutherland, K. P., S. Shaban, J. L. Joyner, J. W. Porter, and E. K. Lipp. 2011. Human pathogen shown to cause disease in the threatened eklhorn coral *Acropora palmata*. *PLoS One* 6 (8):e23468.

Sutton, A. J., T. R. Fisher, and A. B. Gustafson. 2010. Effects of restored stream buffers on water quality in non-tidal streams in the Choptank River Basin. *Water Air Soil Pollut* 208 (1–4):101–118.

Swartz, R. C. 1999. Consensus sediment quality guidelines for polycyclic aromatic hydrocarbon mixtures. *Environ Toxicol Chem* 18 (4):780–787.

Swift, M., N. Troelstrup Jr, N. Detenbeck, and J. Foley. 1993. Large artificial streams in toxicological and ecological research. *J N Am Benthol Soc* 12 (4):359–366.

Taleb, N. N. 2010. *The Black Swan*, 2nd ed. New York: Random House, Inc.

Taper, M. L. and S. R. Lele. 2004. *The Nature of Scientific Evidence: Statistical, Philosophical and Empirical Considerations*. Chicago, IL: University of Chicago Press.

Taylor, F. G., L. K. Mann, R. C. Dahlman, and F. L. Miller. 1975. *Environmental effects of chromium and zinc in cooling-water drift. Cooling Tower Environment—1974*. U.S. Energy Research and Development Administration Symposium, Washington, DC, pp. 408–426.

ter Braak, C. and S. Juggins. 1993. Weighted averaging partial least squares regression (WA-PLS): An improved method for reconstructing environmental variables from species assemblages. *Hydrobiologia* 269/270:485–502.

Tergan, S. O. 2005. Digital concept maps for managing knowledge and information. *Knowledge Information Visualization: Searching Synergies* 3426:185–204.

Tetra Tech, Inc. 2000. *A Stream Condition Index for West Virginia Wadeable Streams*. Report Prepared for U.S. EPA Region 3 Environmental Services Division, and U.S. EPA Office of Science and Technology, Office of Water. Owings Mills, MD: Tetra Tech.

Textor, J., J. Hardt, and S. Knüppel. 2011. DAGitty: A graphical tool for analyzing causal diagrams. *Epidemiology* 22 (5):745.

Thaler, R. H. and C. R. Sunstein. 2009. *Nudge: Improving Decisions About Health, Wealth, and Happiness*. London: Penguin Books.

Thomas, N. J., W. Foreyt, J. Evermann, L. Windberg, and F. Knowlton. 1984. Seroprevalence of canine parvovirus in wild coyotes from Texas, Utah, and Idaho (1972 to 1983). *J Am Vet Med Assoc* 185 (11):1283–1287.

Thompson, D. G., S. B. Holmes, D. G. Pitt, K. R. Solomon, and K. L. Wainio-Keizer. 1994. Applying concentration-response theory to aquatic enclosure studies. In *Aquatic Mesocosm Studies in Ecological Risk Assessment*, edited by R. L. Graney, J. H. Kennedy, and J. H. Rodgers. Boca Raton, FL: CRC Press.

Tolstoy, L. 2007. *War and Peace*. Translated by R. Pevear. Edited by L. Volokhnosy. New York: Vintage Books.

Tomer, M. D. and M. A. Locke. 2011. The challenge of documenting water quality benefits of conservation practices: A review of USDA-ARS's conservation effects assessment project watershed studies. *Water Sci Technol* 64 (1):300–310.

Traas, T. P., J. H. Janse, P. J. Van den Brink, T. Brock, and T. Aldenberg. 2004. A freshwater food web model for the combined effects of nutrients and insecticide stress and subsequent recovery. *Environ Toxicol Chem* 23 (2):521–529.

Tuckerman, S. and B. Zawiski. 2007. Case studies of dam removal and TMDLs: Process and results. *J Great Lakes Res* 33:103–116.

Tullos, D. D., D. L. Penrose, G. D. Jennings, and W. G. Cope. 2009. Analysis of functional traits in reconfigured channels: Implications for the bioassessment and disturbance of river restoration. *J N Am Benthol Soc* 28 (1):80–92.

Tversky, A. and D. Kahneman. 1981. The framing of decisions and the psychology of choice. *Science* 211 (4481):453–458.

Underwood, A. J. 1991. Beyond BACI: Experimental designs for detecting human environmental impacts on temporal variations in natural populations. *Mar Freshw Res* 42 (5):569–587.

Underwood, A. J. 1992. Beyond BACI: The detection of environmental impacts on populations in the real, but variable, world. *J Exp Mar Biol Ecol* 161 (2):145–178.

Underwood, A. J. 1994. On beyond BACI: Sampling designs that might reliably detect environmental disturbances. *Ecol Appl* 4 (1):3–15.

Underwood, A. J. and M. G. Chapman. 2003. Power, precaution, Type II error and sampling design in assessment of environmental impacts. *J Exp Mar Biol Ecol* 296 (1):49–70.

URS Greiner, Inc. and CH2M Hill. 2001. *Remedial investigation report for the Coeur d'Alene basin Remedial Investigation/Feasibility Study*. Seattle, WA: U.S. Environmental Protection Agency, Region 10. URS DCN: 4162500.06200.05.a2.

U.S. Department of Energy (U.S. DOE). 1993. Supplemental Environmental Impact Statement, Petroleum production at maximum efficient rate, Naval Petroleum Reserve No. 1 (Elk Hills), Kern County, California. Tupman, CA, U. S. Department of Energy. DOE/EIS 0158.

U.S. Department of Health Education and Welfare (U.S. DHEW). 1964. *Smoking and Health: Report of the Advisory Committee to the Surgeon General of the Public Health Service*. Washington, DC: Public Health Service, Center for Disease Control. PHS Publication No. 1103.

U.S. Environmental Protection Agency (U.S. EPA). 1986a. *Quality criteria for water (Goldbook)*. Washington, DC: Office of Water, Regulations and Standards. EPA/440/5-76-001.

U.S. Environmental Protection Agency (U.S. EPA). 1986b. *Ambient water quality criteria for dissolved oxygen (freshwater)*. Washington, DC: Office of Research and Development. EPA/440/5-86-003.

U.S. Environmental Protection Agency (U.S. EPA). 1991. *Methods for aquatic toxicity identification evaluations: Phase I. Toxicity characterization procedures*, 2nd ed. Final Report. Duluth, MN: Office of Research and Development. EPA/600/6-91/003.

U.S. Environmental Protection Agency (U.S. EPA). 1992. *Methods for aquatic toxicity identification evaluations: Phase II. Toxicity identification procedures for samples exhibiting acute and chronic toxicity*. Duluth, MN: Office of Research and Development. EPA/600/6-92/080.

U.S. Environmental Protection Agency (U.S. EPA). 1993. *Methods for aquatic toxicity identification evaluations: Phase III. Toxicity confirmation procedures for samples exhibiting acute and chronic toxicity.* Duluth, MN: Office of Research and Development.

U.S. Environmental Protection Agency (U.S. EPA). 1998. *Guidelines for ecological risk assessment.* Washington, DC: Risk Assessment Forum. EPA/630/R-95/002F.

U.S. Environmental Protection Agency (U.S. EPA). 2000a. *Stressor Identification Guidance Document.* Washington, DC: Office of Water. EPA/822/B-00/025.

U.S. Environmental Protection Agency (U.S. EPA). 2000b. *Supplementary guidance for conducting health risk assessment of chemical mixtures.* Washington, DC: Risk Assessment Forum. EPA/630/R-00/002.

U.S. Environmental Protection Agency (U.S. EPA). 2000c. *Nutrient criteria technical guidance manual: Rivers and streams.* Washington, DC: Office of Water, Office of Science and Technology. EPA/822/B-00/002.

U.S. Environmental Protection Agency (U.S. EPA). 2000d. *Ambient water quality criteria recommendations-information supporting the development of state and tribal nutrient criteria: Rivers and streams in nutrient ecoregion XIV.* Washington, DC: Office of Water, Office of Science and Technology. EPA/822/B-00/022.

U.S. Environmental Protection Agency (U.S. EPA). 2000e. *Polynuclear aromatic hydrocarbons (PAHs) analysis in aqueous samples—Casco Bay, ME.* Lexington, MA: U.S. Environmental Protection Agency, Region 1, Office of Environmental Measurement & Evaluation.

U.S. Environmental Protection Agency (U.S. EPA). 2002. *National Recommended Water Quality Criteria: 2002.* Washington, DC: Office of Water, Office of Science and Technology. EPA/822/R-02/047.

U.S. Environmental Protection Agency (U.S. EPA). 2004a. *Long Creek sediment toxicity study 2003.* N. Chelmsford, MA: U.S. EPA Region 1, Office of Environmental Measurement & Evaluation, Ecosystems Assessment Group-Ecology Monitoring Team, Toxicity Testing Laboratory.

U.S. Environmental Protection Agency (U.S. EPA). 2004b. *National recommended water quality criteria.* Washington, DC: Office of Water, Office of Science and Technology.

U.S. Environmental Protection Agency (U.S. EPA). 2006a. *Mid-Atlantic Integrated Assessment MAIA: State of the Flowing Waters Report.* Washington, DC: Office of Research and Development. EPA/620/R-06/001.

U.S. Environmental Protection Agency (U.S. EPA). 2006b. *Estimation and Application of Macroinvertebrate Tolerance Values.* Washington, DC: Office of Research and Development. EPA/600/P-04/116F.

U.S. Environmental Protection Agency (U.S. EPA). 2007a. *Sediment Toxicity Identification Evaluation (TIE) Phases I, II, and II Guidance Document.* Washington, DC: Office of Research and Development. EPA/600/R-07/080.

U.S. Environmental Protection Agency (U.S. EPA). 2007b. *Total Maximum Daily Load Program Needs Better Data and Measure to Demonstrate Environmental Results.* Washington, DC: Office of the Inspector General. Report No. 2007-P-00036.

U.S. Environmental Protection Agency (U.S. EPA). 2011a. *A Field-Based Aquatic Life Benchmark for Conductivity in Central Appalachian Streams.* Cincinnati, OH: Office of Research and Development, National Center for Environmental Assessment. EPA/600/R-10/023F.

U.S. Environmental Protection Agency (U.S. EPA). 2011b. *A National Evaluation of the Clean Water Act Section 319 Program.* Washington, DC: Office of Wetlands, Oceans, and Watersheds.

U.S. Environmental Protection Agency (U.S. EPA). 2011c. *The Effects of Mountaintop Mines and Valley Fills on Aquatic Ecosystems of the Central Appalachian Coalfields.* Washington, DC: Office of Research and Development, National Center for Environmental Assessment. EPA/600/R-09/138F.

U.S. Environmental Protection Agency (U.S. EPA). 2012a. *CADDIS: The Causal Analysis/Diagnosis Decision Information System.* Office of Research and Development. http://www.epa.gov/caddis/index.html (accessed February 1, 2014).

U.S. Environmental Protection Agency (U.S. EPA). 2012b. *CADDIS Volume 2: Sources Stressors and Responses.* Office of Research and Development. http://www.epa.gov/caddis/ssr_home.html (accessed February 1, 2014).

U.S. Environmental Protection Agency (U.S. EPA). 2012c. *Freshwater Biological Traits Database.* Office of Research and Development. http://www.epa.gov/ncea/global/traits/ (accessed January 9, 2014).

U.S. Environmental Protection Agency (U.S. EPA). 2012d. *CADDIS Volume 1: Stressor Identification.* Office of Research and Development. http://www.epa.gov/caddis/si_home.html (accessed March 20, 2013).

U.S. Environmental Protection Agency (U.S. EPA). 2013a. Case Studies-Bioassessment and Enforcement: Using Bioassessment as Evidence of Damage and Recovery Following a Pesticide Spill. Washington, DC: Office of Water. http://water.epa.gov/scitech/swguidance/standards/criteria/aqlife/biocriteria/enforcement.cfm (accessed February 1, 2014)

U.S. Environmental Protection Agency (U.S. EPA). 2013b. *Aquatic Life Ambient Water Quality Criteria for Ammonia—Freshwater.* Washington, DC: Office of Water. EPA 822-R-13-001.

U.S. Environmental Protection Agency (U.S. EPA). 2013c. *Council for Regulatory Environmental Modeling.* Office of the Science Advisor. http://www.epa.gov/crem/ (accessed February 9, 2014).

U.S. Environmental Protection Agency (U.S. EPA). 2013d. BASINS. Office of Water. http://water.epa.gov/scitech/datait/models/basins/ (accessed February 9, 2014).

U.S. Environmental Protection Agency (U.S. EPA). 2013e. AQUATOX. Office of Water. http://water.epa.gov/scitech/datait/models/aquatox/index.cfm (accessed February 9, 2014).

U.S. Environmental Protection Agency (U.S. EPA). 2013f. Causes of Impairment for 303(d) Listed Waters. Office of Water. http://iaspub.epa.gov/waters10/attains_nation_cy.control?p_report_type=T (accessed 15 April, 2013).

U.S. Environmental Protection Agency (U.S. EPA). 2013g. Section 319 Nonpoint Source Success Stories. Office of Water. http://water.epa.gov/polwaste/nps/success319/ (accessed June 6, 2013).

U.S. Environmental Protection Agency (U.S. EPA). 2014a. Ecoregion Maps and GIS Resources. Office of Research and Development. http://www.epa.gov/wed/pages/ecoregions.htm (accessed February 1, 2014).

U.S. Environmental Protection Agency (U.S. EPA). 2014b. Envirofacts. U.S. Environmental Protection Agency. http://www.epa.gov/envirofw/ (accessed February 1, 2014).

U.S. Environmental Protection Agency (U.S. EPA). 2014c. ECOTOX Database. Office of Research and Development, National Health and Environmental Effects Laboratory. http://www.epa.gov/ecotox/ (accessed February 1, 2014).

U.S. Environmental Protection Agency (U.S. EPA), and MPCA. 2004. *Screening Level Causal Analysis and Assessment of and Impaired Reach of the Groundhouse River, Minnesota.* Cincinnati, OH: Office of Research and Development, National Exposure Research Laboratory.

U.S. Geological Survey (U.S. GS). 2008. Blossom Statistical Software. United States Geological Survey. http://www.fort.usgs.gov/products/software/blossom/ (accessed February 9, 2014).

U.S. Geological Survey (U.S. GS). 2014a. StreamStats. U.S. Geological Survey. http://water.usgs.gov/osw/streamstats/(accessed February 9, 2014).

U.S. Geological Survey (U.S. GS). 2014b. Hydrological Simulation Program Fortran. U.S. Geological Survey. http://water.usgs.gov/software/HSPF/ (accessed February 9, 2014).

U.S. Government Accountability Office (U.S. GAO). 2013. *Clean Water Act: Changes Needed If Key EPA Program Is to Help Fulfill the Nation's Water Quality Goals.* Washington, DC: U.S. Government Accountability Office. GAO-14-80.

Vaal, M., J. T. van der Wal, J. Hoekstra, and J. Hermens. 1997. Variation in the sensitivity of aquatic species in relation to the classification of environmental pollutants. *Chemosphere* 35 (6):1311–1327.

van den Berg, M. et al. 1998. Toxic equivalency factors (TEFs) for PCBs, PCDDs, PCDFs for humans and wildlife. *Environ Health Perspect* 106 (12):775–792.

Van den Brink, P. J., N. Blake, T. C. M. Brock, and L. Maltby. 2006a. Predictive value of species sensitivity distributions for effects of herbicides in freshwater ecosystems. *Hum Ecol Risk Assess* 12 (4):645–674.

Van den Brink, P. J., C. D. Brown, and I. G. Dubus. 2006b. Using the expert model PERPEST to translate measured and predicted pesticide exposure data into ecological risks. *Ecol Modell* 191 (1):106–117.

Van Donk, E., H. Prins, H. M. Voogd, S. J. H. Crum, and T. C. Brock. 1995. Effects of nutrient loading and insecticide application on the ecology of Elodea-dominated freshwater microcosms I. Responses of plankton and zooplanktivorous insects. *Arch Hydrobiol* 133:417–439.

Van Metre, P. C. and B. J. Mahler. 2014. PAH concentrations in lake sediment decline following ban on coal-tar-based pavement sealants in Austin, Texas. *Environ Sci Technol* 48:7222–7228.

Van Metre, P. C., B. J. Mahler, and J. T. Wilson. 2009. PAHs underfoot: Contaminated dust from coal-tar sealcoated pavement is widespread in the United States. *Environ Sci Technol* 43 (1):20–25.

Vannote, R., G. Minshall, K. Cummings, J. Sedell, and C. Cushing. 1980. The river continuum concept. *Can J Fish Aquat Sci* 37:130–137.

Van Sickle, J. 2013. Estimating the risks of multiple, covarying stressors in the National Lakes Assessment. *Freshw Sci* 32 (1):204–216.

Van Sickle, J., J. L. Stoddard, S. G. Paulsen, and A. R. Olsen. 2006. Using relative risk to compare the effects of aquatic stressors at a regional scale. *Environ Manage* 38 (6):1020–1030.

Van Wijngaarden, R., J. G. Cuppen, G. H. Arts et al. 2004. Aquatic risk assessment of a realistic exposure to pesticides used in bulb crops: A microcosm study. *Environ Toxicol Chem* 23 (6):1479–1498.

Vekiri, I. 2002. What is the value of graphical displays in learning?. *Edu Psychol Rev* 14 (3):261–312.

Verdonschot, P. F. M., C. K. Feld, E. Tales, A. Melcher, and C. Mielach. 2012. Commonalities and differences in WFD River Basin Management Plans, pp. 185–194 in *WISER Deliverable D5.1-2: Driver-Pressure-Impact and Response-Recovery Chains in European Rivers: Observed and Predicted Effects on BQEs*. http://www.wiser.eu/download/D5.1-2.pdf (accessed February 1, 2014).

Verdonschot, P. F. M., B. M. Spears, C. K. Feld et al. 2013. A comparative review of recovery processes in rivers, lakes, estuarine and coastal waters. *Hydrobiologia* 704 (1):453–474.

Violin, C. R., P. Cada, E. B. Sudduth et al. 2011. Effects of urbanization and urban stream restoration on the physical and biological structure of stream ecosystems. *Ecol Appl* 21 (6):1932–1949.

Wallace, J. B., S. L. Eggert, J. L. Meyer, and J. R. Webster. 1997. Multiple trophic levels of a forest stream linked to terrestrial litter inputs. *Science* 277 (5322):102–104.

Wallace, D. S., S. W. C. West, A. Ware, and D. F. Dansereau. 1998. The effect of knowledge maps that incorporate gestalt principles on learning. *J Exp Edu* 67 (1):5–16.

Waller, L. A. and C. A. Gotway. 2004. *Applied Spatial Statistics for Public Health Data*. New York: Wiley.

Waller, W. T., L. P. Ammann, W. J. Birge et al. 1996. Predicting instream effects from WET tests, Discussion Synopsis. In *Whole Effluent Toxicity Testing: An Evaluation of Methods and Prediction of Receiving System Impacts*, edited by D. R. Grothe, K. L. Dickson, and D. K. Reed-Judkins. Pensacola, FL: CRC Press.

Walsh, C. J., A. H. Roy, J. W. Feminella et al. 2005. The urban stream syndrome: Current knowledge and the search for a cure. *J N Am Benthol Soc* 24 (3):706–723.

Walters, C. 1986. *Adaptive Management of Renewable Resources*. New York: MacMillan.

Walters, C. 1997. Challenges in adaptive managment of riparian and coastal ecosystems. *Conserv Ecol [online]* 1 (2).

Walters, C. and C. Holling. 1990. Large-scale management experiments and learning by doing. *Ecology* 71 (6):2060–2068.

Walters, C., S. J. D. Martell, V. Christensen, and B. Mahmoudi. 2008. An Ecosim model for exploring Gulf of Mexico ecosystem management options: Implications of including multistanza life-history models for policy predictions. *Bull Mar Sci* 83 (1):251–271.

Waters, T. 1995. *Sediment in Streams: Sources, Biological Effects, and Control*. Bethesda, MD: American Fisheries Society.

Webber, E. C., W. G. Deutsch, D. R. Bayne, and W. C. Seesock. 1992. Ecosystem level testing of a synthetic pyrethroid insecticide in aquatic mesocosms. *Environ Toxicol Chem* 11:87–105.

Weed, D. L. 1988. Causal criteria and Popperian refutation. In *Causal Inference*, edited by K. J. Rothman. Chestnut Hill, MA: Epidemiology Resources Inc.

Weed, D. L. 2005. Weight of evidence: A review of concept and methods. *Risk Anal* 25 (6):1545–1557.

Westgate, M. J., G. E. Likens, and D. B. Lindenmayer. 2013. Adaptive management of biological systems: A review. *Biol Conserv* 158:128–139.

West Virginia Department of Environmental Protection (WVDEP). 2006. *Total maximum daily loads for selected streams in the Coal River Watershed, West Virginia: Clear fork appendices*. Charleston, WV: West Virginia Department of Environmental Protection.

West Virginia Department of Environmental Protection (WVDEP). 2008a. *West Virginia integrated water quality monitoring and assessment report*. Charleston, WV: West Virginia Department of Environmental Protection.

West Virginia Department of Environmental Protection (WVDEP). 2008b. *2008 Standard operating procedures*, Vol 1. Charleston, WV: West Virginia Department of Environmental Protection.

West Virginia Department of Environmental Protection (WVDEP). 2013. *Watershed assessment branch 2013 standard operating procedures*. Charleston, WV: West Virginia Department of Environmental Protection.

Wetzel, R. 2001. *Limnology: Lake and River Ecosystems*, 3rd ed. New York: Academic Press.

Wharfe, J., W. Adams, S. E. Apitz et al. 2007. In situ methods of measurement—An important line of evidence in the environmental risk framework. *Integr Environ Assess Manage* 3 (2):268–274.

Whewell, W. 1858. *The Philosophy of the Inductive Sciences, Founded Upon their History*, 2nd ed. London: John W. Parker.

Whittingham, M. J., P. A. Stephens, R. B. Bradbury, and R. P. Freckleton. 2006. Why do we still use stepwise modelling in ecology and behaviour? *J Anim Ecol* 75 (5):1182–1189.

Wickwire, T. and C. Menzie. 2010. The causal analysis framework: Refining approaches and expanding multidisciplinary applications. *Hum Ecol Risk Assess* 16 (1):10–18.

Wiegmann, D. A., D. F. Dansereau, E. C. McCagg, K. L. Rewey, and U. Pitre. 1992. Effects of knowledge map characteristics on information processing. *Contemp Edu Psychol* 17:136–155.

Wilson, E. O. 1998. *Consilience: The Unity of Knowledge*. New York: A. A. Knopf.

Wilson, C. G., R. A. Kuhnle, D. D. Bosch et al. 2008. Quantifying relative contributions from sediment sources in Conservation Effects Assessment Project watersheds. *J Soil Water Conserv* 63 (6):523–532.

Winger, P. V., P. J. Lasier, and K. J. Bogenrieder. 2005. Combined use of rapid bioassessment protocols and sediment quality triad to assess stream quality. *Environ Monit Assess* 100 (1–3):267–295.

Wise, D. R., M. L. Zuroske, K. D. Carpenter, and R. L. Kiesling. 2009. *Assessment of eutrophication in the Lower Yakima River Basin, Washington, 2004–07*. Reston, VA: U.S. Geological Survey. Scientific Investigations Report 2009-5078.

Wiseman, C. D., M. LeMoine, and S. Cormier. 2010a. Assessment of probable causes of reduced aquatic life in the Touchet River, Washington, USA. *Hum Ecol Risk Assess* 16 (1):87–115.

Wiseman, C. D., M. LeMoine, R. Plotnikoff et al. 2010b. *Identification of most probable stressors to aquatic life in the Touchet River, Washington (Final)*. Washington, DC: U. S. Environmental Protection Agency, Office of Research and Development, National Center for Environmental Assessment. EPA/600/R 08/14.

Wolf, P. 2007. Representing causation. *J Exp Psychol* 136:82–111.

Wood, P. J. and P. D. Armitage. 1997. Biological effects of fine sediment in the lotic environment. *Environ Manage* 21 (2):203–217.

Woodman, J. N. and E. B. Cowling. 1987. Airborne chemicals and forest health. *Environ Sci Technol* 21 (2):120–126.

Woods, A., J. Omernik, D. Brown, and C. Kiilsgaard. 1996. *Level III and IV ecoregions of Pennsylvania and the Blue Ridge Mountains, the Ridge and Valley, and*

Central Appalachians of Virginia, West Virginia, and Maryland. Corvallis, OR: U.S. Environmental Protection Agency, Office of Research and Development, National Health and Environmental Effects Research Laboratory. EPA/600/R-96/077.

Woods, A., J. Omernik, and D. Brown. 1999. *Level III and IV ecoregions of Delaware, Maryland, Pennsylvania, Virginia, and West Virginia*. Corvallis, OR: U.S. Environmental Protection Agency, National Health and Environmental Effects Research Laboratory.

Woodward, J. 2003. *Making Things Happen: A Theory of Causal Explanation*. New York: Oxford University Press.

Wright, J. F. 2000. An introduction to RIVPACS. In *Assessing the Biological Quality of Freshwaters: RIVPACS and Other Techniques*, edited by J. F. Wright, D. W. Sutcliffe, and M. T. Furse. Ambleside, UK: Freshwater Biological Association.

Wright, S. 1920. The relative importance of heredity and environment in determining the Piebald pattern of guinea-pigs. *Proc Natl Acad Sci USA* 6 (6):320–32.

Wright, S. 1921. Correlation and causation. *J Agric Res* 20:557–585.

Wrona, F. J., W. Gummer, K. J. Cash, and K. Crutchfield. 1996. *Cumulative impacts within the Northern River Basins*. Northern River Basins Study Report No. 11. Edmonton, Alberta: Environment Canada.

Wrona, F. J., J. Carey, B. Brownlee, and E. McCauley. 2000. Contaminant sources, distribution and fate in the Athabasca, Peace and Slave River Basins, Canada. *J Aquat Ecosyst Stress Recovery* 8 (1):39–51.

Yerushalmy, J. and C. E. Palmer. 1959. On the methodology of investigations of etiologic factors in chronic diseases. *J Chronic Dis* 10 (1):27–40.

Yoder, C. O. and M. T. Barbour. 2009. Critical elements of state bioassessment programs: A process to evaluate program rigor and comparability. *Environ Monitor Assess* 150 (1):31–42.

Yoder, C. O. and J. E. DeShon. 2002. Using biological response signatures within a framework of multiple indicators to assess and diagnose causes and sources of impairments to aquatic assemblages in selected Ohio Rivers and streams. In *Biological Response Signatures*, edited by T. P. Simon. Boca Raton, FL: CRC Press.

Yuan, L. 2010a. Predicting Environmental Conditions from Biological Observations. U.S. Environmental Protection Agency, Office of Research and Development. http://www.epa.gov/caddis/ex_analytical_2.html (accessed February 1, 2014).

Yuan, L. L. 2010b. Estimating the effects of excess nutrients on stream invertebrates from observational data. *Ecol Appl* 20 (1):110–125.

Zehr, S. 2000. Public representations of scientific uncertainty about global climate change. *Pub Understand Sci* 9 (2):85–103.

Ziegler, C. R. 2007. Common candidate cause: Flow alteration. U.S. Environmental Protection Agency, Office of Research and Development. http://www.epa.gov/caddis/ssr_flow_int.html (accessed February 1, 2014).

Ziegler, C. R., J. T. Varricchione, K. Schofield, S. B. Norton, and S. Meidel. 2007a. *Causal analysis of biological impairment in Long Creek: A sandy-bottomed stream in coastal Soutern Maine*. U.S. Environmental Protection Agency, Office of Research and Development, National Center for Environmental Assessment. EPA/600/R-06/065F.

Ziegler, C. R., G. W. Suter II, B. J. Kefford, K. A. Schofield, and G. J. Pond. 2007b. Common candidate cause: Ionic strength. U. S. Environmental Protection Agency, Office of Research and Development. http://www.epa.gov/caddis/ssr_ion_int.html (accessed February 1, 2014).

Zoellick, B. W., T. P. O'Farrell, P. M. McCue, C. E. Harris, and T. T. Kato. 1987. *Reproduction of the San Joaquin kit fox on Naval Petroleum Reserve #1, Elk Hills, California, 1980–1985.* U.S. Department of Energy Topical Report No. 10282-2144. Santa Barbara, CA, EG&G Energy Measurements, Inc.

Zuur, A. F., E. N. Ieno, and C. S. Elphick. 2010. A protocol for data exploration to avoid common statistical problems. *Methods Ecol Evol* 1:3–14.

Zweig, L. D. and C. F. Rabeni. 2001. Biomonitoring for deposited sediment using benthic invertebrates: A test on 4 Missouri streams. *J N Am Benthol Soc* 20 (4):643–657.

Index

A

Abundance-based similarity index, 371–372
Acid mine drainage (AMD), 358, 370, 373
  discrimination of sites, 372
  Stonecoal Branch, 377, 378
Adjustment sets, *see* Deconfounding sets
Agent causation, 18, 48
Alteration, 56, 57
  evidence, 370–372
  habitat, 359–360
Ambient waters, 219–221
AMD, *see* Acid mine drainage (AMD)
Analogy, 41–42
Analysis of covariance, 197
Antecedence, 49, 56–58
AQUATOX model, 247
Arsenic
  evidence of exposure, 420
  levels in foxes, 400
  stressor–response relationships, 425
  temporal sequence, 419
Artificial intelligence, 42–43
  research in, 182
Assessment, 273; *see also* Causal assessment; Ecological causal assessment
  sequences, 284–286
Association, 139; *see also* Case-specific observations
  covariation quantification, 147–150
  difference quantification, 140–147
  improbability of observation, 141, 142, 143
  interpretation, 150–153
  magnitude of difference, 141–147
  between variables, 139
Autocorrelation, 163–164
Automated variable selection techniques, 194–195

B

BACI designs, *see* Before-After-Control-Impact designs (BACI designs)
BACIP, *see* Before-After-Control-Impact-Pairs (BACIP)
Bank stability, 359
Barite ($BaSO_4$), 419
Barium, 400, 428, 431
  evidence of exposure, 420
  median concentrations, 423
  temporal sequence, 419
Bayes' theorem, 36
Before-After-Control-Impact designs (BACI designs), 38, 205, 206
  for cause–effect linkage investigation, 210
Before-After-Control-Impact-Pairs (BACIP), 205
Benthic fish, 390
Benthic invertebrates, 117, 389
  assemblages shift, 312
  mayflies in, 171
  recovery, 288
Best management practices (BMPs), 296
Biological response
  to candidate stressors, 180, 181
  for causal assessment, 178
  information, 159
  "plausibility" regions of, 367, 369
  of primary interest, 128
Biomarkers, 234–235
  in causal assessments, 238–241
Biotic ligand model (BLM), 245
BKME, *see* Bleached kraft mill effluent (BKME)
Bleached kraft, 388
Bleached kraft mill effluent (BKME), 388
Bleaching, 388
BLM, *see* Biotic ligand model (BLM)
BMPs, *see* Best management practices (BMPs)

Printed and bound by CPI Group (UK) Ltd, Croydon, CR0 4YY

18/10/2024

01776267-0010